Signal Measurement and Estimation Techniques for Micro and Nanotechnology

Signal Measurement and Estimation Techniques for Micro and Nanotechnology

Cédric Clévy • Micky Rakotondrabe
Nicolas Chaillet
Editors

Signal Measurement and Estimation Techniques for Micro and Nanotechnology

Editors
Cédric Clévy
AS2M Department
FEMTO-ST Institute
UMR CNRS
6174 - UFC
ENSMM/UTBM
rue Alain Savary 24
25000 Besançon
France
cclevy@femto-st.fr

Micky Rakotondrabe
FEMTO-ST Institute
UMR CNRS
6174 - UFC
ENSMM / UTBM
rue Alain Savary 24
25000 Besançon
France
mrakoton@femto-st.fr

Nicolas Chaillet
FEMTO-ST Institute
rue Alain Savary 24
25000 Besançon
France
nicolas.chaillet@femto-st.fr

ISBN 978-1-4899-9039-6 ISBN 978-1-4419-9946-7 (eBook)
DOI 10.1007/978-1-4419-9946-7
Springer New York Dordrecht Heidelberg London

Printed on acid-free paper

Springer is part of Springer Science+Business Media (www.springer.com)

Preface

Micro- and nanotechnologies are recent fields of interest that offer high innovation potential. These technologies lead to small dimension multifunctional systems that are of great interest in various domains: biomedical, aerospace, military, automotive, small factories, etc.

The technologies initially developed for the microscale have brought an information revolution through the capability of integrating an incredible number of microscopic transistors in a small surface. More recently, MEMS technologies were developed to introduce mechanical, optical, and/or thermal functions to integrate microactuators and microsensors. Inkjet printer heads, accelerometers, micro-mirrors, micro-relays, and pressure sensors are among the most known and widespread devices that open cost-effective and highly integrated solutions to the car industry, aeronautics, medicine, biology, energy, and telecommunication domains. One step further, nanotechnologies deal with the technology at the nanoscale.

All developed devices aim at sensing or acting at a very small scale, so small that new phenomena negligible at the macroscale become very influent. These scale effects are generally not fully mastered. They include for instance adhesion forces and high noise-to-signal ratio. Furthermore, microsystems dedicated to micro/nanopositioning or micro/nanomanipulation require micrometric and submicrometric accuracy for motions and micronewtons for forces. Such severe characteristics require the need of integrated systems able to perform motions in a controlled way, which requires proprioceptive measurement. Actually, while many solutions exist for actuation, sensing at the micro/nanoscale for measuring their motions is a great challenge. Several projects have emerged during these last years to develop new solutions able to efficiently measure at the micro/nanoscale in an integrated way. The editors of this book have organized a scientific workshop during the IEEE-International Conference on Robotics and Automation (ICRA) held in May 2010 in Anchorage Alaska USA to consider and discuss these issues. The workshop has brought researchers and engineers together to present, discuss, and exchange ideas on this new topic: "Measurement at the micro/nano-scale." The exciting discussions and exchanges between the speakers of the workshop and the audience, composed of engineers, researchers, and students, have resulted the necessity to make a perennial archive available for a large public of the interesting presentations and discussions. This is the motivation of this book.

The book is made of eight chapters.

Chapter 1 introduces the specificities of the micro/nanoscales. The scale effects as well as some typical corresponding magnitudes are presented. The difficulty to get internal measurement is particularly highlighted. The interesting contribution of the control theory (observers, estimators) to that issue ends the chapter.

Chapter 2 is a deep presentation of the so-called self-sensing technique. This technique, allowing the use of an actuator also for the measurement of position and/or force, is presented in this chapter as an efficient approach for piezoelectric-based microactuators. Recent developments are therefore detailed and discussed along the chapter.

Chapter 3 is concerned with the efficient use of the well-known Kalman filtering to reduce noises in signals measured at the micro/nanoscale. The powerful of this filter makes it very complementary with existing microsensors that measures micronewtons forces and micrometers displacements during micromanipulation or microassembly tasks.

In Chap. 4, the design and development of a new generation of microforces sensors based on a capacitive principle are detailed. The singular properties of these sensors make them recognized in various applications at the micro/nanoscale such as micromanipulation, microassembly, and small objects characterizations.

Chapter 5 deals with the particular application of manipulation and characterization of biological cells. The mechanical characterization of oocytes is detailed using computer vision including microscopy and a new polymeric device.

In Chap. 6, the authors describe innovative techniques to characterize thin-film nanostructures, notably helical nanobelts. These nanostructures can be further used as NEMS, tools for nanohandling. Their mechanical properties, such as stiffness, are therefore investigated in the chapter.

Chapter 7 is dedicated to mechanism approaches to enhance the performances of MEMS. The chapter is particularly focused on the dynamics enhancement of sensors based on optical MEMS. Both theory and design cases are included.

Finally, Chap. 8 discusses the state-observer approach to estimate signals in scanning probe microscopes. This chapter introduces the notion of parameter amplification that allows the enhancement of the measurement accuracy and that is demonstrated through an illustrative example.

We express our deep thanks to the authors who describe new results in a very didactic way. Most of them originally participated to the workshop mentioned above.

We are also very grateful to Steve Elliot and Andrew Leigh from Springer USA for their support.

Besançon, France

Cédric Clévy
Micky Rakotondrabe
Nicolas Chaillet

Contents

Contributors

Juan Camilo Acosta Institut des Systémes Intelligents et de Robotique, Université Pierre et Marie Curie, Place Jussieu, 75252 Paris Cedex, France, Acosta@isir.upmc.fr

Gildas Besançon Control Systems Department, GIPSA-lab, Grenoble Institute of Technology and Institut Universitaire de France, Ense[3] BP 46, 38402 Saint-Martin d'Hères, France, gildas.besancon@grenoble-inp.fr

Felix Beyeler Institute of Robotics and Intelligent Systems, ETH, Zurich, Switzerland, fbey@ethz.ch

Mokrane Boudaoud FEMTO-ST Institute, 32 avenue de l'Observatoire, 25004 Besançon, France, mokrane.boudaoud@femto-st.fr

Robert F. Casper Samuel Lunenfeld Research Institute, Toronto Mount Sinai Hospital, 600 University Ave, Toronto, Ontario, Canada, M51 1X5, rfcasper@aol.com

Nicolas Chaillet FEMTO-ST Institute, rue Alain Savary 24, 25000 Besançon, France, nicolas.chaillet@femto-st.fr

Cédric Clévy AS2M Department, FEMTO-ST Institute, UMR CNRS, 6174 - UFC, ENSMM/UTBM, rue Alain Savary 24, 25000 Besançon, France, cclevy@femto-st.fr

Roxanne Fernandes Samuel Lunenfeld Research Institute, Toronto Mount Sinai Hospital, 600 University Ave, Toronto, Ontario, Canada, M51 1X5

Yassine Haddab AS2M Department, FEMTO-ST Institute, 32 avenue de l'Observatoire, 25004 Besançon, France, yassine.haddab@femto-st.fr

Hideki Hashimoto Institute of Industrial Science, The University of Tokyo, 4-6-1 Komaba, Meguro-ku, 153-8505 Tokyo, Japan, hashimoto@iis.u-tokyo.ac.jp

Gilgueng Hwang Laboratory for Photonics and Nanostructures, CNRS, route de Nozay, 91460 Marcoussis, France, gilgueng.hwang@lpn.cnrs.fr

Ioan Alexandru Ivan AS2M Department, FEMTO-ST Institute, 24 rue Alain Savary, 25000 Besançon, France, alex.ivan@femto-st.fr

Andrea Jurisicova Samuel Lunenfeld Research Institute, Toronto Mount Sinai Hospital, 600 University Ave, Toronto, Ontario, Canada, M51 1X5

Yann Le Gorrec AS2M Department, FEMTO-ST Institute, 32 avenue de l'Observatoire, 25004 Besançon, France, legorrec@femto-st.fr

Xinyu Liu Department of Mechanical and Industrial Engineering, University of Toronto, 5 King's College Road, Toronto, ON, Canada, M5S 3G8

Institute of Biomaterials and Biomedical Engineering, University of Toronto, Toronto, ON, Canada

Department of Chemistry and Chemical Biology, Harvard University, 12 Oxford Street, Cambridge, MA 02138, USA, xliu@gmwgroup.harvard.edu

Philippe Lutz AS2M Department, FEMTO-ST Institute, 24 rue Alain Savary, 25000 Besançon, France, plutz@femto-st.fr

Simon D. Muntwyler Institute of Robotics and Intelligent Systems, Tannenstrasse 3, 8092 Zurich, Switzerland, msimon@ethz.ch

Bradley J. Nelson Institute of Robotics and Intelligent Systems, ETH, Zurich, Switzerland, bnelson@ethz.ch

Micky Rakotondrabe FEMTO-ST Institute, UMR CNRS, 6174 - UFC, ENSMM/UTBM, rue Alain Savary 24, 25000 Besançon, France, mrakoton@femto-st.fr

Stephane Regnier Institut des Systémes Intelligents et de Robotique, Université Pierre et Marie Curie, Place Jussieu, 75252 Paris Cedex, France, stephane.regnier@upmc.fr

Gustavo A. Roman Department of Mechanical and Aerospace Engineering, University of Florida, Gainesville, FL 32611-6250, USA, jabronie@ufl.edu

Yu Sun Department of Mechanical and Industrial Engineering, University of Toronto, 5 King's College Road, Toronto, ON, Canada, M5S 3G8

Institute of Biomaterials and Biomedical Engineering, University of Toronto, Toronto, ON, Canada, sun@mie.utoronto.ca

Alina Voda Control Systems Department, GIPSA-lab, Grenoble University, Ense[3] BP 46, 38402 Saint-Martin d'Hères, France, alina.voda@grenoble-inp.fr

Gloria J. Wiens Department of Mechanical and Aerospace Engineering, University of Florida, Gainesville, FL 32611-6250, USA, gwiens@ufl.edu

Chapter 1
Microscale Specificities

Cédric Clévy and Micky Rakotondrabe

Abstract Within these last decades, we assist the spectacular emergence of the micro and nanotechnology. These technologies are tools derived from several complementary fields like micro/nanorobotics, control, biology, mechanics, physics and chemistry. Considered scale is between some tens of nanometers to some hundreds of micrometers. At this scale, manipulating objects, measuring signals or controlling systems constitute a great challenge because of the non-classical specificities that exist. In particular, since several years, researchers and engineers attempt to develop convenient measurement techniques or sensors (technology) for the micro/nanoscale. The spirit was that these techniques or technology could provide essential information and signals during the positioning, manipulation or assembly (position and force signals) of micro/nano-objects or on their characteristics (stiffness, etc.) with the required resolution, accuracy, range and bandwidth. These last years, additionally to the performances, several projects also highlight the packageability aspects.

This chapter reminds the specificities of the micro/nanoworld. After introducing the chapter, dimensions, ranges and order of magnitudes that typify this world will be given. Thanks to the literature survey, it will be demonstrated that the signals measurement at the microworld is a new and open field due to the lack of sensors presenting at the same time the embeddability, the required precision and large range, and low cost. Finally, we present the possible contribution of observer and estimator theory to the measurement at the micro/nano-world in order to complete the measurement provided by existing sensors.

C. Clévy (✉)
Université de Franche-Comté, FEMTO-ST Institute, Département AS2M – Automatique et Systèmes Micro-Mécatroniques, UMR CNRS 6174 – ENSMM / UFC / UTBM, 24, rue Alain Savary, 25000 Besançon, France
e-mail: cclevy@femto-st.fr

C. Clévy et al. (eds.), *Signal Measurement and Estimation Techniques for Micro and Nanotechnology*, DOI 10.1007/978-1-4419-9946-7_1,

Keywords Micro/nano-world • Specificities • Micrometric dimensions • Signals • Sensors

1.1 Introduction

Micro and nanotechnology, i.e. technology related to micro/nano-domain, is today a growing field that presents a great interest at various levels. It covers various applications such as microfabrication, micro/nano-assembly of NEMS /MEMS/MOEMS, micro/nano-manipulation and characterization of biological or artificial objects, micro/nano-positioning, surface scanning, medical treatment, sensing, material and surface characterization, etc. [1]. Micro/nanotechnology is therefore considered as feature for several domains: military, automotive, medical, aerospace, micro/nanorobotics, telecommunication, computer, etc.

First, in the biology field, a lot of studies are done to characterize or manipulate biological cells like oocytes. Microtechnology has been used for that. It was established that the success of these studies, in particular IVF (In vitro fertilization), are closely dependent on the use of:

- Suitable handling (micro)tools with appropriate design
- Accurate enough motion generators (microrobots or micromanipulators) and adequate motion strategies
- (micro)Force sensors able to measure interactions between biological cells and tools (handling, injection)

Among these requirements, one of the main challenges is to have suitable sensors providing the suitable range, resolution, size and bandwidth. For such applications, the studied cells are placed in liquid environment. When only the tip part of the force sensor is placed inside the liquid environment, the main difficulty relies in splitting the interaction forces (between the tip part of the force sensor and biological cell) and forces induced by the meniscus (liquid/air interface). Reversely, when the whole sensor is placed inside the liquid environment, size, bio and electrical compatibility of the sensor integration must be strictly ensured.

Plenty of products result from miniaturization such as with our cellphones, flat-faced screen televisions, GPS (Global Positioning System), medical devices and several other examples that also integrate several complex functions in reduced volumes. Assembling such devices enables the integration of several components that are made of different materials and that have different functionalities (sensor, actuator, telecommunication, mechanics...) to fabricate one complex product. It opens to new products because it improves the integration possibilities or can be cost effective solutions for them [2–4]. At the macro scale (let's say human size), complex assembly tasks are often manually done for costs reasons. At this scale, only simplest and most repetitive tasks are done automatically using robots or manipulation tools. At the microscale, for complex but also for simple tasks, human dexterity is limited for the challenge and is source of success and quality loss. At this scale, robotization has to be generalized for both simple and complex tasks.

Among them, it is required to take into account surface forces that predominate volume forces (contrary to the macro size)[5]. To ensure good micromanipulation dexterity, force and position sensors are required to obtain reliable information about the manipulated micro-object behavior. The sensors have so to be placed the closer possible to the object, as a consequence, they must be as small as possible. In addition, low level control has to be placed the closer possible to the sensitive part of the sensor in order to obtain good quality signal and reduce the influence of external perturbations (mechanical noise, thermal drift, electrostatic influence...).

These two examples (study of biological cells and microsystems) illustrate the needs of sensors for micro and nano technology fields. They show that sensors have to:

- Be integrated in the experimental set-up
- Be as small as possible
- Furnish the necessary performances (sub-micrometric accuracy, range, sub-nanometric resolution, high bandwidth, etc.)
- Take into account of environmental variations

Sensors that meet all of these requirements are still under development in research laboratories and in some companies. Their development addresses great challenges such as compact design, high performances, microfabrication and packaging. This chapter reminds the main characteristics of the micro/nano-scale that makes very severe the performances that sensors should provide, and that makes very hard the development of appropriate ones.

The chapter is organized as follows. In Sect. 1.2 we give some dimensions, ranges and order of magnitude that typify the micro/nano-scale. Afterwards, we survey in Sect. 6.3 the different physical characteristics that make different this scale relative to the macro-scale. Section 1.4 is dedicated to remind some existing sensors and measurement techniques that are actually used. Finally, Sect. 1.5 introduces the plan of this book.

1.2 Dimensions, Ranges, and Order of Magnitudes at the Micro/Nano-Scale

1.2.1 Definitions

Before giving the dimensions, ranges and order of magnitudes typifying the micro-/nano-scale, let us first give definitions of that are related to this scale.

There are several papers that define microsystems. Despite that, all definitions highlight the notion of miniaturized systems, the difference rely on the limit of the sizes to be considered as microsystems and on the presence of functionality or not in them. These definitions varies according if we are in Europe, in Asia or in America (US and Canada), and if they are given by companies or research institutes. The definition given here tries to match with several of them. It is noticed that the term

Microsystems is sometimes with capital in the literature as well as in dictionary. However, along this book, according to the context and to the authors, expressions Microsystems and microsystems are alternately used. Finally, the definition reported here is given by trying to match with those given in the literature.

Definition 1.2.1. **Microsystems** are systems whose the largest dimension is smaller than some centimeters (intuitively less than 2 cm) and that possess at least two of the following functions: sensing, signal transmission or treatment, actuation, power supply or signal transmission. However, beyond the reduced dimensions, one can actually classify as Microsystems when the miniaturized systems perform micro or sub-micrometric resolution and/or accuracy in their functionality.

Sometimes, microsystems are confused with MEMS. In fact, MEMS (Micro Electro Mechanical Systems) are particular case of them.

Definition 1.2.2. **MEMS** are defined as microsystems having dimensions strictly below $1\times1\times1$ mm.

Remark 1.2.1. From Definition 1.2.1 and Definition 1.2.2, it can be deduced that microsystems and MEMS are functional systems. The name of MEMS that includes electric and mechanics confirms itself the presence of functionalities in these systems. Thus, any miniaturized structures and objects, in particular with dimensions less than $1\times1\times1$ mm, without functionalities are given the name of **microstructures** and **micro-objects** in the literature.

Remark 1.2.2. Instead of using MEMS, the literature also utilizes the expression **MOEMS** (Micro Opto Electro Mechanical Systems) when the considered MEMS includes an optical functionality, such as for orienting beam or for spectrometry. When the dimensions of the functional systems go under $1 \times 1 \times 1\,\mu$m, they are given the name of **NEMS** (nano electro mechanical systems).

Remark 1.2.3. If any robots and sensors have characteristics that match with Definition 1.2.1, they are called **microrobots** and **microsensors**. The term microsensors is however less used.

Definition 1.2.3. **Microcomponents** is the expression used for micro-objects, microstructures, MEMS or relatively small microsystems that have to be assembled in order to derive a miniaturized product.

Definition 1.2.4. **Micromanipulation** is the task consisting to manipulate micrometric objects (micro-objects). Objectives of micromanipulation lies on the precise positioning or characterization of these objects. The same definition holds for **nanomanipulation** except the considered objects which in this case are nanometric.

Definition 1.2.5. **Microassembly** is a set of tasks consisting to assemble micro components (which themselves are micro-objects, microstructures, MEMS, or relatively small microsystems) in order to derive a more complex microstructure, MEMS/MOEMS or Microsystems. The same definition holds for **nanoassembly** except the considered components which in this case ranges between the nanometer and some micrometers.

To perform micro/nanoassembly, micro/nanomanipulation, characterization or tasks in general at the micro/nano-scale, different tools could be involved: microactuators, sensors, microrobots, etc. In order to be convenient with the micro/nano-scale, these tools should provide some performances that are required at this scale. The following definitions (see for e.g. [13]) provide some common terminology often used to characterize sensors and actuators.

Definition 1.2.6. The resolution of a sensor is the smallest change it can detect in the quantity that it is measuring.

Definition 1.2.7. The resolution of an actuator is the smallest repeatable motion that can be made.

Definition 1.2.8. The accuracy of a sensor is the maximum difference that will exist between the actual value (which must be measured by a primary or good secondary standard) and the indicated value at the output of the sensor. Again, the accuracy can be expressed either as a percentage of full scale or in absolute terms.

Definition 1.2.9. The range of a sensor is the maximum and minimum values of applied parameter that can be measured.

Definition 1.2.10. The range of an actuator is the maximum and minimum values of parameters that it can performed.

Definition 1.2.11. The bandwidth of a sensor or an actuator corresponds to the measure of signals frequency that can pass without significant phase-lag and amplitude attenuation.

1.2.2 Characteristic Values at the Micro/Nano-Scale

1.2.2.1 Distance and Dimensions

The volume of microsystems is less than some cubic centimeters according to Definition 1.2.1. Dimensions of microcomponents that constitute them are identified between the micron and some millimeters. This range corresponds to the range of micromanipulation and microassembly completed with the range of meso. The term **meso** designates the interface between the macro and the micro-scale [14]. The accuracy required for the positioning during tasks at the micro/nano-scale should often be better than some micrometers. Furthermore, the range of positioning required during these tasks vary between 10 μm (one-tenth of the minimal length of a micro-object) up a tens of centimeter (slightly superior to the maximal dimension of a microsystem) [15]. Considering the nanomanipulation, nanoassembly and tasks that cover the nano-scale, the same calculation is used, one has just to replace *micro* by *nano*. Figure 1.1 resumes the dimensions and distances that typify the micro/nano-scale.

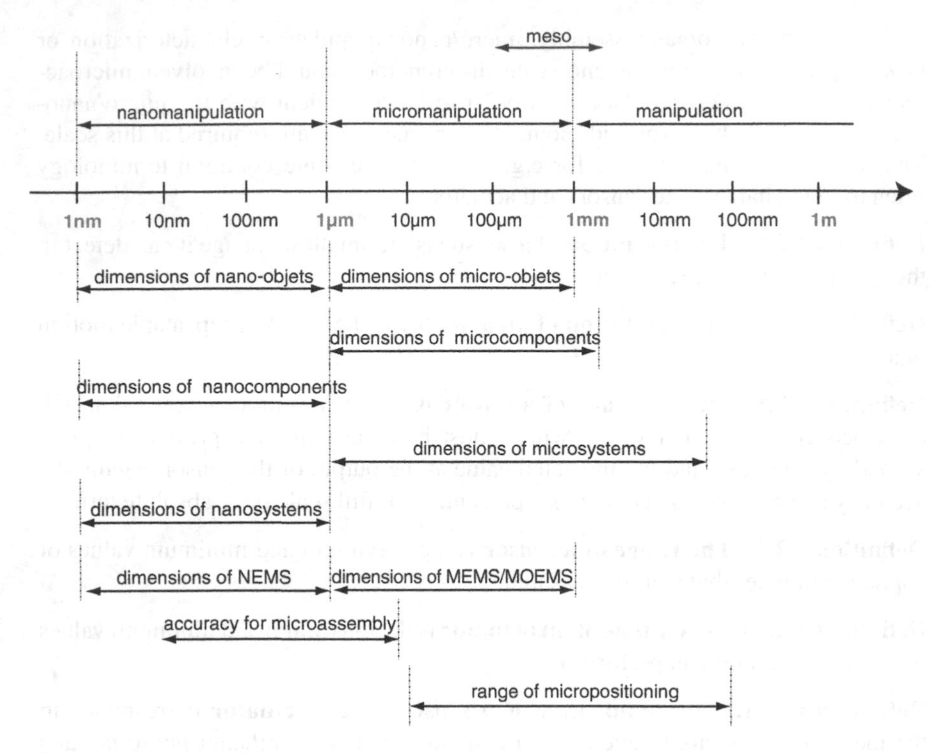

Fig. 1.1 Some dimensions and distances that typify the micro/nano-scale

1.2.2.2 Micro and Sub-Micro Forces

The well known forces that characterize the micro/nano-scale are the adhesion forces. These forces are called surface forces because they arise from the interaction of two or several surfaces close enough, contrary to volumic forces (weight, etc.) that are the interaction between volumes. Adhesion forces exist for any objects interaction but they become apparent when the volumes are relatively weak. This is the case for objects and structures at the micro/nano-scale. There are three adhesion forces classically cited in the literature: the van der Waals forces, the capillary force and the electrostatic force. To have an idea of magnitudes, we consider a silicone material based micro-sphere of 10 µm in diameter. Among the adhesion forces, the capillary force is the most dominant when the weight of the micro-sphere becomes less than 10 mN [5]. If the diameter of the micro-sphere is 1 µm (the minimal size of a micro-object), the electrostatic force is the most weak. Its intensity is nearly 100 fN. Generally, adhesion forces range between some femtoNewton to some hundreds of nanoNewton [16].

To compute the weights involved at the micro/nano-scale, we still consider the silicone material, the density of silicone being 2,300 kg/m^3. The weight of micro-objects calculated with a cubic structured silicone having a side between 1 μm and 1 mm varies between 20 fN and 20 μN. For microcomponents, the weight calculated with a cubic with a side between 1 μm and 8 mm ranges between 20 fN and 12 mN. The weight of microsystems calculated with a cubic of minimal size equals to 10 μm varies between 20 pN and can reach 50 mN. Regarding nano-objects, nanocomponents and nanosystems, their weights are below that of their counterpart at the micro.

While there are no identified values of manipulation force in nanomanipulation, the micromanipulation and microassembly contexts arise maniplation forces ranging generally between 10 and 350 mN with a resolution between 10 and 400 μN [17].

Figure 1.2 resumes the typical values of forces at the micro/nano-scale.

1.3 Physical Characteristics and Main Issues at the Micro/Nano-Scale

Scaling down to micro and nano scale brings a lot of specificities that have to be taken into account when designing, fabricating, or controlling micro and nano-devices. These specificities are also valuable when designing sensors or measurement techniques for the micro/nano-scale. Among all, the most important specificities are the surface forces (adhesion forces), the low signal to noise ratio at this scale, the small available space and the high influence of the environment to small systems in general. These items are detailed in this section.

1.3.1 Surface Forces and Their Influences

Nature provides very interesting examples of micro-nano scale specificities. For instance, everybody already wondered about how insects can walk on Walls, ceilings or on water surface or how ants can bring so heavy parts in regard with their own weight. Such an example illustrates the fact that surface forces (van der Waals, capillary, electrostatic) are dominant relative to volume forces (weight for example) when the dimensions are small, contrary to human sizes devices. According to the context, surface forces may be very strong. For example, they enable Gecko lizards walking along Waals despite their important weight (can reach 100 g) to support. Recently, powerful Scanning Electron Microscope (SEM) has been used to enable the observation and explanation of this phenomena [6, 7]. In fact, Gecko feet are covered by billion hairs having at their tip spatulas. These spatulas are very small (100–200 nm of diameter) and the hairs are extremely flexible making easy to adapt to the contact between the feet and the support. The aim is to have a very close and large surface contact for any roughness of the support surface. Hence, this closeness

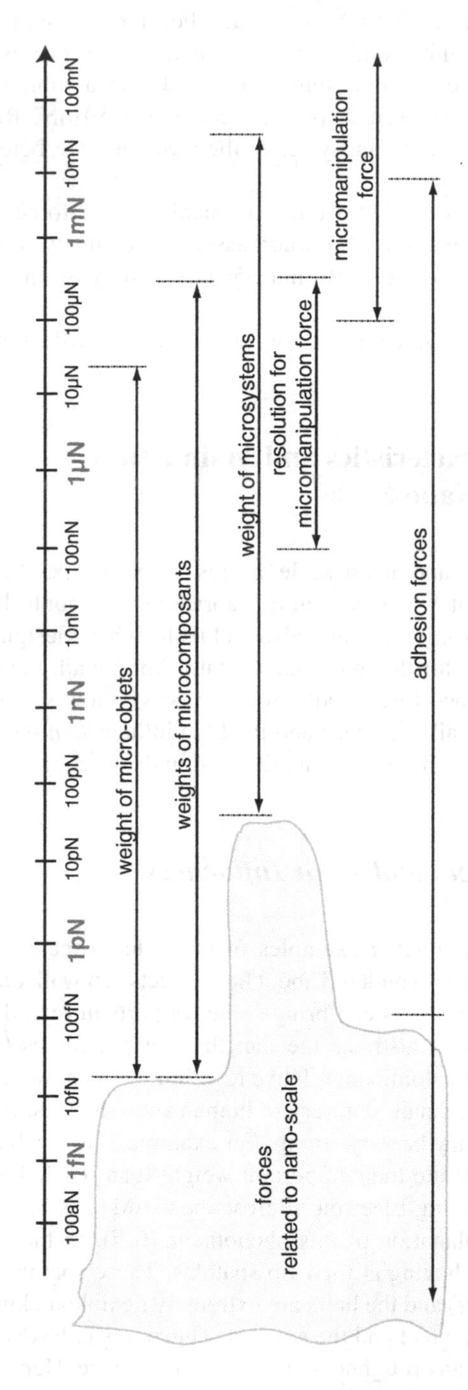

Fig. 1.2 Some typical values of force at the micro/nano-scale

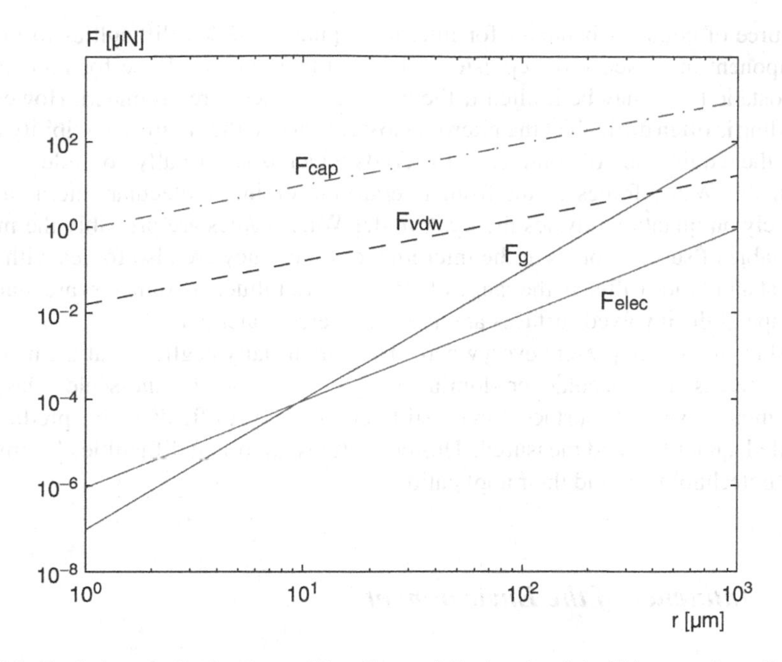

Fig. 1.3 Evolution of adhesion forces (F_{cap}, F_{elec} and F_{vdw}) and weight (F_g) versus the radius of a silicone ball

enables to take advantage from van der Waals forces (interaction forces that strongly depends on the distance between two surfaces). Scientists are trying to develop bio-inspired synthesized surfaces based on this principle [8].

As already cited, surface forces are mainly composed of the capillary force, the electrostatic force and van der Waals forces [9]. Figure 1.3 resumes the evolution of these forces when the radius of a silicone micro-sphere is below 1 mm. It clearly shows that the weight becomes insignificant relative to adhesion forces when the object is too small.

The capillary force is induced by the presence of liquid (often water due to hygrometry) between two surfaces. Influent parameters are the quantity of liquid at the interface, materials, shapes and relative orientation of both surfaces. The amplitude of capillary force may vary a lot from an experimental setup to another one due to the strong influence of hygrometry. Indeed, a dry (small hygrometry rate) environment makes capillary force nearly negligible. Conversely, when present, the capillary force is often the strongest surface forces and may reach several milli-Newtons.

The electrostatic force is mainly due to Coulomb interactions between objects with the presence of charged particles or when charges appear due to triboelectrification. Influent parameters are materials (notably their conductivity) of both surfaces in regards, their shape and the distance between them. The electrostatic force thus acts at distance and may be source of strong attraction or repulsion. Both cases are source of complex behavior for micromanipulation tasks: difficulties to grasp a component in presence of repulsion forces, difficulties to release for repulsion. Electrostatic force may be limited if the involved surfaces are grounded. However, grounding is often difficult at the micro/nano-scale due to the small accessibility and due to the requirement of conductive materials which is not usually possible.

Van der Waals forces result from interatomic or intermolecular interactions. They rely on quantum physics theory. Van der Waals forces are probably the most predictable of surface forces at the micro/nano-scale. They are also forces with the smallest amplitude (often in the range of 10–100 nN). Influent parameters are mainly the shape of the involved surfaces and the considered materials.

Surface forces are present everywhere. They are usually negligible at the macro-scale whereas they become predominant at the micro and nano-scale. Despite the numerous studies, surface forces still remain very difficult to be predicted, modelled, quantified and measured. This constitutes important difficulties for micro and nanotechnologies and their applications.

1.3.2 Influence of the Environment

Surface forces but also the behavior of micro and nano scale devices (or devices acting at the micro or nanoscales) are greatly influenced by environmental conditions like temperature, hygrometry, charges in presence, light sources, etc. For example, temperature variations within 10°C can be source of several tens of microns of deflexions in piezocantilevers used as actuators in microgrippers or mobile stage devices [11]. Such unwanted deflections result from a combination of thermal expansion and variation of piezoelectric constants in the piezoelectric actuators. All materials used for experimental set-up also suffer from dilatation and their combination is source of uncertainties in the control of the tasks to be performed. For example a 10-cm aluminum bar constituting an experimental set-up used to study micro or nanoscale devices is source of several tens of micrometers deviation. Hygrometry also strongly influences the behavior of passive and active materials used to compose micro and nano-scale devices or devices acting at this scale. Finally, dusts are often neglected in several studies and literature, however they must also be taken into account. Indeed, their dimensions are quite similar to those of the manipulated micro-objects and therefore they can be confused or interfer with the latter. To partially solve all these problems, it is required to perform measurements to quantify the effects of environmental effects on the studied devices and then to control them. Nevertheless, controlled environments are very expensive and bring constraints like free space or vibration isolation making them sometimes difficult to use. Several control techniques can also be applied, among them robust and optimal control techniques enables to consider parameters variations as disturbance to be rejected [10, 11].

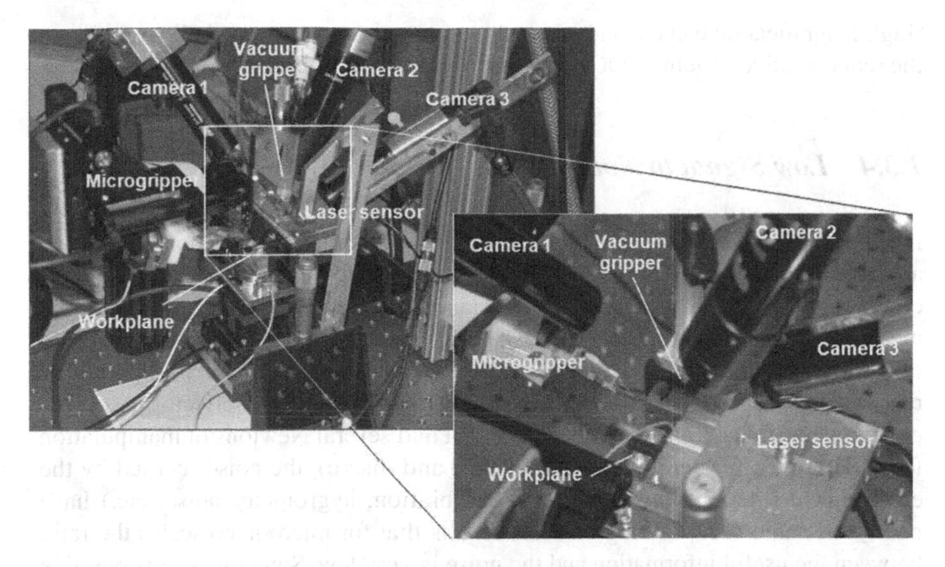

Fig. 1.4 Example of microassembly station developed at FEMTO-ST Institute (Besançon France) including three high magnification cameras, one piezoelectric microgripper mounted on a 4-DOF microgripper, one vacuum gripper mounted on a 3-DOF micropositionner, one triangulation laser sensor and one workplane mounted on a 5-DOF microrobot

1.3.3 Difficulties to Direct Sense

Micro and nanotechnology implementation often requires several measurements with suitable range, resolution and bandwidth and in different directions (i.e. with multiple DOF: degrees of freedom). This can be done either directly (the required information is directly quantified by a sensitive component) or indirectly (the required information is deduced from an intermediate measure signal). However, direct sensing is preferred relative to indirect sensing at the micro and nanoscales because of the following advantages that it can offer:

- Direct sensing may enable to take more precisely into account surface and contact forces
- It facilitates to prevent from or to limit the influence of environmental conditions
- The signal to noise ratio can be drastically improved

Despite all of these positive aspects, direct sensing remains extremely challenging at the micro/nano-scale because it requires very small sensitive parts and integrated sensors that have the required performances (accuracy, bandwidth, resolution), and eventually in multi-DOF measurement. Such sensors are missing today and several research and projects (at the international level) are focusing on such specific field in order to further allow the well characterization, packaging devices and control at the micro/nano-scale. In Fig. 1.4, a microassembly station developed at FEMTO-ST is illustrated. The figure shows that the used measurement systems (high magnification camera and triangulation laser sensor) is very bulky relative to the real operative system (4-DOF microgripper, 8-DOF microrobot and workspace).

1.3.4 Low Signal to Noise Ratio

Displacement and force signals are often the essential information in several applications at the micro/nano-scale. These signals are very weak relative to their counterparts at the macro-scale. For instance, the range of displacement to be measured during a micropositioning task performed by a microgripper is in the order of 40 μm. The micromanipulation force performed by the same microgripper may reach 30 mN. At the macro-scale, a classic gripper used as end-effector in a robot performs several centimeters of displacement and several Newtons of manipulation force. However, for both scales (micro/nano and macro), the noises caused by the environment (thermal noise, mechanical vibration, hygrometry noise, etc.) have nearly the same order of magnitude. It results that for micro/nano scales the ratio between the useful information and the noise is very low. Sometimes, it is possible to bypass the noise issues by using sensors with high performances filtering systems embedded. For instance, displacement sensors, often based on optical principle or interferometry, offer a high reduction of the noise by maintaining a relatively high bandwidth and accuracy. However, these sensors are very expensive and very bulky. On the other hand, displacement sensors that are packageable and much more cheap such as strain gages are very sensitive to noises. Figure 1.5 gives the comparison of the displacement of a piezoelectric microgripper measured by optical sensor and a strain gauge sensor. This figure shows that the noise generated by the optical sensor (± 0.1 μm peak-to-peak) is nearly ten times reduced than that generated by the strain gauge (± 2 μm peak-to-peak).

1.4 Some Sensors and Measurement Techniques at the Micro/Nano-Scale

Obtaining suitable and reliable experimental data requires the use of sensors. For several applications at the micro/nano-scale, measuring forces between two objects and measuring displacements (relative or absolute) signals are the most widespread. In many techniques, the measures of forces are indirectly derived by using the measures of displacements. The main reason is that there is a real lack of force sensors with the appropriate performances and dimensions for the micro/nano-scale. Despite the recent development of capacitive based force sensors[1], covering the whole range of force at this scale is not yet possible nowadays. Among the forces measurement techniques, those without contact between the sensor and the studied

[1] http://www.femtotools.com/

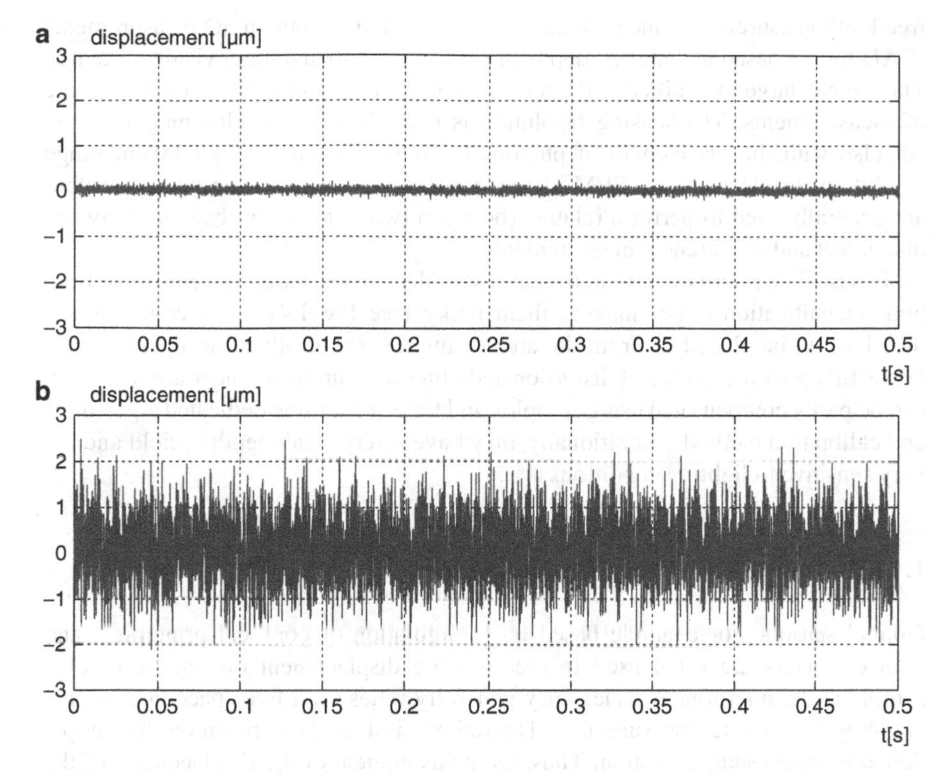

Fig. 1.5 Noises obtained with two displacemlent sensors: (**a**) with an optical sensor, (**b**) with a strain gauge sensor

sample or device are more preferred than those with contact. Such principles enable to bypass (or at least limit) some inconveniences like sensors integration difficulty or like the influence of the sensor on the studied device behavior. Conversely, other applications require the establishment of a contact between the sensor and the devices. For instance, in some micromanipulation and microassembly tasks, the force sensor is used as gripping tool at the same time.

Existing sensors can be classified as exteroceptive and proprioceptive (embedded sensors) sensors.

1.4.1 Exteroceptive Sensors

1.4.1.1 Vision-Based Systems

Vision-based systems enable position, speed or force (through deformation) measurements [5, 12]. They can be, for instance, high resolution cameras, high magnification optical microscopes and SEM. Each of them enables planar (i.e. 2 degrees of freedom) measurements and combinations can be done to obtain 3D measurements (CAD model-based techniques, depth and shape from focus, stereovision systems). They have large magnification ranges making them applicable to large range of measurements. The sensing resolution is often linked with this magnification but also with the wavelength of photons for optical microscopes (100 nm range resolution) or electrons for SEM (10 nm resolution). Image processing algorithm are generally used to perform relative (between two objects) or absolute (between one object and a reference) measurements.

To visualize micro or sub-micro scales samples, vision-based systems must have high magnification optics making them bulky (see Fig. 1.4). As a consequence, they have to be placed far from the area of interest (generally more than 10 mm). Powerful optics are source of distortion and other measurement uncertainties which can be partly compensated using complex and time consuming dedicated algorithms and calibration methods. Additionally, they have a very small depth of field and are very sensitive to lighting conditions.

1.4.1.2 Optical Sensors and Interferometers

Optical sensors (for example based on triangulation or confocal principles) and interferometers are often used to measure the displacement on one point of a sample at the micro/nano-scale. They generally relies on a free space laser signal enabling non contact measurement. The reflexion of the laser beam on the sample depends on the sample motion. Thus, the measurement of the displacement of the reflected signal enables to deduce the static and the dynamic point displacement. Both devices offers 10 nm resolution measurements, but, in this case, optical sensor measuring range is limited to about 500 μm whereas interferometers offer more than 10 mm with a resolution up to 1 or 2 nm. Both device have bulky sensing heads for micro/nano-scale characterization (about $50 \times 50 \times 20$ mm sizes). So far, optical sensors and interferometers offer the best performances (resolution, accuracy, range, bandwidth) among the sensors at this scale. However, their main inconveniences are the large sizes and the costs which are very expensive. Figure 1.6, which presents an optical sensor used to measure the linear motion of the TRING-module microrobot developed in [18], illustrates the bulky sizes of the sensor face to the microrobot itself.

1.4.1.3 Eddy Current Sensors

Eddy current sensors comprise a probe that emits an alternating magnetic field. When the sensor probe approaches a surface, small currents are induced inside it. They are source of a repulsive magnetic field. The interactions between these two magnetic fields mainly depends on the distance between the sensor head and the surface. Thus, this distance is deduced by measuring the evolution of the magnetic fields interaction. The range of measurement directly depends on the size

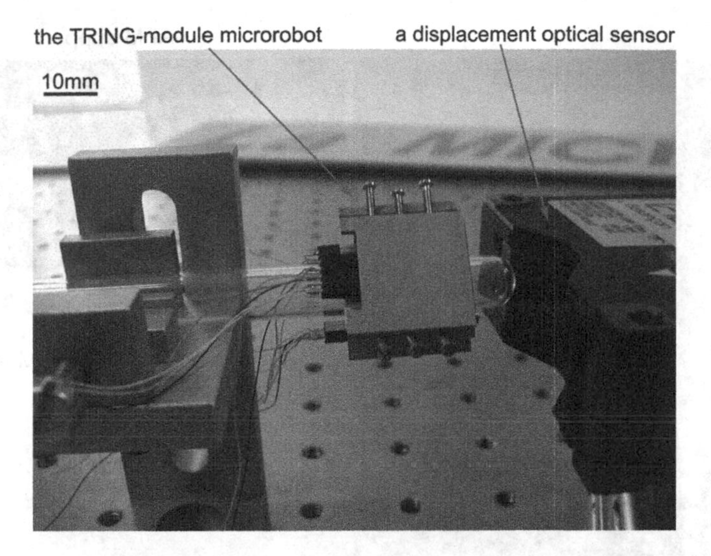

Fig. 1.6 An optical sensor used to measure the linear motion of the TRING-module microrobot developed in [18]

of the sensor probe. Smallest commercially available sensor probes having 3 mm in diameter can measure about 0.5 mm of displacement range with 30 nm in resolution and 1 μm in linearity. To obtain better quality of measurements, the surface to be measured has to be three times bigger than the sensing probe.

1.4.1.4 Capacitive Sensors

Capacitive sensors principle relies on the capacitance changes that happen between two electrodes (sensor probe and target surface). Once the supplied electric field and surface electrodes fixed, only the distance between both electrodes influence the capacitance changes. Smaller commercially available sensor probes (about 5 mm in diameter) induce smaller measurement ranges (lower than 500 μm). Resolution versus the range of measurement is linear and often corresponds to some thousandths of percents making capacitive sensor resolution extremely good. Hence, the obtained resolution can reach 0.03 nm with 0.1 μm linearity and 50 μm of range. To obtain a good quality of measurements, the target surface has to be 1.5 times bigger than the sensing probe.

1.4.1.5 AFM Microscopy Based Force Measurement

AFM (Atomic Force Microscopy) has been used not only to characterize surfaces and their roughness but also to measure and characterize adhesion forces [16]. The principle, presented in Fig. 1.7, is as follows. A piezotube actuator moves a silicone

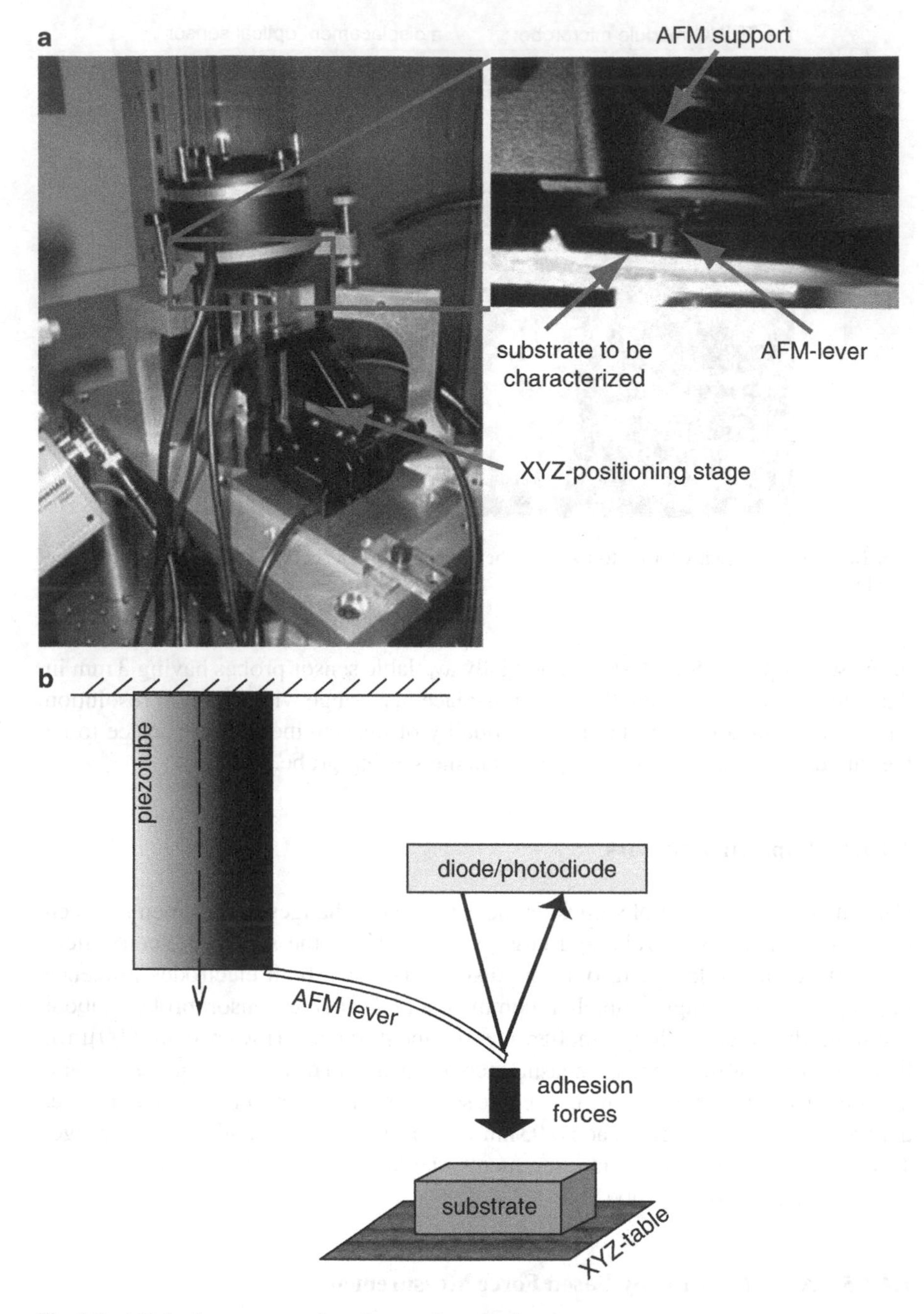

Fig. 1.7 AFM microscopy used to characterize adhesion forces

cantilever (AFM lever) backwards or towards the sample to be studied. According to the relative distance between the AFM-lever and the sample or according if there is a contact or not, interaction forces with varying amplitude appears between them. The stiffness of the AFM-lever being known, it is possible to characterize these interaction forces, mostly composed of adhesion forces, by measuring its deflection thanks to a diode/photodiode measurement system. The main advantage of the use of AFM-microscopy is the high resolution and accuracy that it can provide if the piezotube actuator is conveniently controlled [19].

1.4.2 Proprioceptive Sensors

This section introduces several measurement principles and sensors that can be embedded onto microscale devices. These sensors are particularly embeddable thanks to their physical principles favorable to miniaturization. They can be either structured independently and afterward be integrated in the working devices or directly embedded during the design and development of the devices. In the latter case, most of these embedded sensors are structured through microfabrication techniques.

1.4.2.1 Strain Gage

Strain gages are based on the elastic strain of electrically conductive materials. Strain generates a modification of the electrical resistance of the stretched material. A whetstone bridge often enables to measure this resistance change and then to deduce the strain evolution. To maximize its sensitivity, the sensitive part of the strain gage is often constituted of long and thin conductive material. For size reduction reasons, this material is stripped along a zig-zag pattern. Strain gage can be fabricated on thin films that have to be glued on the surface of the element whose deformation has to be measured. They can also be directly fabricated on the surface thanks to microfabrication techniques. In particular, materials with very good gage factors (sensibility factor) can be processed with microfabrication techniques. While strain gage offer a good embeddability, their main disadvantages are their fragility and in some gage sensors the provided noisy signals. Figure 1.8 pictures a piezoactuator with two kinds of displacement sensors: an embedded strain gauge sensor (proprioceptive) and an external optical sensor (exteroceptive). As we can see, while the optical sensor is very large relative to the sizes of the device (piezoactuator and support), the strain gauge offers a good packageability desirable for batch fabrication and industrialization.

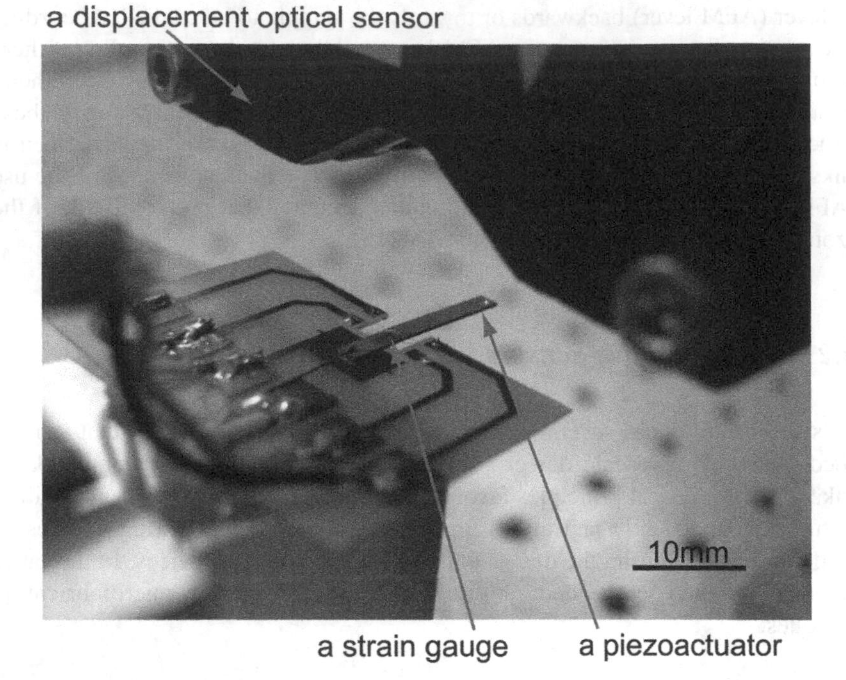

Fig. 1.8 A piezoactuator with strain gauge sensor (proprioceptive) and an optical sensor (exteroceptive)

1.4.2.2 Capacitive Sensors

Capacitive sensors principle can also be miniaturized using microfabrication techniques. Contrary to the macro-scale where electrodes are usually two planar surfaces, a micro-scale capacitive sensor often consists in interdigited comb-like electrode arrays. This principle enables to obtain a good resolution because the sensitivity of the sensor directly depends on the electrodes surfaces (which results from the sum of elementary comb surfaces). Section 1.4 is particularly dedicated to the capacitive sensors for micro/nano-scale.

1.4.2.3 Piezoelectric Sensors

Piezoelectric sensors make use of the direct effect of piezoelectric materials (generation of electrical charges from an applied strain). Quantifying these charges permits to deduce the applied force or displacement. Piezoelectric sensors enable very high measurement resolution. These sensors provide a high bandwidth [20]. However, they are not adapted to static measurement because of the drift (creep) characteristics [21]. Piezoelectric sensors can be based on classical piezoelectric materials such as PZT [22, 23] or PVDF (PolyVinylidine DiFluoride) [24, 25].

Table 1.1 Sensors used at the micro/nano-scale

Sensor types	Strength	Weakness
Optical sensors and interferometers	Resolution, accuracy, bandwidth	Expensive, bulky, one dof and pinpoint measurement
Vision based measurement	2 or more dof measurement possible, large range	Expensive, bulky and limited bandwidth
Eddy current sensors	Good sizes-cost-range-resolution compromise	One dof measurement, depends on the quality of the target surface
Strain gage	Reduced sizes, cheap, 2-dof possible	Very fragile, noisy, long calibration and preparation before use
Capacitive sensors (external)	Large bandwidth, accuracy	Non-linear, limited working distance, limited range
Capacitive sensors (embedded)	Large bandwidth, accuracy, embeddable	Non-linear, limited range
Piezoelectric sensors	Large bandwidth, embeddable, self-sensing possible	Difficult for static measurements,
Piezomagnetic sensors		Non-linear, non-embeddable, one dof measurement

There are several possibilities to obtain embedded piezoelectric sensors in micro-devices and microactuators. The first consists in using inserting a piezoelectric layer in the device. When the latter performs a deformation, the piezolayer also undergoes a deformation which finally results charges on its surfaces. The deformation and/or the force can afterward be estimated from these charges. The second concerns the self-sensing which consists in using the (piezoelectric) actuator as a sensor at the same time. More precisely, the electrodes used for the piezoactuators are also used to collect the charges during the deformation. A convenient electronic circuit as well as an observer should be therefore employed in order to estimate the deformation or force. Section 1.2 will be focused on self-sensing techniques and their applications at the micro/nano-scale.

1.4.3 Conclusion About Sensors

Table 1.1 resumes the above cited sensors added with other existing ones used at the micro/nano-scale.

The integration of sensors with suitable performances (high bandwidth, very high accuracy and convenient size, integration ability) remains a great challenge for the micro/nano-scale. In fact, it appears that the lack of such sensors is the main limitation to successfully perform the control of robots in the micro/nano world and to push back the limits of automation, as for example required in rapid and precise microassembly. The lack of embeddable high performances sensors also prevents industry from the batch fabrication and development of industrializable micro-products. These last years, the technological obstacles have led researchers to the design of a new generation of integrated sensors (Silicon/PZT, etc.), self-sensing methods in active materials and advanced signal estimation coming from control theory.

1.5 Control Theory as a Complement Tool to Measure at the Micro/Nano-Scale

1.5.1 Presentation and Modeling of the Illustrative Example

On the one hand, high accuracy sensors are expensive and bulky (optical, interferometer, etc.). On the other hand, small sensors are fragile and very sensitive to noises (strain gages, etc.). Furthermore, some applications require the measurement of both displacement and force during the tasks, as example pick-and-place tasks in micromanipulation and microassembly applications.

In order to gain space and to go to the packageability of microsystems, two approaches were proposed: (1) the use of small and embeddable sensors (for example strain gages) and the rejection of the noises using the Kalman filtering, (2) the use of reduced number of sensors and the application of observers to complete the measurement. Another advantage of using observers is that other information can be estimated from the measured signal (displacement). According to the level of missing information (for example force) and to the available one (for example, displacement), but also according to the final use (for feedback control, for characterization, etc.), there are different kinds of observer techniques that can be used. More explanations will be displayed through an example: we consider cantilever structured piezoactuators usually used in microgripper. As presented in Fig. 1.9a, two piezoactuators (piezocantilevers) constitute a microgripper. While one actuator is controlled on position, the second one is controlled on force. In order to improve the performances of the two actuators, feedback control must be used (see Fig. 1.9b, where δ_r is the position and F_r the force, subscript r meaning reference). This particularly allows to reject the effect of disturbances due to the environment (manipulated objects, thermal variation, etc.). Embedding optical sensors which offer the best performances in these actuators is not possible. This is why strain gage are used. Furthermore, strain gage furnish only displacement information and one should use an observer to estimate the force.

Consider one of the two piezocantilevers that constitute the microgripper. The two actuators being made by the same materials and have the same dimensions, their behaviors and models are similar and the analysis are identical for both. The model linking the input voltage U, the force F applied at the tip and the output deflection δ, in the linear case, is:

$$\delta = d_{\mathrm{p}} \cdot U \cdot D(s) + s_{\mathrm{p}} \cdot F \cdot D(s), \tag{1.1}$$

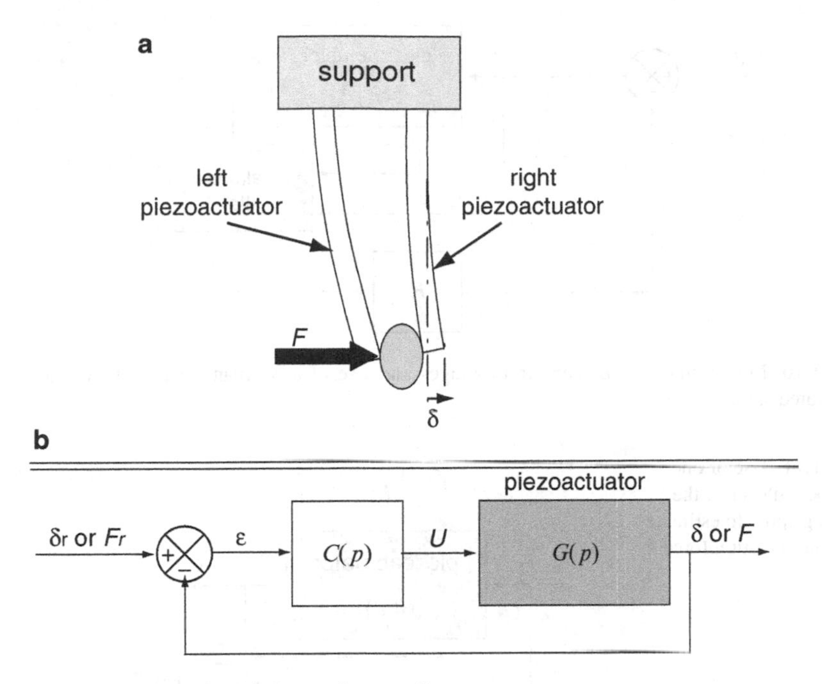

Fig. 1.9 A microgripper based on two piezocantilevers

where d_{p} and s_{p} are the piezoelectric and the elastic coefficients, respectively, and $D(s)$ (with $D(0) = 1$) is the dynamic part.

When the applied electrical field – through the voltage U – is high, the nonlinearities behavior of the piezoelectric materials becomes significant. These nonlinearities are the hysteresis and the creep and need to be taken into account when applying an observer. The nonlinear model of the actuator is therefore [26]:

$$\delta = H(U) \cdot D(s) + C_r(s) \cdot U + s_{\mathrm{p}} \cdot F \cdot D(s), \tag{1.2}$$

where $H(\cdot)$ is an operator that describes the (static) hysteresis and $C_r(s)$ is a linear approximation of the creep [27].

1.5.2 Strain Gages Sensors, Kalman Filtering and State Feedback Control

In [28], strain gage were used to measure the deflection δ of the piezocantilever with a view to reduce the sizes of the whole microsystems (actuators and sensors). To reduce the noises of the measured signals, the authors apply a Kalman filtering computed with the linear model (1.1). In addition to the noises rejection, the technique allows the estimation of the states of the system and therefore allows the use of state feedback control techniques (Fig. 1.10).

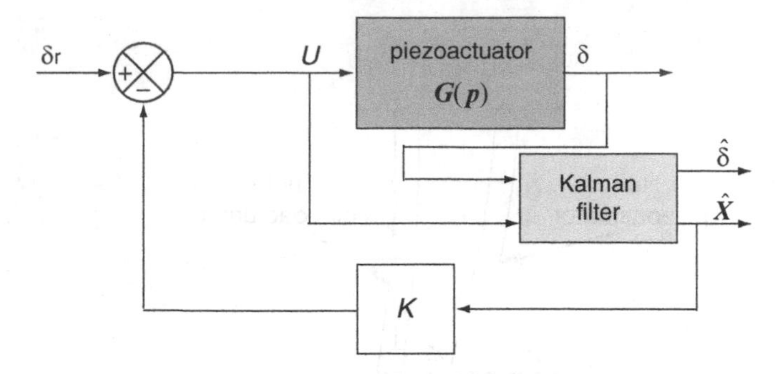

Fig. 1.10 Measurement of δ with strain gauges and use of a Kalman filter (state variables X, estimated signal ?)

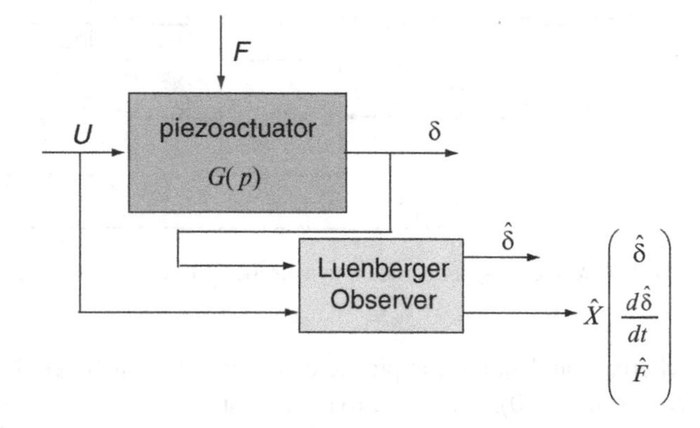

Fig. 1.11 Use of one piezocantilever of the microgripper to estimate the manipulation force F

1.5.3 Force Estimation Using the Luenberger Observer

In [29], the second piezocantilever of the microgripper is used only to estimate the manipulation force. For that, the force has been considered as a state of the system. Because a derivative is required in the state equation, the author considers the condition $\frac{dF}{dt} = 0$. As a result, the state vector is composed of the deflection δ, its derivative $\frac{d\delta}{dt}$ and the force F. Based on the model in (1.11), a Luenberger observer has been applied (Fig. 1.11).

1.5.4 Force Estimation and Unknown Input Observer Technique

When the force is considered as a state to be estimated, it requires that the dynamics model of the force is known. In the previous case, the derivative of F has been considered to be null and the estimation was only valuable for static case.

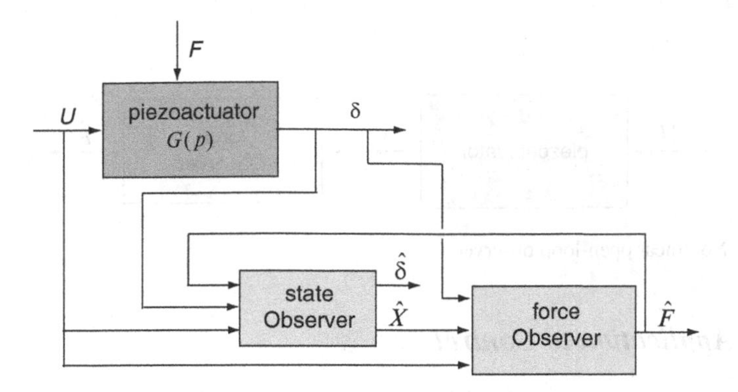

Fig. 1.12 Inverse-Dynamics-Based UIO technique to estimate the unknown input force

It has also been demonstrated that the dynamics of the force in piezocantilevers always depend on the characteristics of the manipulated object [30]. Therefore, considering the force as a state is not convenient if the estimate will be used in a control purpose. This is why the Unknown Input Observer technique (UIO) has been proposed recently [31]. The Inverse Dynamics Based UIO technique [32] was especially applied. In this, we consider the force as an unknown input. A classical observer is first employed to estimate the state vector (composed of the deflection and its derivative). Afterward, a second observer is applied to estimate the force (Fig. 1.12).

1.5.5 Force Estimation in the Nonlinear Case, Open Loop Observer

In [33], an other approach was proposed to estimate the force. It consists in using the nonlinear model in (1.2) and directly deducing the force:

$$\hat{F} = \frac{1}{s_{\mathrm{p}} \cdot D(s)} (\delta - H(U) \cdot D(s) - C_r(s) \cdot U), \tag{1.3}$$

where the hysteresis $H(\cdot)$ was modeled by the Bouc–Wen approach.

This method requires the bistability (the (direct) model and its inverse are both stable) of $D(s)$ as its inverse is used in the (open-loop) observer. If the system is linear, the method can also be applied. Indeed, the term $d_{\mathrm{p}}U$ of (1.1) is a linear approximation of the hysteresis term $H(\cdot)$ in (1.2), the creep $C_r(s)$ being set to zero.

As presented in Fig. 1.13, the observer has an open-loop structure, and therefore is sensitive to model uncertainties.

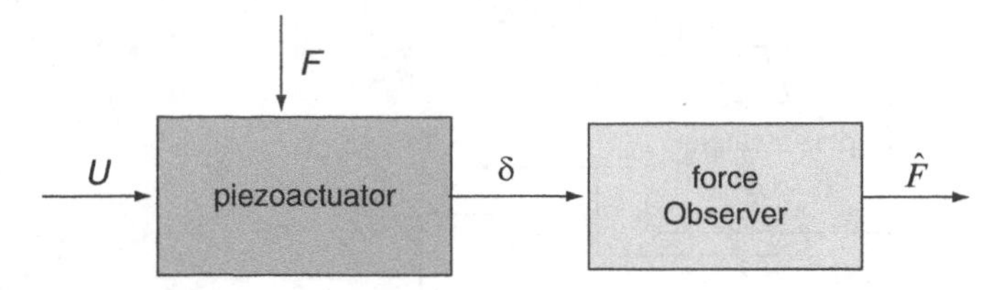

Fig. 1.13 Nonlinear open-loop observer

1.5.6 Application to Control

Controlling systems is essential to improve their performances and to satisfy certain requirements of the applications. Open-loop control techniques (feedforward) have proved their efficiency for piezoelectric cantilevers. Their main advantage is the embeddability of the whole systems because no sensor is used. However, they are compromised by the presence of external disturbances or eventual model uncertainties. Closed-loop techniques (feedback) are therefore recognized as long as the systems are used in conditions where disturbances are present. The lack of precise, large bandwidth, and embeddable sensors for micro/nano-scale invites researchers/engineers to use or to complement the available measures with other techniques of measurement. In this spirit, possible complementary techniques of measurement are observers. For instance, the Luenberger observer for the estimation of δ and $\frac{d\delta}{dt}$ of the piezoactuator was successfully used in a state feedback control law of the displacement [29]. References [30,34,35] used successfully the nonlinear open-loop observer to estimate the force and to apply H_∞ based controllers. These combinations of observers and feedback control allowed the piezoactuators to perform micrometric accuracy, to maintain several hundreds of Hz of bandwidth and to efficiently reject the effects of external disturbances such as thermal variation or unwanted environmental vibration.

1.5.7 Self-Sensing Based Measurement Combined with Observer

As shown above, accurate sensors are often bulky and expensive while integrable ones are fragile and not robust. Thus, instead of using the displacement measurement and afterward estimating the missing information (force, velocity, etc.), one can measure charges appearing on the surfaces of the piezocantilevers and derive all information and signals necessary for the feedback. Such technique, called self-sensing[2] is presented in Fig. 1.14, bypasses the use of any displacement sensors. However, it requires to utilize a well-designed electronic circuit followed by a

[2]Self sensing may also be applied to other kinds of active materials than piezoelectric ones.

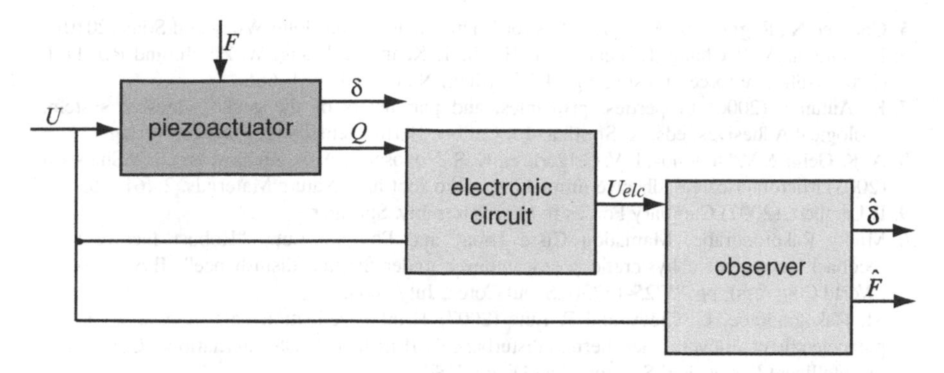

Fig. 1.14 Principle scheme of the self-sensing technique to estimate displacement and/or force

convenient observer. The electronic circuit will uses the available signals (charges Q and input voltage U) to derive the exploitable voltage U_{elc}. Then, from the new available information U_{elc} and U, an observer will give an estimate $\hat{\delta}$ and $\hat{F}$ of the displacement δ and of the force F respectively. This technique will be detailed in Sect. 1.2.

1.6 Conclusion

This chapter presented some issues and specificities at the micro/nano-scale. These specificities make very difficult the development of high performances and packageable sensors or measurement systems for this scale. Some existing sensors were afterward presented and compared. It follows that the sensors that can really offer the required resolution, accuracy and bandwidth are often bulky and expensive. In the last part of the chapter, we shown that observer techniques, from control theory, could be good complementary tools to sensors in order to measure signals at the micro/nano-scale with some miniaturization and packageability features.

References

1. J. Peirs, (2001) Design of micromechatronic systems: scalelaws, technologies, and medical applications. PhD thesis, K.U.Leuven Dept. of Mech. Eng., Leuven, Belgium.
2. H. Van Brussel, J. Peirs, D. Reynaerts, A. Delchambre, G. Reinhart, N. Roth,M. Weck, and E. Zussman, (2000) Assembly of microsystems. Annals of the CIRP, 49(2):451–472.
3. D. O. Popa, R. Murthy et A. Das, (2009) M3- deterministic, multiscale, multirobot platform for microsystems packaging : design and quasi-static precision evaluation. IEEE Transaction on Automation Science and Engineering (TASE), 6:345–361.
4. S. Bargiel, K. Rabenorosoa, C. Clévy, C. Gorecki and P. Lutz, (2010) Towards Micro-Assembly of Hybrid MOEMS Components on Reconfigurable Silicon Free-Space Micro-Optical Bench, Journal of Micromechanics and Microengineering, 20(4).

5. Chaillet, N., Regnier, S.: Microrobotics for Micromanipulation. John Wiley and Sons (2010)
6. K. Autumn, W.P. Chang, R. Fearing, T. Hsieh, T. Kenny, L. Liang, W. Zesch, and R.J. Full. (2000) Adhesive force of a single gecko foot-hair. Nature, 405:681–685.
7. K. Autumn, (2006) Properties, principles, and parameters of the gecko adhesive system. Biological Adhesives, eds. A. Smith and J. Callow, Berlin Heidelberg: Springer Verlag.
8. A. K. Geim, S.V. Dubonos, I. V. Grigorieva, K. S. Novoselov, A. A. Zhukov, and S. Y.Shapoval (2003) Microfabricated adhesive mimicking gecko foot-hair. Nature Materials, 2:461–463.
9. P. Lambert, (2007) Capillary Forces in Microassembly, Springer.
10. Micky Rakotondrabe, Mamadou Cissé Diouf and Philippe Lutz, "Robust feedforward-feedback control of a hysteretic piezocantilever under thermal disturbance", IFAC - WC , (World Congress), pp:13725-13730, Seoul Corea, July 2008.
11. M. Rakotondrabe, C. Clévy and P. Lutz (2007) H-inf. "Deflection control of a unimorph piezoelectric cantilever under thermal disturbance", IEEE/RSJ IROS International Conference on Intelligent Robots and Systems, San Diego, USA.
12. G. Yang, J. A. Gaines, and B. J. Nelson, (2005) Optomechatronic Design of Microassembly Systems for Manufacturing Hybrid Microsystems. IEEE Trans. On Inductrial Electronics, Vol. 52, N. 4.
13. Joseph J. Carr and John M. Brown, "Introduction to Biomedical Equipment Technology", Prentice Hall, Third Edition, ISBN: 0-13-849431-2, April 2010.
14. Pierre Lambert, "A contribution to microassembly: a study of capillary forces as a gripping principle", PhD thesis, Université Libre de Bruxelles, 2005.
15. Micky Rakotondrabe, "Développement et commande modulaire d'une station de microassemblage", PhD thesis, Université de Franche-Comté at Besançon, 2006.
16. M. Gauthier, S. Régnier, P. Rougeot and N. Chaillet, "Forces analysis for micromanipulation in dry and liquid media", Journal of MicroMechatronics, Vol.3, pp.389–413, 2006.
17. Cédric Clévy, "Contribution à la micromanipulation robotisée : un système de changement d'outils automatique pour le micro-assemblage", PhD thesis, Université de Franche-Comté at Besançon, 2005.
18. Micky Rakotondrabe, Yassine Haddab and Philippe Lutz, "Development, Modeling, and Control of a Micro-/Nanopositioning 2-DOF Stick-Slip Device", IEEE - Transactions on Mechatronics (T-mech), Vol.14, Issue 6, pp:733–745, December 2009.
19. Micky Rakotondrabe and Patrick Rougeot, 'Presentation and improvement of an AFM-based system for the measurement of adhesion forces', IEEE - CASE, (International Conference on Automation Science and Engineering), pp:585–590, Bangalore India, August 2009.
20. M. Motamed and J. Yan, 'A review of biological, biomimetic and miniature force sensing for microflight', IEEE IROS, 2005.
21. C. K. M. Fung, I. Elhajj, W. J. Li and N. Xi, 'A 2-d pvdf force sensing system for micromanipulation and microassembly', IEEE ICRA, 2002.
22. D. Campolo, R. Sahai and R. S. Fearing, 'Development of piezoelectric bending actuators with embedded piezoelectric sensors for micromechanical flapping mechanisms', IEEE ICRA, 2003.
23. K. Motoo, F. Arai, Y. Yamada, T. Fukuda, T. Matsuno and H. Matsura, 'Novel force sensor using vibrating piezoelectric element', IEEE ICRA, 2005.
24. D-H. Kim, B. Kim, S-M. Kim and H. Kang, 'Development of a piezoelectric polymer-based sensorized microgripper for microassembly and micromanipulation', IEEE IROS, 2003.
25. A. Shen, N. Xi, C. A. Poumeroy, U. C. Wejinya and W. J. Li, 'An active micro-force sensing system with piezoelectric servomechanism', IEEE IROS, 2005.
26. M. Rakotondrabe, Y. Haddab and P. Lutz, 'Quadrilateral modelling and robust control of a nonlinear piezoelectric cantilever', IEEE - Transactions on Control Systems Technology (T-CST), Vol.17, Issue 3, pp:528–539, May 2009.
27. M. Rakotondrabe, C. Clévy and P. Lutz, Complete open loop control of hysteretic, creeped and oscillating piezoelectric cantilever, IEEE - Transactions on Automation Science and Engineering (T-ASE), 7(3), pp. 440–450, 2010.

28. Y. Haddab, Q. Chen and P. Lutz, 'Improvement of strain gauges micro-forces measurement using Kalman optimal filtering.', IFAC Mechatronics, 19(4), 2009.
29. Y. Haddab. 'Conception et réalisation d'un système de micromanipulation contrôlé en effort et en position pour la manipulation d'objets de taille micrométrique'. PhD dissertation (in French), University of Franche-Comté, Laboratoire d'automatique de Besancon, 2000.
30. Micky Rakotondrabe, Yassine Haddab and Philippe Lutz, 'Modelling and H-inf force control of a nonlinear piezoelectric cantilever', IEEE/RSJ - IROS, (International Conference on Intelligent Robots and Systems), pp:3131–3136, San Diego CA USA, Oct-Nov 2007.
31. M. Rakotondrabe and P. Lutz, 'Force estimation in a piezoelectric cantilever using the inverse-dynamics-based UIO technique', IEEE - ICRA, (International Conference on Robotics and Automation), pp:2205–2210, Kobe Japan, May 2009.
32. C-S. Liu and H. Peng, 'Inverse-dynamics based state and disturbance observers for linear time-invariant systems', ASME Journal of Dynamics Systems, Measurement and Control, vol.124, pp.375–381, September 2002.
33. M. Rakotondrabe, I. A. Ivan, S. Khadraoui, C. Clevy, P. Lutz, and N.Chaillet, 'Dynamic Displacement Self-Sensing and Robust Control of Cantilevered Piezoelectric Actuators Dedicated to Microassembly Tasks', Submitted in IEEE/ASME AIM (Advanced Intelligent Mechatronics), 2010.
34. M. Rakotondrabe, C. Clévy and P. Lutz, 'Modelling and robust position/force control of a piezoelectric microgripper', IEEE - CASE, (International Conference on Automation Science and Engineering), pp:39–44, Scottsdale AZ USA, Sept 2007.
35. M. Rakotondrabe and Y. Le Gorrec, 'Force control in piezoelectric microactuators using self scheduled H_∞ technique', ASME Dynamic Systems and Control Conference and IFAC Symposium on Mechatronic Systems, MIT 2010.

Chapter 2
Self-Sensing Measurement in Piezoelectric Cantilevered Actuators for Micromanipulation and Microassembly Contexts

Ioan Alexandru Ivan, Micky Rakotondrabe, Philippe Lutz, and Nicolas Chaillet

Abstract This chapter aims to develop a self-sensing technique to measure the displacement and the force in piezoelectric microactuators dedicated to micromanipulation and microassembly contexts. In order to answer the requirements in these contexts, the developed self-sensing should perform a long duration measurement of constant signals (displacement and force) as well as a high precision. Furthermore, we propose to consider the dynamics in the displacement self-sensing measurement such that a positioning feedback is possible and therefore a high micro/nanopositioning accuracy is obtained. The experimental results validate the proposed technique and demonstrate its conveniency for micromanipulation and microassembly contexts.

Keywords Self-sensing • Force and displacement signals • Cantilevered piezoelectric actuators • Micro and nano-scale

2.1 Introduction

Piezoelectric materials are very prized to develop systems acting at the micro/nanoscale. This recognition is due to the high resolution, high bandwidth and high force density that they can offer. They can be divided into two main categories as actuators: continuous (benders, stack, tube, etc.) or discrete (stick-slip, inch-worm etc.). One of the applications of piezoelectric materials is the piezogripper (Fig. 2.1a, b) often integrated in cells (Fig. 2.1c) in order to perform micromanipulation/microassembly of small parts, such as biological objects or microstructures.

M. Rakotondrabe (✉)
FEMTO-ST Institute, UMR CNRS, 6174 - UFC, ENSMM/UTBM, rue Alain Savary 24, 25000 Besançon, France
e-mail: mrakoton@femto-st.fr

C. Clévy et al. (eds.), *Signal Measurement and Estimation Techniques for Micro and Nanotechnology*, DOI 10.1007/978-1-4419-9946-7_2,

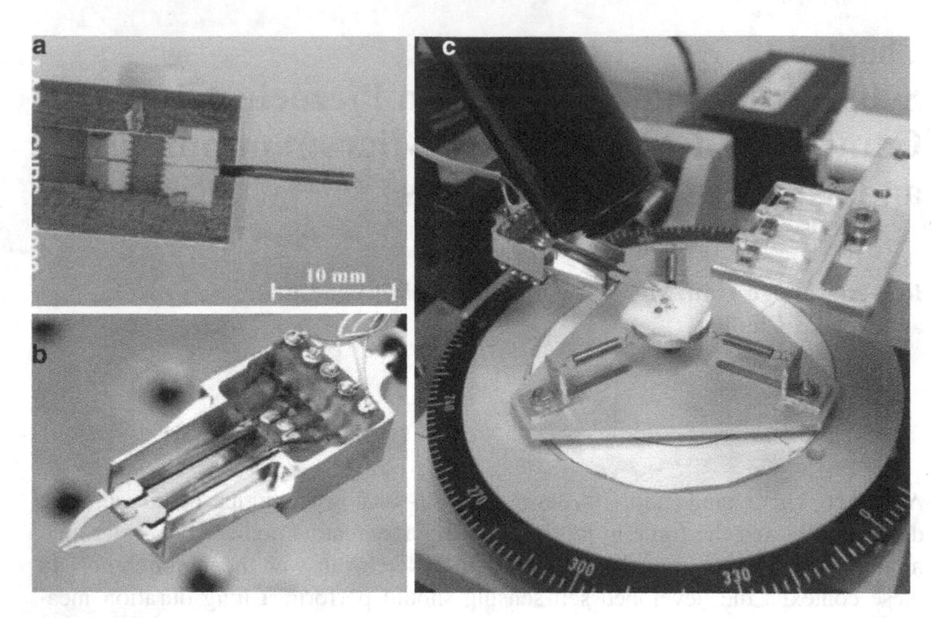

Fig. 2.1 Piezoelectric devices and systems, courtesy of AS2M Department, FEMTO-ST Institute (Besanon FR). (**a**) Double-unimorph gripper. (**b**) Duo-Bimorph gripper called MOC (Microrobot On Chip). (**c**) Working table with piezoelectric gripper

A piezogripper is made up of two piezoelectric cantilevers (piezocantilevers) (Fig. 2.2a). When the cantilevers are subjected to electric voltage, they bend and can maintain and/or position a manipulated micro-object. When properly controlled, piezogrippers can pick, transport and place micro-objects with micrometric or submicrometric positioning accuracy. While one cantilever is often controlled on position, the second one is controlled on force in order to avoid the destruction of the handled object (Fig. 2.2b). It is obvious that the successful of such control requires the measurement or estimation of the displacement and force signals. Furthermore, the availability of these signals also allow to characterize the manipulated objects if necessary.

The control and automation of piezogrippers as well as the study of the control of each elementary piezocantilever are still limited in laboratory [1, 2]. This is mainly due to the lack of embeddable and precise sensors, both for displacement and force, that can be used for packaged and feedback controlled systems. On the one hand, embeddable sensors such as strain gages are very noisy and fragile. On the other hand, very precise and large range displacement sensors (capacitive or magnetic displacement readers, optical sensors, interferometers, etc.) are bulky and expensive. Another solution can be the micromachined piezoelectric systems (Piezo-MEMS) incorporating piezoresistive or capacitive sensing elements, but these systems are not yet fully developed and available on the market. A possible solution to bypass all these limitations is the utilization of open-loop control techniques where no sensors are required [3], but they do not account the external disturbances

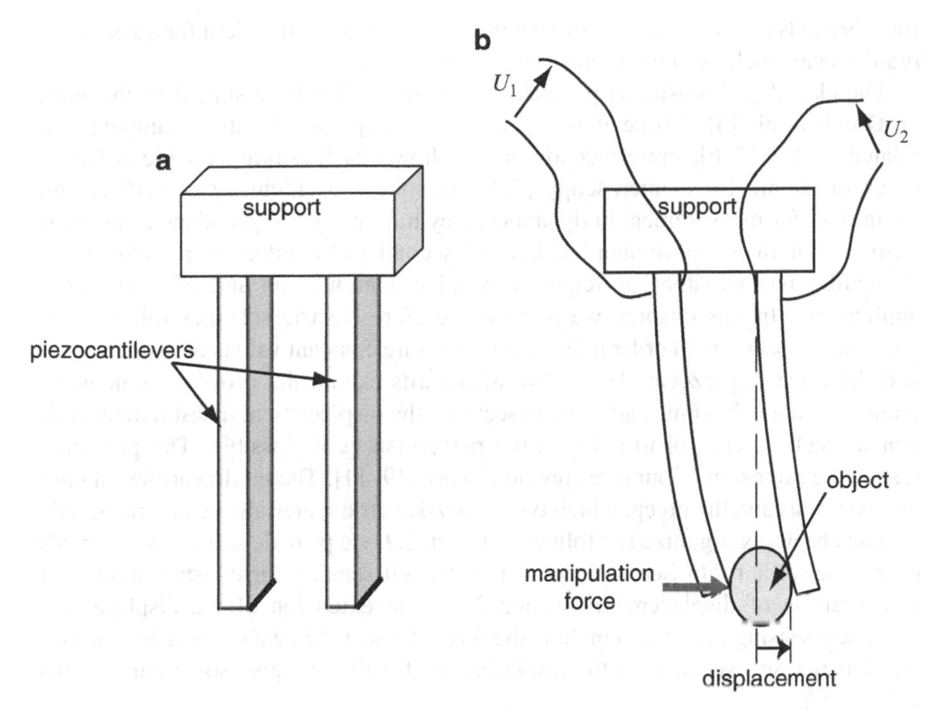

Fig. 2.2 (**a**) Principle of a piezogripper. (**b**) Manipulation of a micro-object

(such as thermal variation) as well as the internal uncertainties of the actuators models making therefore the loss of the accuracy. To sum up, it appears that existing methods in piezogrippers control do not answer the requirement of high performances and the packageability and that newer approach should be proposed for that.

In this chapter, we propose to apply the self-sensing approach to measure the signals in piezocantilevers of a piezogripper. Self-sensing has the advantages to be directly packaged and to offer a feedback control possibility if well designed. It is, therefore, suitable for the developement of embedded and high performances piezogrippers dedicated to micromanipulation and microassembly cells.

Self-sensing consists of using at the same time an actuator as also a sensor. This is particularly achievable for piezoelectric actuators thanks to the direct and converse piezoelectric effects, subjected that convenient electric circuit and observer are used. This simple and cost-effective up-grade solution for existing piezoelectric grippers is based on current integration, that allows displacement and/or force measurement. Charge measurement is often rejected on false idea that PZT is a very bad isolating material. In fact, not the leaking resistivity, which is high enough to preserve charges for hundreds of seconds, but ferroelectric material non-linearities (hysteresis, creep) as well as temperature influence put challenge on charge-based self-sensing. Existing literatures in self-sensing methods do not account both hysteresis and creep compensations. In addition, they are only intended for vibrations or short-term duration; hence, they are not convenient for quasi-static requirements such as in micromanipulation/microassembly.

The idea of self-sensing in piezoelectric cantilever has been started by the work of Dosch et al. [4]. While it is not a new concept on vibration damping and related control [5, 6], more recently, it has shown its feasibility for piezoelectric tubes of atomic force microscopy [7, 8]. Despite the efficiency of self-sensing techniques for measurement in dynamics, they have not yet been adapted for long-term measurement (more than hundreds of seconds) of displacement and/or force in cantilevered actuators, as required by micromanipulation and microassembly applications. In this chapter, we propose therefore electric schemes followed by convenient observers in order to be able to measure constant values of displacement and the force in piezocantilevers. We afterwards extend the proposed scheme to include both the dynamic and static apsects for the displacement measurement such that a feedback control to enhance the performances is possible. The presented results are extension of our three previous works [9–11]. The nonlinearities, mainly the hysteresis and the creep, which typify piezoelectric materials are compensated.

The chapter is organized as follows. In Sect. 2.2, we provide some fundamentals on piezoelectricity. In Sect. 2.3, we detail the self-sensing for constant long-term measurement of displacement. Section 2.4 is an extension of the displacement static self-sensing in order to include the force. In Sect. 2.5, we propose to consider the dynamic self-sensing for the displacement. Finally, we give some conclusions in Sect. 2.6.

2.2 Fundamentals of Piezoelectric Actuators and Self-Sensing

2.2.1 Piezoelectric Cantilevered Actuators

In piezogrippers, according to the geometry of the piezocantilevered actuators, displacement scale ranges from micrometer to millimeter, and usual blocking force from 100 μN to 100 mN. Micro-position accuracy can go down to a couple nanometers and response time is in the order of milliseconds. In the sequel, the terms *piezoelectric cantilever* and *piezocantilever* are alternately employed.

Piezocantilevers are usually made of piezoelectric PZT ceramics ($Pb[Zr_x Ti_{1-x}]O_3$) and amorphous (mettalic) materials. Polymers such as PVDF (polyvinylidene fluoride) are rarely employed. Unimorph cantilevers are made of a single layer of piezoelectric material deposited or stuck on an amorphous base material (cross-section shown in Fig. 2.3a). When an external electric field is applied, the piezoelectric layer tends to contract/expand making the whole structure to bend (Fig. 2.4) accordingly to well-stated constitutive equations [12] and depicted briefly in Sect. 2.3. Bimorph actuators (Fig. 2.3b, c) are made up of two piezoelectric layers that are to be driven in an opposite manner (depending on polarization direction). Main advantage with respect to unimorphs is the increase of displacement range, blocking force and a better temperature effects rejection.

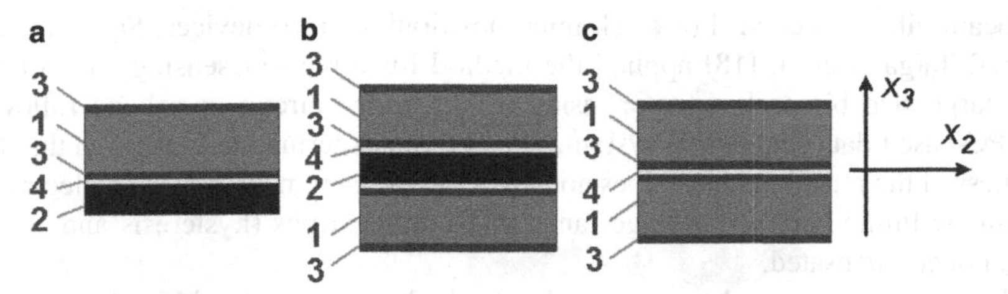

Fig. 2.3 Cross-section of different piezoelectric bender structures: (**a**) unimorph, (**b**) three-layer bimorph, (**c**) two-layer bimorph

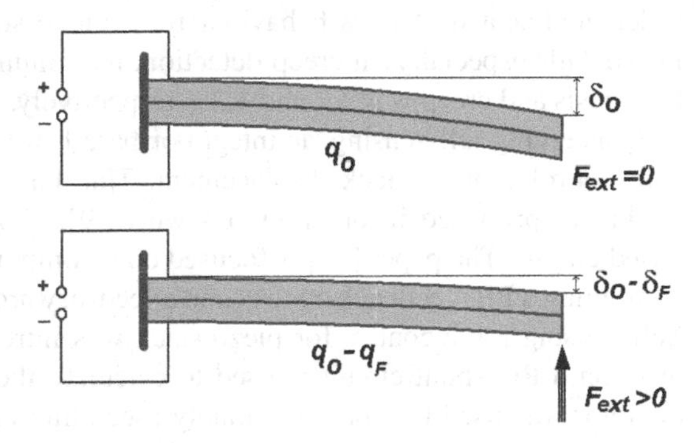

Fig. 2.4 Electric charge in a piezoelectric cantilever bending under external voltage

With the advent of multiple screen-printing and sol-gel techniques, multi-layer piezocantilevers applications are also raising [12, 13].

Given the continuous interest in piezoelectric actuators over the past decades, the literature is rich in studying and modeling these devices behaviour, static, and dynamic nonlinearities as well as their limits [14–17] for providing closed or open loop control solutions.

2.2.2 Self-Sensing: Methods, Challenges, and Applications

Self-Sensing allows actuating a piezoelectric device while sensing its displacement (strain) and/or applied force (stress). The intrinsic technique may be further used in a closed-loop system without needing external sensors.

The first use of "self-sensing" term dates back to 1992 when Dosch et al. [4] while his work consisted in successfully damping the vibration of a piezoelectric beam without using the aid of external sensors. Voltage drop provided from a capacitive bridge was processed in an analog circuit, amplified and returned back to the piezoelectric element. Sooner, several independent applications began to emerge for beam vibration control or stack micropositioning piezo devices. Several years later T. Taigami et al. [18] applied the method for force self-sensing and control of a large size bimorph actuator, using a half-bridge circuit, a voltage follower and PC-based data acquisition system. The authors experimentally verified that the stiffness of the grasping object does not affect the measurement; however, electronic schematic limited applied voltage range and nonlinearities (hysteresis and creep) were not compensated.

Creep compensation by shaping the applied step signal with a logarithmic function allowed a reduction of open loop control error by a factor of 5–10 [19]. Issues related to compensation of hysteresis and creep with inverse systems came also into discussion [20]. In [21], hysteresis and creep models were inverted using Newton iterative procedure, reported displacement open loop control error being 2% (not specified if on full actuator range). In [22], adaptive identification based on an unscented Kalman filter was proposed to make dynamic prediction of a first order nonlinear hysteresis behaviour for a piezo stack actuator. Without entering into details especially on creep detection, maximum reported percentage error for hysteresis and creep was 3.2 and 1.2% respectively.

Concerning self-sensing, an integrator-based circuit was introduced [23] to sense and control a piezo stack displacement. This basic principle of circuit is closed to that of proposed in our previous work [9], showing advantages over bridge-based circuit. The paper [23] is focused on a compounding control of displacement combining a PID feedback control and a feedforward control of hysteretic behavior. Self-sensing force control for piezo stack was introduced in [24, 25]. The voltage drop on a RC shunt circuit is used to calculate the charges. A hyperbola-shaped operator was used to model separately ascending and descending branches of the hysteresis. Force is controlled classically with a PID feedback under different loading conditions. In paper [26], a generalized Maxwell-slip hysteresis operator was used. Static displacement error is reported to 2–4% while static force error is 2–6%.

The issues of these papers focus on short-term (less than 1 s) force self-sensing control or vibration damping applications. For instance in [5], a modified bridge circuit with adaptable gain was intended for vibration control under structural deformation, in [6] self-sensing actuators ameliorated positioning and vibration in hard-disk drives, in [27] self-sensing microdispensing system demonstrated better positioning performance than with external sensors.

Some sensorless methods neighbored to self-sensing concept consist in shaping several electrodes on actuator and dedicating them separately for actuating or sensing [28]. In this way, nonlinear effects are avoided and signals well separated. However, a fraction of actuating capability is lost. Recently, a SPM piezo-tube scanner with a new electrode pattern allowed self-measurement of nanometer resolution with improved transfer function [7, 8].

Summarizing, self-sensing first advantage is that system will be more flexible in terms of space occupation, allowing better miniaturization and dexterity in terms of DoF (degrees of freedom). Also, actuator's dynamics will no longer be affected by mechanically attached sensors (e.g. strain gages). The number of connecting cables will be reduced. The system will be more cost-effective.

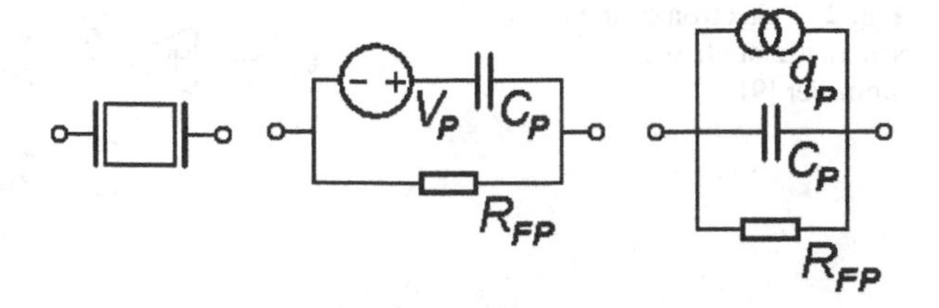

Fig. 2.5 Symbol and equivalent electrical schematics of piezo actuators

Disadvantages consist in adding a supplementary electronic circuit, but of reduced complexity. On the other hand, provided measurement is not fully-static. Electronic circuit is based on a capacitive bridge (or divider) or a current integrator. Specific applications with dedicated electrodes for sensing may measure directly strain-induced voltage. Attention has to be paid for preserving the charge as long as possible. Non-linearities such as hysteresis and creep due to ferroelectric domain relaxation put the most challenge on self-sensing technique, limiting its accuracy, especially in force sensing. Identifying these nonlinear effects requires an extra procedure. Finally, there are working temperature constraints that should be taken into account to prevent loss of accuracy.

Very few publications deal with self-sensing techniques intended for automated or tele-operated micromanipulation/microassembly cells. For these systems, the required periods of time for preserving the static values are between dozens to several hundreds of seconds. A quasi-static method for deriving displacement and force from actuated uni- and bi-morph cantilevers is therefore proposed in the sequel. The nonlinearities (hysteresis and creep) are especially taken into account.

2.2.3 *Electronic Circuit and Experimental Setup*

In Fig. 2.5, static equivalent schematic is shown. C_P is the piezo element capacitance, R_{FP} is the PZT leaking resistance (in the $G\Omega$ range) and V_P and q_P are the piezoelectrically generated voltage or charge, respectively.

The electronic schematic as in Fig. 2.6 integrates the current across the actuator. It provides the main signal for the self-sensing observer. An optional fixed value C_R capacitor in anti-parallel with C_P improves signal range by keeping the same sensitivity.

Output voltage of the circuit is:

$$V_{out} = -\frac{1}{C}\int_0^T i(t)\mathrm{d}t = -\frac{1}{C}Q. \tag{2.1}$$

Where the charge Q varies with the applied voltage and is proportional with the capacitance difference $(C_P - C_R)$. In reality, as will be seen, nonlinear ferroelectric

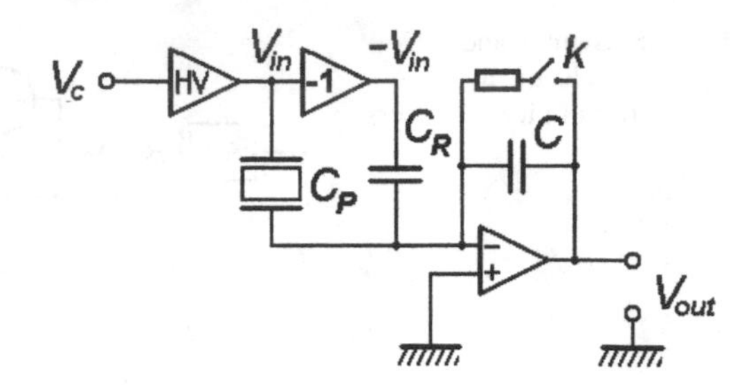

Fig. 2.6 Electronic circuit schematic of charge amplifier [9]

effects and slow charge leaks in the piezoelectric material will require several compensation terms. Supplementary practical details of the circuit design and how to improve its parameters are presented in [9].

The experimental setup is depicted in Fig. 2.7a, b. Several rectangular actuators (made of PZT bonded on a *Cu* or *Ni* substrate) were tested, of length between 10 and 15 mm, width between 1 and 2 mm, and total thickness of 0.27–0.45 mm. The FemtoTools force sensor and the optical displacement reader type Keyence LC-2420 were used for only intermediate tests. Also a SIOS SP-120 miniature plane-mirror interferometer was employed if higher resolution was necessary. The high voltage (HV) amplifier allowed supplying the actuator with a voltage up to 150 V. The DSpace DS1103 real-time controller board served to implement under Matlab Simulink the required observers, as will be detailed in the next chapters.

2.3 Static Displacement Self-Sensing

2.3.1 Charge Calculation

The developed self-sensing systems can be divided into three main parts, as in Fig. 2.8. In the figure, V_{in} is the applied input voltage, δ is the deflection of the cantilever while $\hat{\delta}_s$ is the estimate from the self-sensing. Q is the charge appearing on the surfaces of the electrodes and V_{out} is the voltage from the electronic circuit. The piezoelectric actuators and the charge amplifier (current integrator) were previously discussed. Acquired V_{out} signal is processed to provide detected displacement. For this purpose a data processing system should implement an observer (estimator).

In the absence of external force, we have a theoretically linear relation between displacement and applied voltage [29]:

$$\delta = -\frac{3d_{31}}{1+\frac{d_{31}^2}{4s_{11}^E\varepsilon_{33}^S}}\frac{L^2}{h^2}V_{\text{in}}, \tag{2.2}$$

where s_{11}^E is the compliance coefficient along the beam, ε_{33}^S and d_{31} are dielectric and piezoelectric material coefficients.

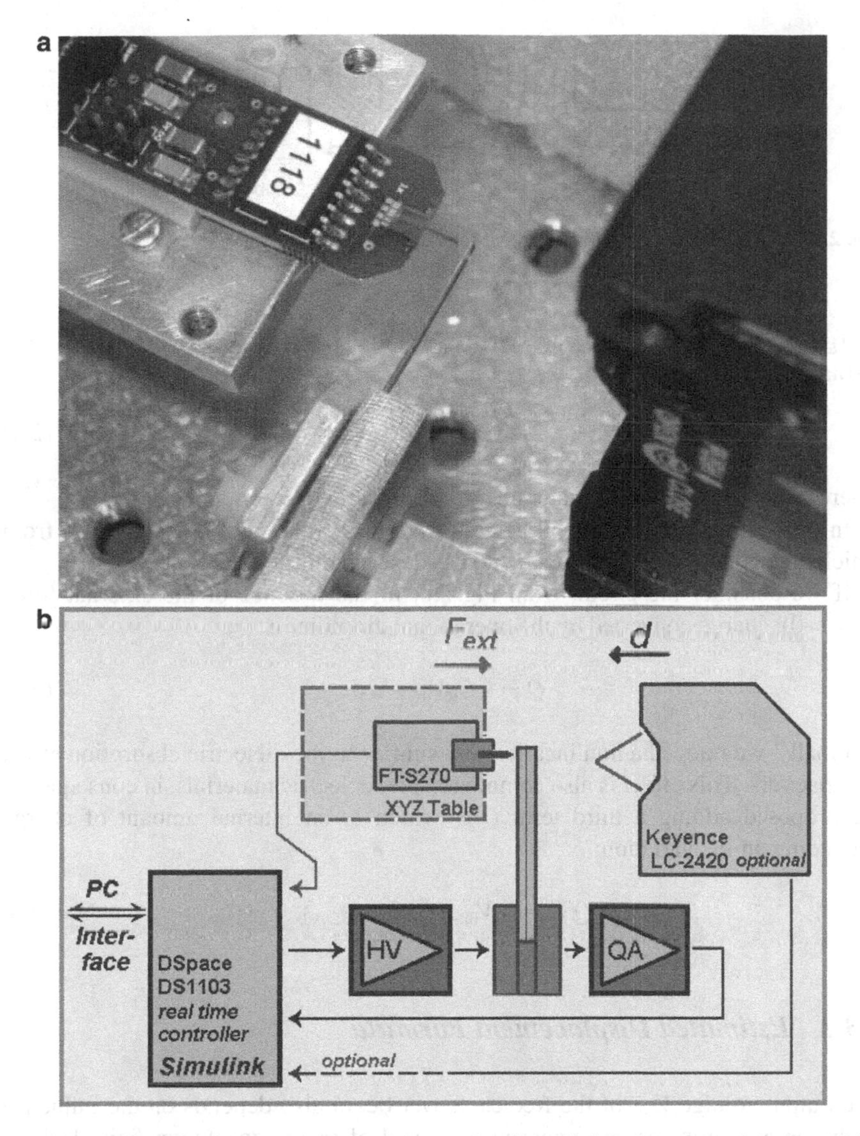

Fig. 2.7 (**a**) Photo of experimental set-up. A force sensor is on the *left*, the actuator is in the *middle* and an optical displacement reader is located in the *right side* of the image. (**b**) Schematic block diagram used for experiments, consisting of the piezoactuator, force and displacement sensors, high voltage amplifier (HV), charge amplifier (QA) and real-time controller. Directions for positive force and displacement are figured

Using the relation between the applied voltage and the capacitance for bimorph piezoelectric cantilever beam

$$Q = \frac{4wL\varepsilon_{33}^{S}}{h}V_{\text{in}} \tag{2.3}$$

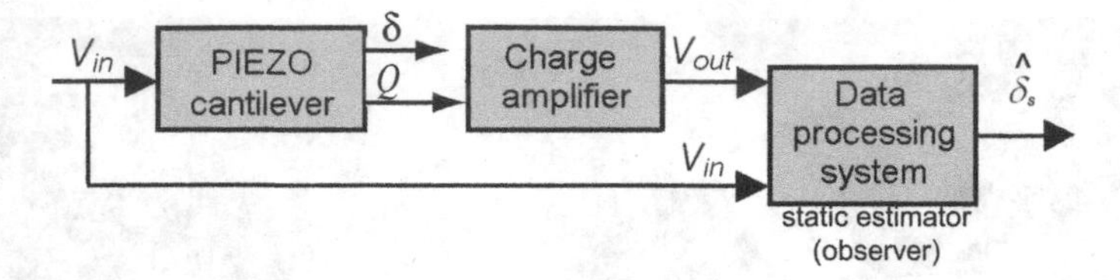

Fig. 2.8 Principle of the static displacement self-sensing

charge directly results and, as stated previously, is quasi-proportional to free displacement δ:

$$Q = \frac{4wh\varepsilon_{33}^S}{3d_{31}L}\left(1 + \frac{d_{31}^2}{4s_{11}^E\varepsilon_{33}^S}\right)\delta = \alpha\delta, \tag{2.4}$$

where α is denoted as an actuator charge-displacement coefficient.

In the sequel, this charge will be converted into a measurable voltage V_{out} from which the deflection δ will be estimated as described in Fig. 2.8.

If we consider the circuit from Fig. 2.6, in the absence of the external force ($F_{\mathrm{ext}} = 0$), charge converted by the operational amplifier is:

$$Q = -C_{\mathrm{R}}V_{\mathrm{in}} + \alpha\delta. \tag{2.5}$$

Finally, we noticed a nonlinearity very similar to the dielectric absorption effect in capacitors. This effect is also common in piezoelectric materials, in consequence we proposed adding a third term Q_{DA} which is an internal amount of charge depending on ε_{33} variation:

$$Q = -C_{\mathrm{R}}V_{\mathrm{in}} + (\alpha\delta + Q_{\mathrm{DA}}). \tag{2.6}$$

2.3.2 Estimated Displacement Formula

The output voltage V_{out} of the free cantilever beam also depends on the influence of the op-amp non-zero bias current i_{BIAS} and of the piezoactuator finite leaking resistance R_{FP}:

$$V_{\mathrm{out}} = \frac{C_{\mathrm{R}}}{C}V_{\mathrm{in}} - \frac{\alpha\delta + Q_{\mathrm{DA}}}{C} - \frac{1}{C}\int \frac{V_{\mathrm{in}}(t)}{R_{\mathrm{FP}}}\mathrm{d}t - \frac{1}{C}\int i_{\mathrm{BIAS}}(t)\mathrm{d}t. \tag{2.7}$$

The estimated displacement $\hat{\delta}_s$, may be computed as follows:

$$\hat{\delta}_s = -\frac{C}{\alpha}V_{\mathrm{out}} - \frac{Q_{\mathrm{DA}}(V_{\mathrm{in}},t)}{\alpha} + \frac{C_{\mathrm{R}}}{\alpha}V_{\mathrm{in}} - \frac{1}{R_{\mathrm{FP}}\alpha}\int V_{\mathrm{in}}(t)\mathrm{d}t - \frac{1}{\alpha}\int i_{\mathrm{BIAS}}(t)\mathrm{d}t. \tag{2.8}$$

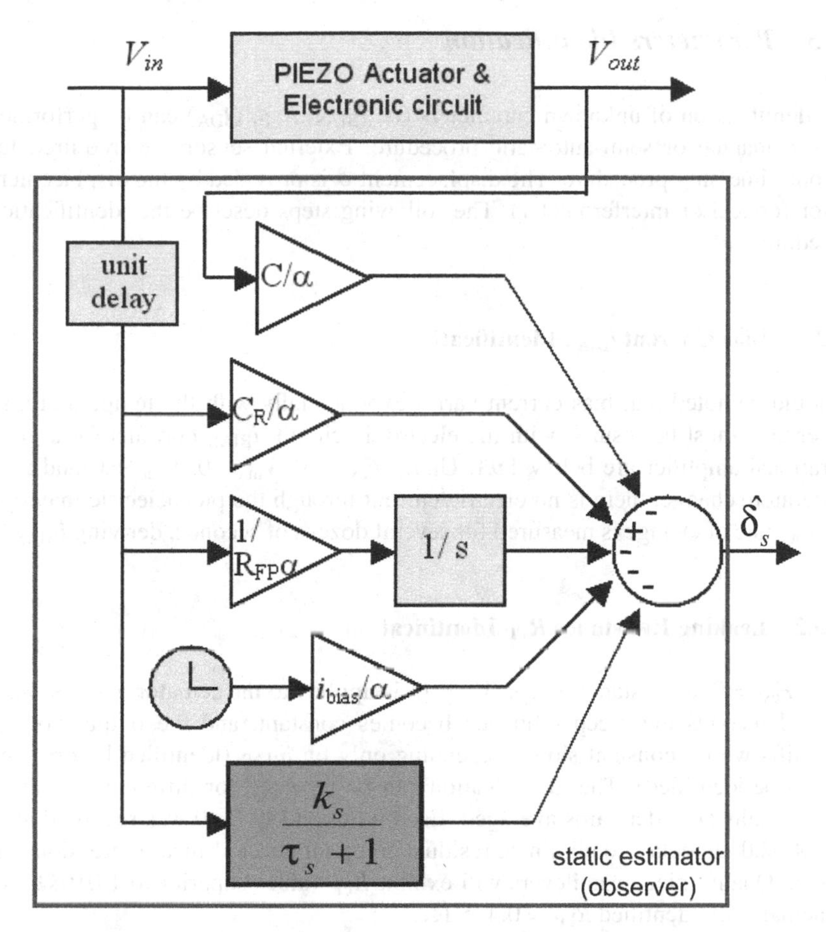

Fig. 2.9 Block-scheme (in Simulink) of the observer of the static displacement self-sensing. Reprinted with permission, Copyright 2009 American Institute of Physics [9]

Regarding the dielectric absorption term, we will consider a simple relaxation effect described by a simple first-order transfer function:

$$Q^*_{\mathrm{DA}}(V_{\mathrm{in}}, s) = \frac{Q_{\mathrm{DA}}(V_{\mathrm{in}}, s)}{\alpha} = \frac{k_{\mathrm{s}}}{\tau s + 1} V_{\mathrm{in}}, \tag{2.9}$$

where k_{s} is a static gain and τ a time constant.

Figure 2.9 presents the final observer bloc-scheme. Some parameters of the expression (8) are known while others require identification. Their calculation will be presented in the next section.

2.3.3 *Parameters Identification*

The identification of unknown parameters (α, i_{BIAS}, R_{FP}, Q_{DA}) can be performed under a manual or semi-automatic procedure. External sensors are required for this one-time only procedure. The displacement δ is provided by the displacement sensor (optical or interferometer). The following steps describe the identification procedure.

2.3.3.1 Bias Current i_{BIAS} Identification

It should be noted that bias current varies exponentially with the temperature, so proper care must be insured with the electronic circuit. i_{BIAS} currents for a good operational amplifier are below $1pA$. Under $F_{ext} = 0$, $V_{in} = 0$, $V_{out} \cong 0$ and zero temperature change, there is no electric current through the piezoelectric material. The V_{out} rate of change is measured for several dozens of seconds, deriving i_{BIAS}.

2.3.3.2 Leaking Resistance R_{FP} Identification

Under $F_{ext} = 0$, a constant voltage ($V_{in} \neq 0$) is applied to the actuator. After several hundred seconds the creep influence becomes constant, and the output voltage V_{out} shifts with a constant slope, depending only on i_{BIAS} (identified before) and R_{FP} (to be identified). The identification can be repeated for different V_{in} values whose results are afterwards averaged. Each point on Fig. 2.10 was recorded after a 1,000–2,000 s delay, to eliminate residual creep influence. Linear regression was applied. Quality piezo cantilevers will exhibit R_{FP} values superior to 1,010 Ω. For our actuator we identified $R_{FP} = 0.435\,\mathrm{T}\Omega$.

2.3.3.3 Displacement Coefficient α Identification

A step signal is applied on the free actuator. To avoid oscillatory of the actuator, the step signal is shaped with ramp of around 20 V/s (Fig. 2.11). Measured values of δ and V_{out} immediately after the step signal V_{in} will serve to compute α:

$$\alpha = (-CV_{out} + C_R V_{in})/\delta. \tag{2.10}$$

An alternate method for deriving α is to apply one or several sinusoidal signals as in Fig. 2.12 and use amplitude values in (2.10), i.e. use magnitude at low frequency.

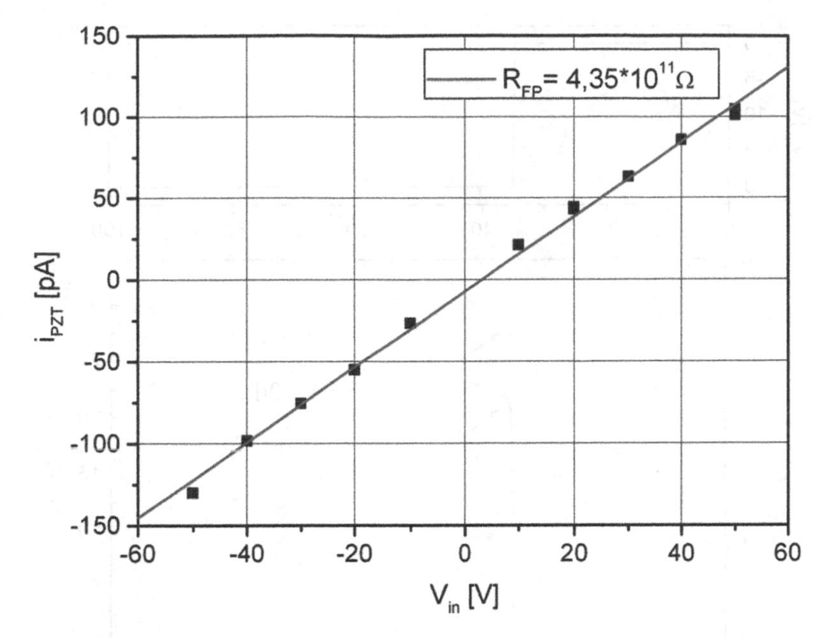

Fig. 2.10 Leakage current of PZT actuator measured under constant DC voltage values. Calculated insulation resistance is 0.435 TΩ. Reprinted with permission, Copyright 2009 American Institute of Physics [9]

2.3.3.4 Identification of Dielectric Absorption Transfer Function

The last part to be identified in displacement formula (2.8) is the dielectric absorption $Q_{\mathrm{DA}}(V_{\mathrm{in}},t)$ of the piezoelectric material. We have:

$$\Delta\hat{\delta}_s(s) = Q^*_{\mathrm{DA}}(s)V_{\mathrm{in}}(s), \tag{2.11}$$

where $\Delta\hat{\delta}_s = \hat{\delta}_s - \delta$ is the difference between estimated (using already identified parameters) and measured tip displacement.

Identification of k_{s} and τ is performed on a step response, allowing the calculation of the static gain and time constant easily (Fig. 2.13).

2.3.4 *Displacement Self-Sensing Results*

In Fig. 2.14, an input signal V_{in} was applied in several steps between +20 and −25 V, under null external force. Data was recorded for 1,020 s – sufficient for most micromanipulation and microassembly applications involving piezoelectric actuators. A very good agreement is found: measured and detected displacement curves almost superpose.

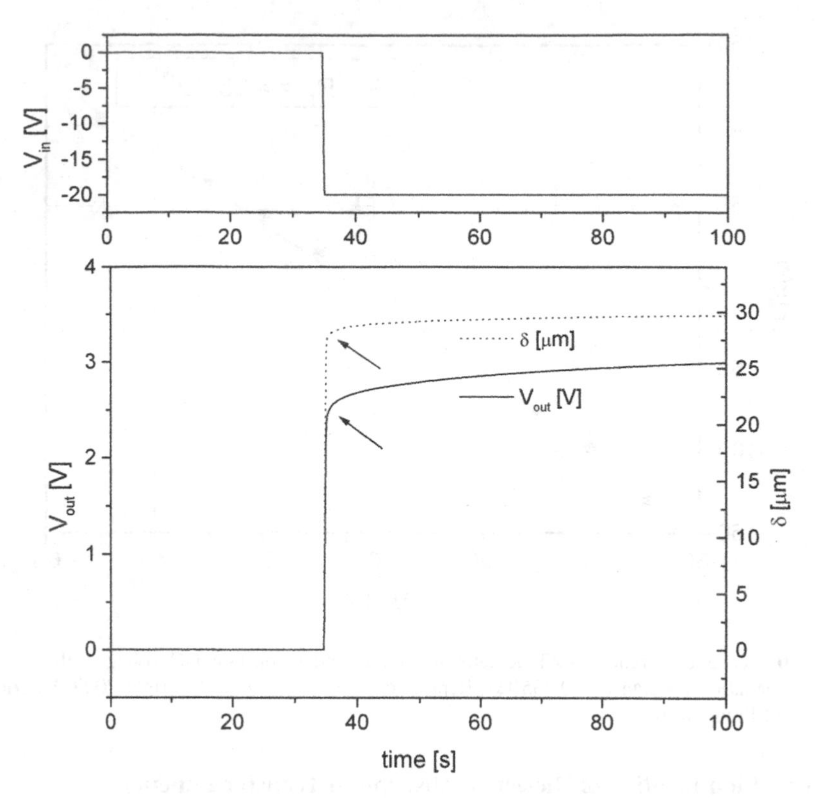

Fig. 2.11 Identification of α coefficient from -20 V step input voltage. Reprinted with permission, Copyright 2009 American Institute of Physics [9]

A comparative representation of displacement errors is made as follows. Three graphs are traced (Figs. 2.15–2.17), from uncompensated to fully compensated with respect to leaking resistance and dielectric absorption. Measurement with Keyence optical displacement reader provided a poorer linearity than self-sensing signal, making impossible accurate error evaluation, SIOS interferometer was eventually employed. Our constraint on the utilized interferometer is that data is only available offline. Vertical error lines in the figures can be neglected and are due to the linear interpolation and sampling period mismatch between the two data sets (acquired at sampling rates of 10 and 16.11 Hz). As seen in the Fig. 2.15, peak-to-peak error of uncompensated signal is 2.75 μm. Compensation of R_{FP} leaking resistance allowed a reduction of maximum error to 1.05 μm (Fig. 2.16). Adding the compensation of dielectric absorption (Fig. 2.17) provided a 0.38 μm (0.55%) peak-to-peak error.

Unaveraged measured self-sensing signal noise in displacement (Fig. 2.18) is of only 1.6 nm (RMS), being ten times less noisy than that of filtered Keyence LC-2420 sensor (16.7 nm rms noise on 4096 averaged samples). However, as expected, SIOS SP 120 interferometer showed best results: 0.5 nm rms noise.

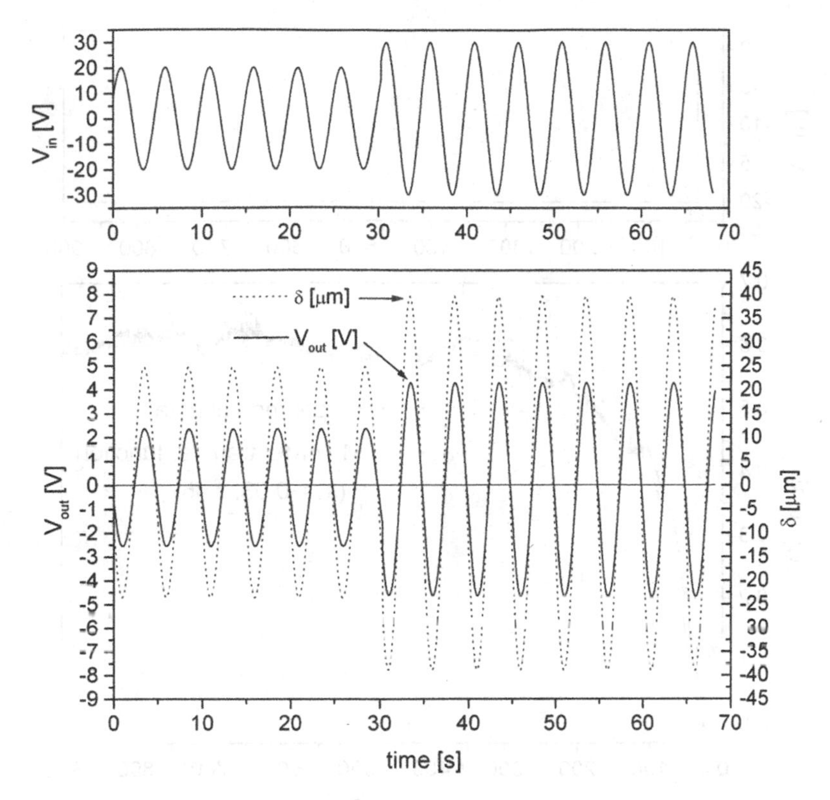

Fig. 2.12 Identification of α coefficient from sinusoidal signals. Reprinted with permission, Copyright 2009 American Institute of Physics [9]

2.4 Static Displacement and Force Self-Sensing

2.4.1 Force-Displacement Self-Sensing Detection

A self-sensing system for displacement and force detection can be divided into three main parts, as in Fig. 2.19. Piezoelectric actuator is submitted to a known input voltage V_{in} and to an unknown external force F_{ext}. Resulted charge Q is converted by an electronic amplifier to a measurable voltage V_{out}. Acquired signal is processed to provide detected displacement and/or force.

Observer algorithm runs on a PC or is deployed into a real-time processor or microcontroller. External signals can be provided to improve self-sensing accuracy, for instance temperature variation compensation. As the charge cannot be kept indefinitely, external resetting before each measurement could offset large parts of the error. Externally provided close-contact information could also help observers in ameliorating force sensing.

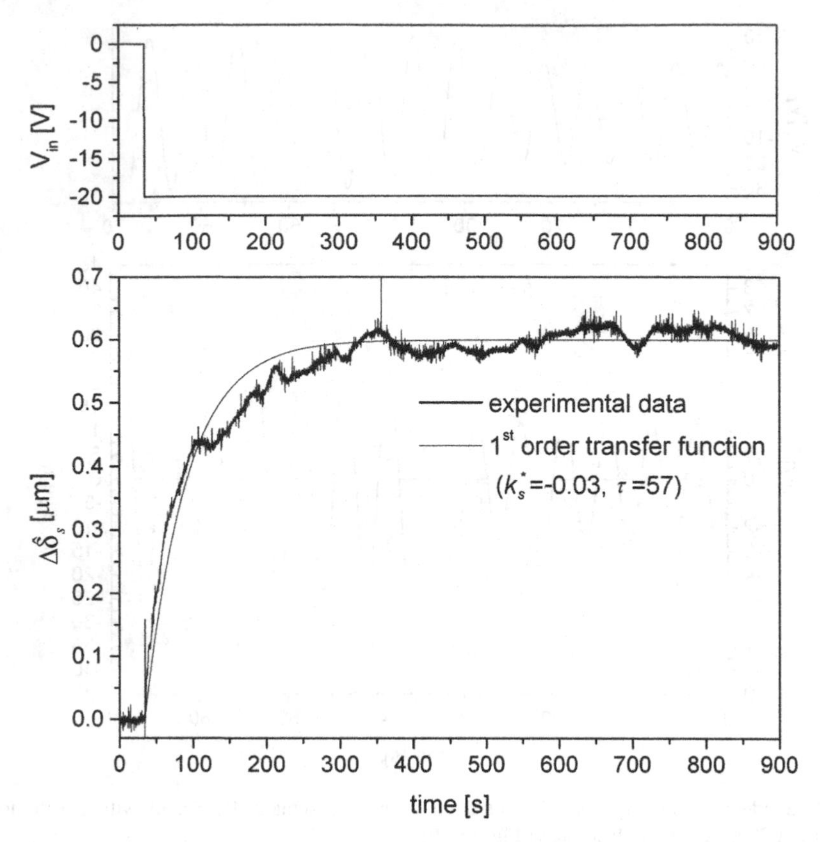

Fig. 2.13 Identification of dielectric absorption $Q_{\mathrm{DA}}^*(s)$ transfer function. Reprinted with permission, Copyright 2009 American Institute of Physics [9]

2.4.2 Piezoelectic Bimorphs: Theoretical Approach

The force equilibrium of a cantilever can be simplified as:

$$\mathbf{F}_{\mathrm{ext}} + \mathbf{F}_{\mathrm{el}} + \mathbf{F}_{\mathrm{piezo}} = 0, \tag{2.12}$$

where F_{ext} is the external force (to evaluate), $F_{\mathrm{el}} \cong k_{\mathrm{beam}}\delta$ is the elastic bending force (proportional to the beam flexure δ) and F_{piezo} is the reverse piezoelectric effect generated force.

When an external electrical field E_3 is applied along z-axis (poling direction), stress along x-axis T_1 and electrical displacement along z-axis D_3 are:

$$T_1 = c_{11}S_1 - e_{31}E_3, \tag{2.13}$$

$$D_3 = \varepsilon_{33}^S E_3 + e_{31}S_1, \tag{2.14}$$

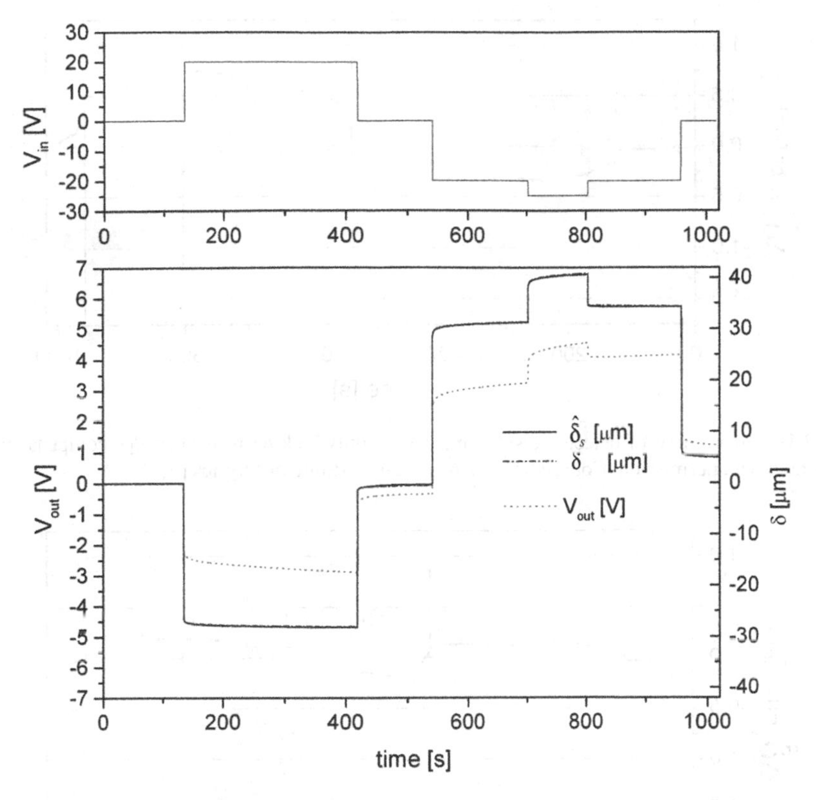

Fig. 2.14 Measured and detected displacement for an arbitrary V_{in} input signal ($F_{\text{ext}} = 0$). Reprinted with permission. Copyright 2009 American Institute of Physics [9]

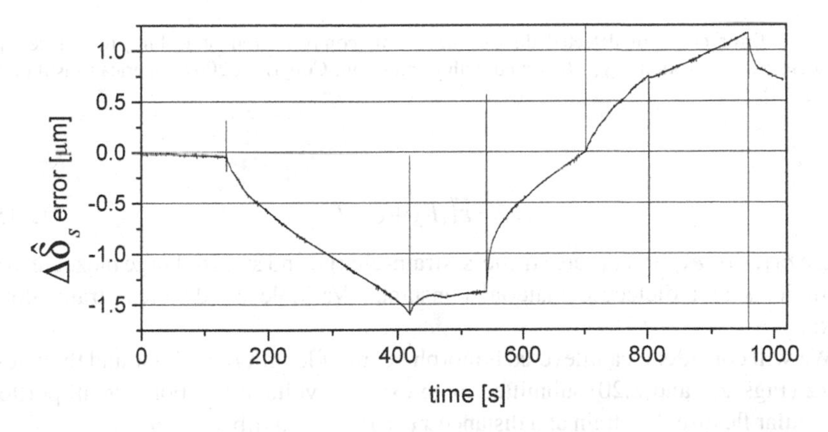

Fig. 2.15 Error curve of detected displacement with no leaking resistance R_{FP} and dielectric absorption Q_{DA} compensation. Reprinted with permission, Copyright 2009 American Institute of Physics [9]

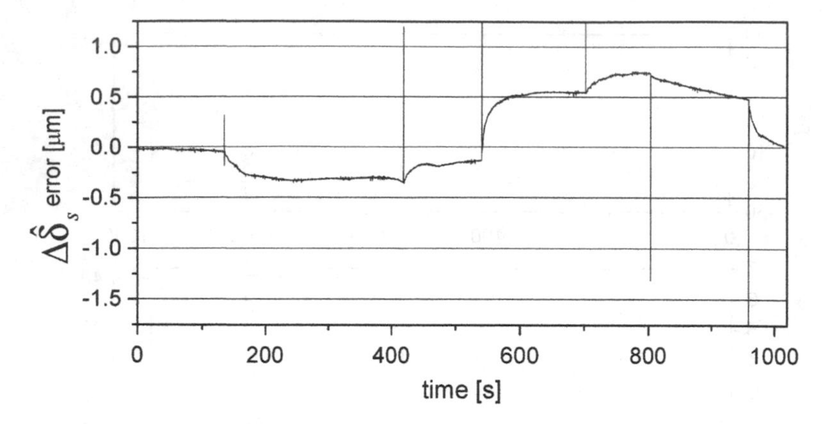

Fig. 2.16 Error curve of detected displacement with only leaking resistance R_{FP} compensation. Reprinted with permission, Copyright 2009 American Institute of Physics [9]

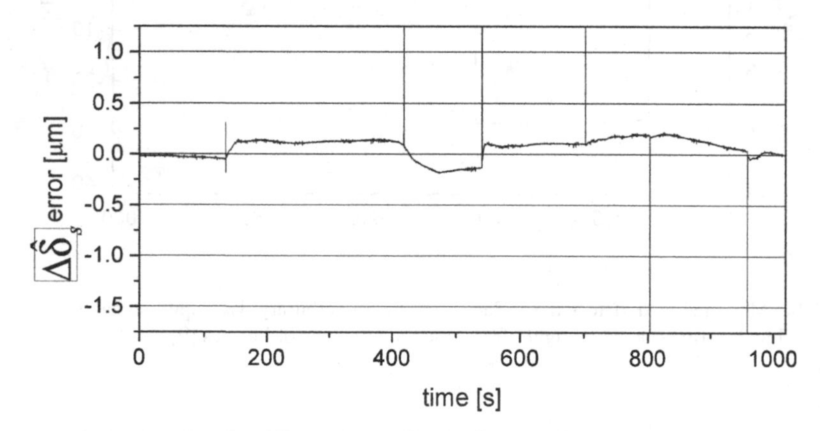

Fig. 2.17 Error curve of detected displacement with compensation of leaking resistance R_{FP} and dielectric absorption Q_{DA}. Reprinted with permission, Copyright 2009 American Institute of Physics [9]

or

$$D_3 = \varepsilon_{33}^T E_3 + d_{31} T_1, \tag{2.15}$$

where c_{11}, d_{31}, e_{31} and ε_{33} are stiffness, strain-charge and stress-charge piezoelectric coefficients and dielectric material constants. Variable S_1 denotes strain along x-axis.

We will consider a cantilevered bimorph beam of length L, width w and thickness $\pm h/2$ (Figs. 2.3 and 2.20) submitted to an external voltage V_{in}. For a small portion in circular flexure the strain at a distance x_3 from neutral fiber is:

$$S_1 = \frac{\mathrm{d}S_1 - \mathrm{d}S_0}{\mathrm{d}S_0} = \frac{(R + x_3 - R)\mathrm{d}\alpha}{R\mathrm{d}\alpha} = \frac{x_3}{R}. \tag{2.16}$$

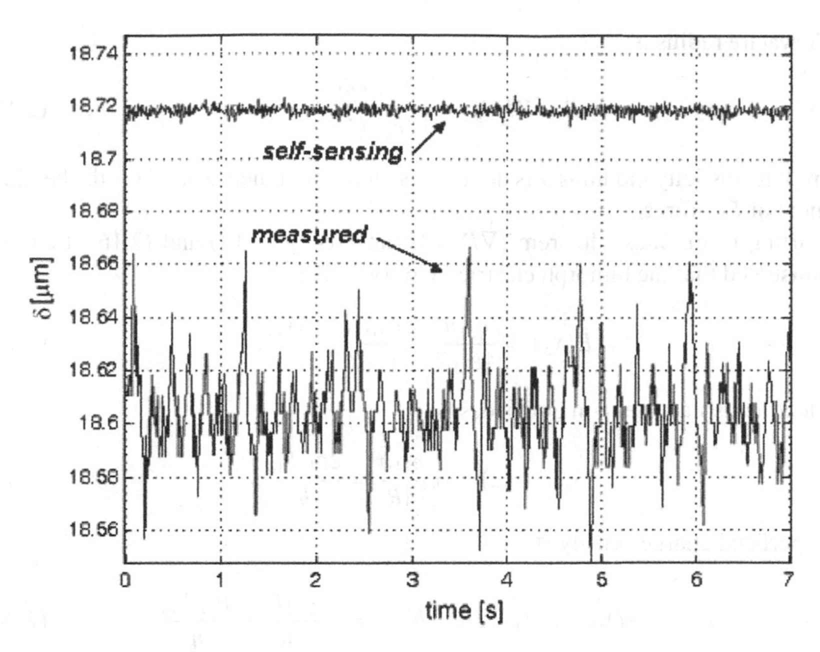

Fig. 2.18 Zoomed measured (Keyence LC2420) and detected displacement for noise evaluation. Reprinted with permission, Copyright 2009 American Institute of Physics [9]

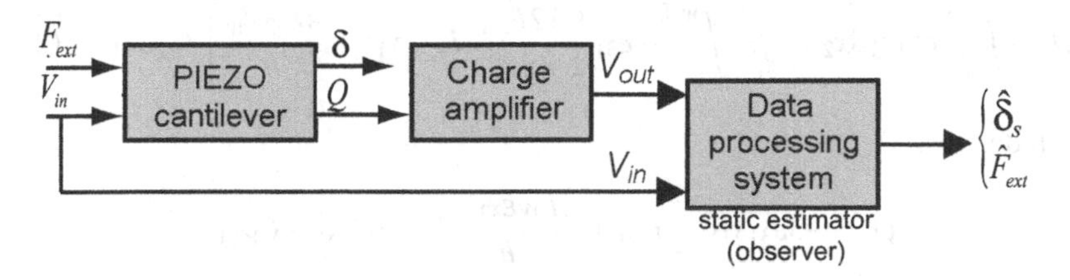

Fig. 2.19 Principle of the static displacement and force self-sensing

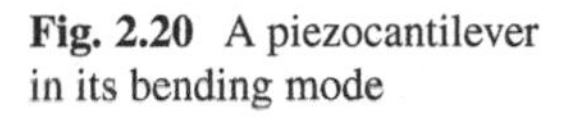
Fig. 2.20 A piezocantilever in its bending mode

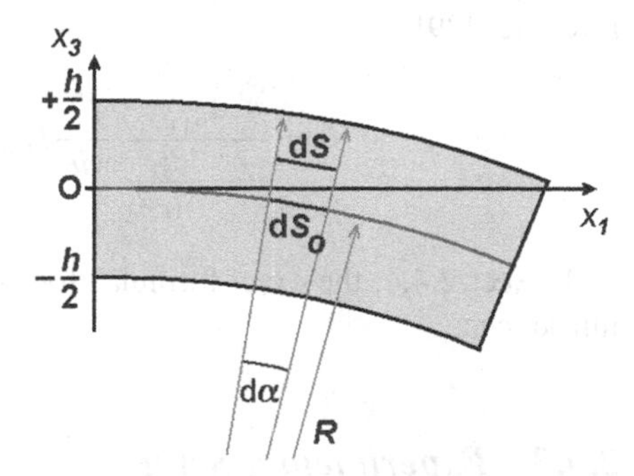

Curvature radius is:

$$R = YI/M = \frac{Ywh^3}{12F_{ext}(L - x_1)}, \tag{2.17}$$

where Y is elasticity modulus, I is the x_1-axis moment of inertia and M is the bending moment of F_{ext} force.

Starting from Gauss theorem ($\nabla D = 0$) and using (2.14) and (2.16) it can be demonstrated that the bimorph electric field formula is:

$$E(x_3) = \frac{e_{31}h}{4\varepsilon_{33}R} - \frac{e_{31}x_3}{\varepsilon_{33}R} - \frac{2V_{in}}{h}. \tag{2.18}$$

Electric displacement on both sides is:

$$D_3(\pm h/2) = \pm\frac{e_{31}h}{4R} \mp \frac{2\varepsilon_{33}V_{in}}{h}. \tag{2.19}$$

Superficial charge density σ is:

$$\sigma = -D_3(+h/2) + D_3(-h/2) = -\frac{e_{31}h}{2R} + \frac{4\varepsilon_{33}V_{in}}{h}. \tag{2.20}$$

Charge due to applied voltage and external force is:

$$Q = \int\int_A \sigma \mathrm{d}x_1 \mathrm{d}x_2 = \int_0^L \int_0^w \left[-e_{31}\frac{h}{2}\frac{12F_{ext}}{Ywh^3}(L - x_1) + \frac{4\varepsilon_{33}V_{in}}{h} \right] \mathrm{d}x_2 \mathrm{d}x_1. \tag{2.21}$$

which derives:

$$Q = -3e_{31}s_{11}\frac{L^2}{h^2}F_{ext} + \frac{4Lw\varepsilon_{33}V_{in}}{h} = \beta F_{ext} + C_P V_{in}, \tag{2.22}$$

where β is the force sensitivity coefficient and C_P is the actuator capacitance.

Displacement of the beam submitted to external voltage and force is derived from reference [29]:

$$\delta = \frac{4s_{11}^E}{1 + \frac{d_{31}^2}{4s_{11}^E\varepsilon_{33}^S}} \frac{L^3}{wh^3} F_{ext} - \frac{3d_{31}}{1 + \frac{d_{31}^2}{4s_{11}^E\varepsilon_{33}^S}} \frac{L^2}{h^2} V_{in}. \tag{2.23}$$

In Sect. 2.4.4, the exact formulae for displacement and force estimation will be introduced.

2.4.3 Experimental Setup

The setup schematic overview is depicted in Fig. 2.7. Reference force is measured with a FT-S270 micromachined capacitive sensor (from *FEMTO-tools* company) mounted on a XYZ micro-translation table for close-contact adjustment. The displacement is measured by a Keyence LC-2420 optical head. Both sensors are employed as reference in identification and error evaluation tests, their presence is of course finally not necessary when the self-sensing method is used.

Recall the equation of the output voltage already presented in (2.1):

$$V_{out} = -\frac{1}{C}\int_0^T i(t)\mathrm{d}t = -\frac{1}{C}Q, \tag{2.24}$$

where, if we ignore all non-linear piezoelectric effects, charge is proportional to applied voltage V_{in} and external force F_{ext} like in (2.22):

$$Q = (C_P - C_R)V_{in} + \beta F_{ext}. \tag{2.25}$$

2.4.4 Detected Displacement and Force Formulae

Formulae (2.24)–(2.25) require further compensation against op-amp bias current i_{BIAS} and piezoelectric actuator leaking resistance especially when working in quasi-static (more than a couple seconds of integration time). When we are in force detection mode, the expression (2.24) becomes:

$$V_{out} = \frac{C_R - C_P}{C}V_{in} - \frac{\beta F_{ext}}{C} - \frac{1}{C}\int \frac{V_{in}(t)}{R_{FP}}\mathrm{d}t - \frac{1}{C}\int i_{BIAS}(t)\mathrm{d}t. \tag{2.26}$$

Hence, estimated external force is:

$$\hat{F}_{ext} = -\frac{C}{\beta}V_{out} + \frac{C_R - C_P}{\beta}V_{in} - \frac{1}{R_{FP}\beta}\int V_{in}(t)\mathrm{d}t - \frac{1}{\beta}\int i_{BIAS}(t)\mathrm{d}t. \tag{2.27}$$

All above expressions do not take into account nonlinear nature of piezoelectric ceramics that introduce large parts of uncertainty. For instance, measurements on a unimorph beam of 15×1×0.2 mm showed $C_P = 1.74$ nC/V and $\beta = 1.03$ nC/mN. Blocking force of such a beam was only 0.07 mN/V. This indicates that 1% of error (nonlinearity) in charge-to-applied voltage characteristic introduces 24% uncertainty in the estimation of force. Given that ferroelectric behaviour of the PZT material shows even 15% of nonlinearity, a compensation of these unwanted effects is unavoidable.

Hence, we will replace $\frac{-C_P}{C}V_{in}$ term from (2.26) with two models for hysteresis and creep. The Vout expression becoming:

$$V_{out} = (V_{hyst} + V_{creep}) + \frac{C_R}{C}V_{in} - \frac{\beta F_{ext}}{C} - \frac{1}{C}\int \frac{V_{in}(t)}{R_{FP}}\mathrm{d}t - \frac{1}{C}\int i_{BIAS}(t)\mathrm{d}t. \tag{2.28}$$

Final formula for estimated force will include two additional operators for nonlinear effects: $F_C(s)$ for creep and $F_H(s)$ for hysteresis compensation:

$$\hat{F}_{\text{ext}} = -\frac{C}{\beta}V_{\text{out}} + \frac{C_R}{\beta}V_{\text{in}} - \frac{1}{R_{FP}\beta}\int V_{\text{in}}(t)\mathrm{d}t - \frac{1}{\beta}\int i_{\text{BIAS}}(t)\mathrm{d}t$$
$$-\frac{1}{\beta}F_C(s)\cdot V_{\text{in}} - \frac{1}{\beta}F_H(s)\cdot V_{\text{in}}. \quad (2.29)$$

Formula for free ($F_{\text{ext}} = 0$) piezoelectric beam displacement is taken from [9]:

$$\hat{\delta}_{\text{free\_s}} = -\frac{C}{\alpha}V_{\text{out}} + \frac{C_R}{\alpha}V_{\text{in}} - \frac{1}{R_{FP}\alpha}\int V_{\text{in}}(t)\mathrm{d}t - \frac{1}{\alpha}\int i_{\text{BIAS}}(t)\mathrm{d}t, \quad (2.30)$$

where α is the displacement coefficient.

When submitted to both external voltage and force, theoretical expression (2.23) is prone to nonlinearities, hence it is better to estimate the displacement with the following formula:

$$\hat{\delta}_s = \hat{\delta}_{\text{free\_s}} - \hat{F}_{\text{ext}}/k_Z, \quad (2.31)$$

where k_Z is transverse mechanical stiffness of the beam.

The Simulink detection model of force and displacement implemented into a dSPACE real time controller sums several terms from expressions (2.29), (2.30) and (2.31) (see Fig. 2.21). As seen, a supplementary third-order low-pass Butterworth filter was introduced to cancel the noise of $\hat{F}_{\text{ext}}/k_Z$ term from (2.21) which is far superior to $\hat{\delta}_{\text{free\_s}}$.

2.4.5 Self-Sensing Parameter Identification

Identification for displacement self-sensing method was already discussed in the Sect. 2.3. The parameters were bias current i_{BIAS}, leaking resistance R_{FP} and displacement coefficient α. For simplicity reasons, we suppressed dielectric absorption compensation (Q_{DA}) but that can be introduced from [9] if more accuracy is required. We will have to add a supplementary series of parameters intended for force self-sensing, such as: force sensitivity β, transverse stiffness k_Z, hysteresis operator F_H, and creep transfer function F_C.

Identification procedure may be done in the following possible manner.

2.4.5.1 Force Sensitivity β and Transverse Stiffness k_Z Identification or Estimation

Actuator is short-circuited (in fact $V_{\text{in}} = 0$) and an external force step $F_{\text{ext}} \neq 0$ is applied. In our case force was applied by slightly moving the force sensor

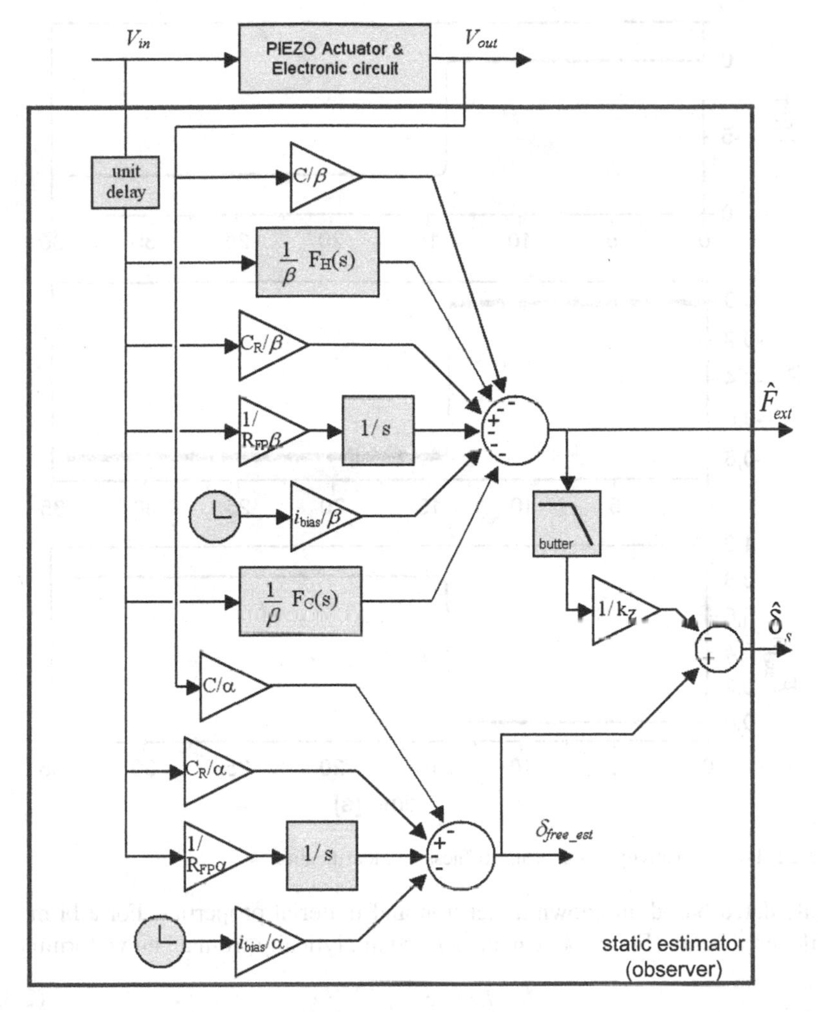

Fig. 2.21 Force and displacement detection observer implemented under Matlab Simulink. Reprinted with permission, Copyright 2009 American Institute of Physics [10]

(FT-S270) against the fixed actuator and therefore the applied force F_{ext} is known. Displacement, output voltage and force are recorded (Fig. 2.22). Force sensitivity β and beam stiffness k_Z are identified:

$$\begin{cases} k_Z = F_{\text{ext}}/(-\delta) \\ \beta = V_{\text{out}}/F_{\text{ext}}. \end{cases} \tag{2.32}$$

If a force sensor is not available, β parameter can be alternately calculated with only the laser displacement sensor aid. Transverse beam stiffness k_Z will

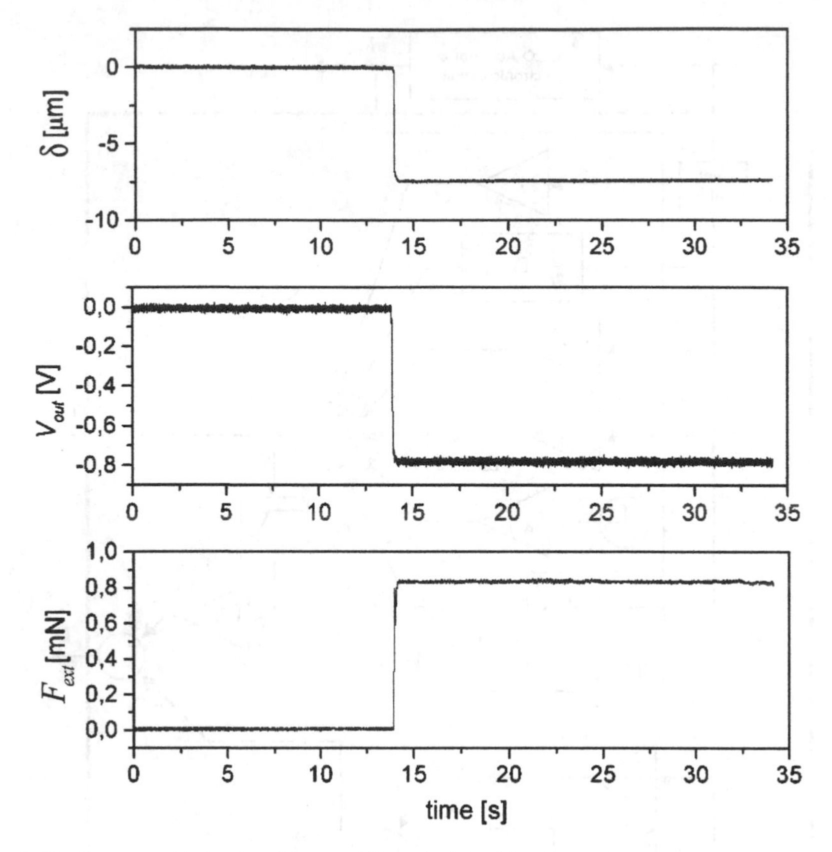

Fig. 2.22 Force sensitivity β and beam stiffness k_Z identification

be calculated based on known dimension and material properties. For a bi-morph cantilevered beam (Fig. 5), k_Z can be derived analytically from a known formula:

$$k_Z = 3YI/L^3 \tag{2.33}$$

where I is the bending moment of inertia around x_2.

Afterwards force sensitivity β will be estimated from:

$$\beta = V_{\text{out}}/(-\delta_{\text{meas}} k_Z). \tag{2.34}$$

2.4.5.2 Bias Current i_{BIAS} Identification

Bias current is known to vary exponentially with temperature. However, for a good amplifier and with some care regarding temperature variation, its influence will be small enough to consider i_{BIAS} constant [9]. Identification of i_{BIAS} is done under $F_{\text{ext}} = 0$, $V_{\text{in}} = 0$, $V_{\text{out}} \cong 0$ and stable environmental temperature, when current through the piezoelectric material is supposed null. The slight and linear V_{out} rate of change gives i_{BIAS} value, as in Fig. 2.23 (upper part).

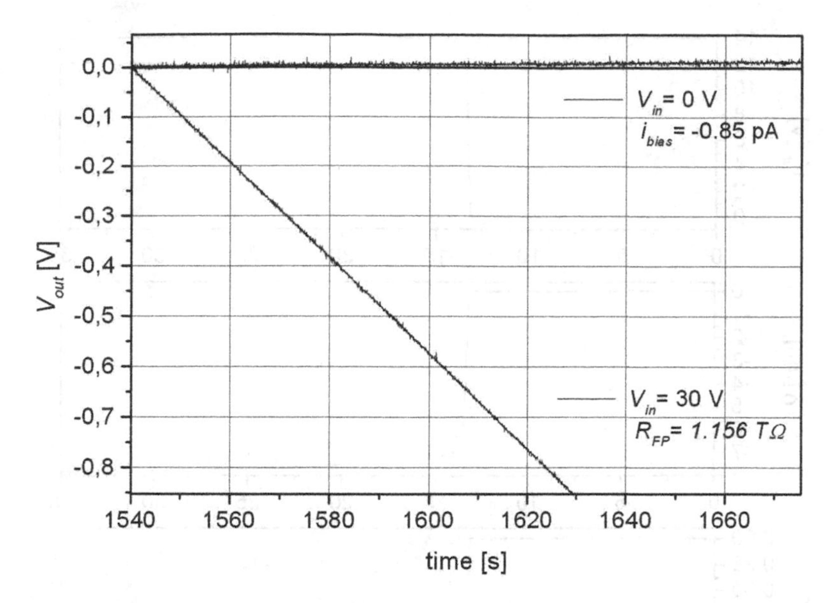

Fig. 2.23 Identification of bias and leakage current (resistance) of PZT actuator performed for input voltages of zero and +30 V, respectively

2.4.5.3 Leaking Resistance R_{FP} Identification

The procedure is similar to that presented in [9], under zero force $F_{\text{ext}} = 0$ and a constant applied voltage $V_{\text{in}} = 30\,\text{V}$ (see Fig. 2.23). After more than 1,000 s the piezoelectric material relaxation (creep) becomes negligible and V_{out} rate of change only depends on bias current (already identified) and on leaking resistance to be identified from (2.26) or (2.27).

As seen from Fig. 2.23, for a good circuit the influence of leaking resistance at high values of V_{in} is more significant than that of bias current. For some applications i_{BIAS} can even be neglected.

2.4.5.4 Displacement Coefficient α Identification

A ramped step signal is applied to the actuator submitted to null external force ($F_{\text{ext}} = 0$). The ramped signal of around 20 V/s suppresses transition oscillations. Displacement δ, input and output voltages V_{in} and V_{out} are recorded, as in Fig. 2.24. Coefficient α that links output voltage to displacement is computed easily with values immediately after the voltage step (that occurs at 13 s in the Fig. 2.24).

In our case, identified displacement coefficient was $\alpha = 3.435 \times 10^{-3}$C/m.

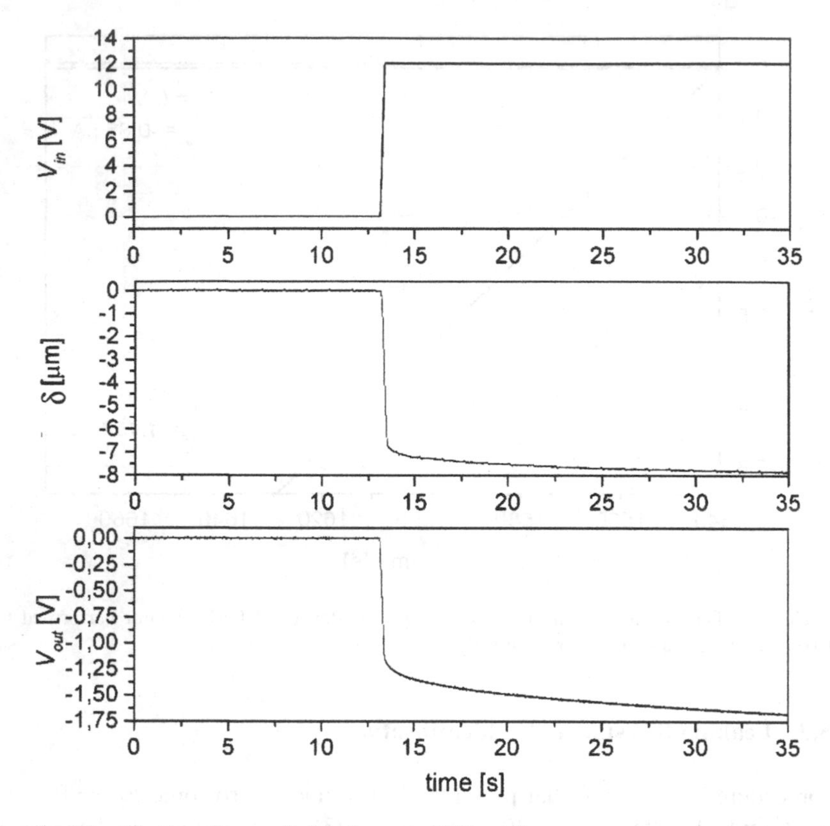

Fig. 2.24 Identification of α coefficient from a +12 V step input

2.4.5.5 Prandtl–Ishlinskii Hysteresis Operator Identification

It is well known that ferroelectric materials are characterized by hysteresis nonlinearity between applied voltage V_{in} and free actuator bending δ, as in Fig. 2.25.

There exist several hysteresis models for linearizing piezoelectric systems. The principle of the usually linearization method is the implemention of their inverse models (Fig. 2.26a), also known as feedforward compensation technique.

Cantilever bending δ is proportional with the charge [8,9]. In our case, there will be an inherent hysteresis between applied voltage V_{in} and output voltage V_{out}. Force self-sensing accuracy strongly depends on proper compensation of hysteresis. There are two equally possible options:

- Use of the feedforward compensation technique by cascading an inverse hysteresis operator, as in Fig. 2.26a
- Or substracting output signal with that of the direct operator (model). A principle scheme is given in Fig. 2.26b

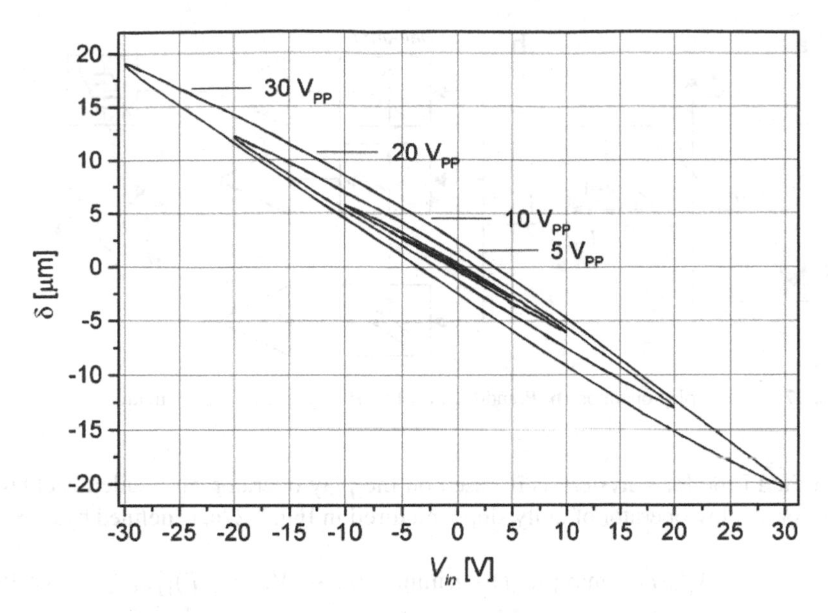

Fig. 2.25 Typical hysteretic behaviour of piezo actuators

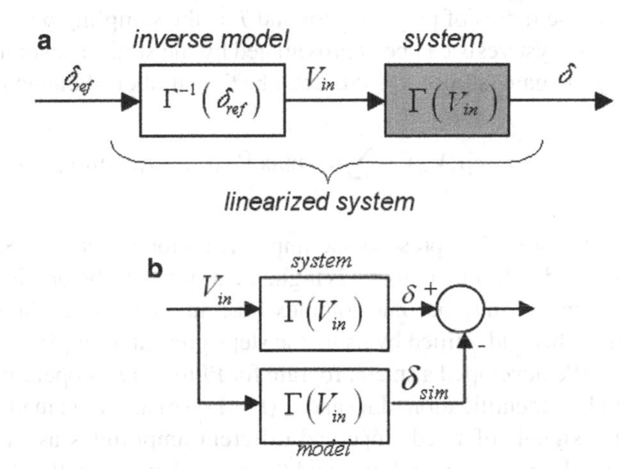

Fig. 2.26 (**a**) Nonlinear compensation of displacement by cascading the system with its inverse model. (**b**) Parallel compensation of nonlinear system. Reprinted with permission, Copyright 2009 American Institute of Physics [10]

The latter option (Fig. 2.26b) was chosen given the particularity that V_{in} applied signal can thus remains totally external to the self-sensing system (sometimes required by applications).

Hysteresis can be estimated/compensated with several methods: the Preisach model [30], the Prandtl–Ishlinskii model (PI) [3, 31, 32], the Bouc–Wen model [33] or asymmetric hyperbola shaped operator [34]. However, the PI model is especially appreciated for its ease of implementation, ease of obtaining the direct/inverse model and its accuracy. It was particularly discussed in previous papers [3].

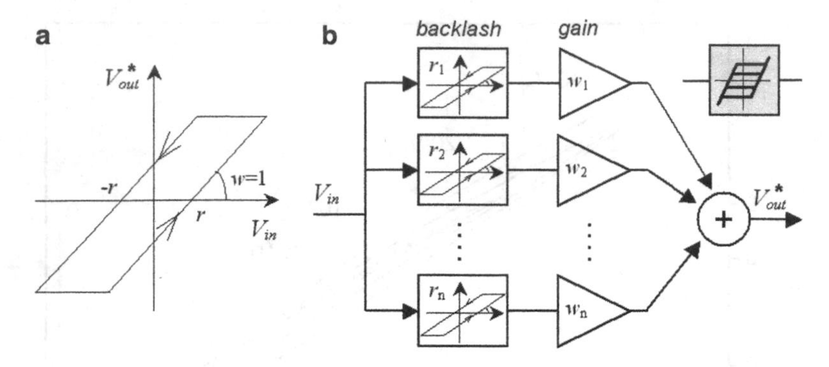

Fig. 2.27 (**a**) The play operator. (**b**) Prandtl–Ishlinskii (PI) hysteresis approximation

In the PI model, a hysteresis is based on the play operator, also called backlash operator. A play operator of unity slope, pictured in Fig. 2.27a, is defined by:

$$V^*_{\text{out}}(t) = \max\{V_{\text{in}}(t) - r, \min[V_{\text{in}}(t) + r, V^*_{\text{out}}(t - T)]\}, \tag{2.35}$$

where $V^*_{\text{out}}(t)$ is the output voltage compensated against bias and leaking currents, r is the radius of play operator and T is the sampling period.

A hysteresis can be approximated by the sum of several play operators weighted by the gain (slope) w_i [35]. Let n be the number of elements, so we have:

$$V^*_{\text{out}}(t) = F_{\text{H}}(V_{\text{in}}) = \sum_{i=1}^{n} w_i \cdot \max\{V_{\text{in}}(t) - r_i, \min[V_{\text{in}}(t) - r_i, V^*_{\text{out}}(t - T)]\}. \tag{2.36}$$

Figure 2.27b presents the implementation scheme in Simulink of the hysteresis PI model. If the number n is high, the model will be precise. However, it is too high, the model may become complex for implementation. Parameters $bw_i = 2r_i$ and w_i have been identified by using the steps presented in [3].

We developed a small program for PI hysteresis operator identification, as shown in [3]. Identification of w_i and r_i $(i = 1 \cdots n)$ arrays is made with a couple triangular V_{in} signals of fixed slope and different amplitudes as in Fig. 2.28a, b. The slope should be low enough to avoid the phase-lag due to the dynamics of the cantilever, which therefore would modify the hysteresis curve. However, it should not be too low in order to avoid the influence of the creep on the same hysteresis curve. In our case the slope was chosen at ±20 V/s. The maximum amplitude is chosen to be 30 V peak-to-peak, corresponding to the range use. The identification is performed with the signal of the largest amplitude. Smaller amplitude signals are further used for identified operator error evaluation. If errors are too high consider changing the signal slope or reduce the voltage interval.

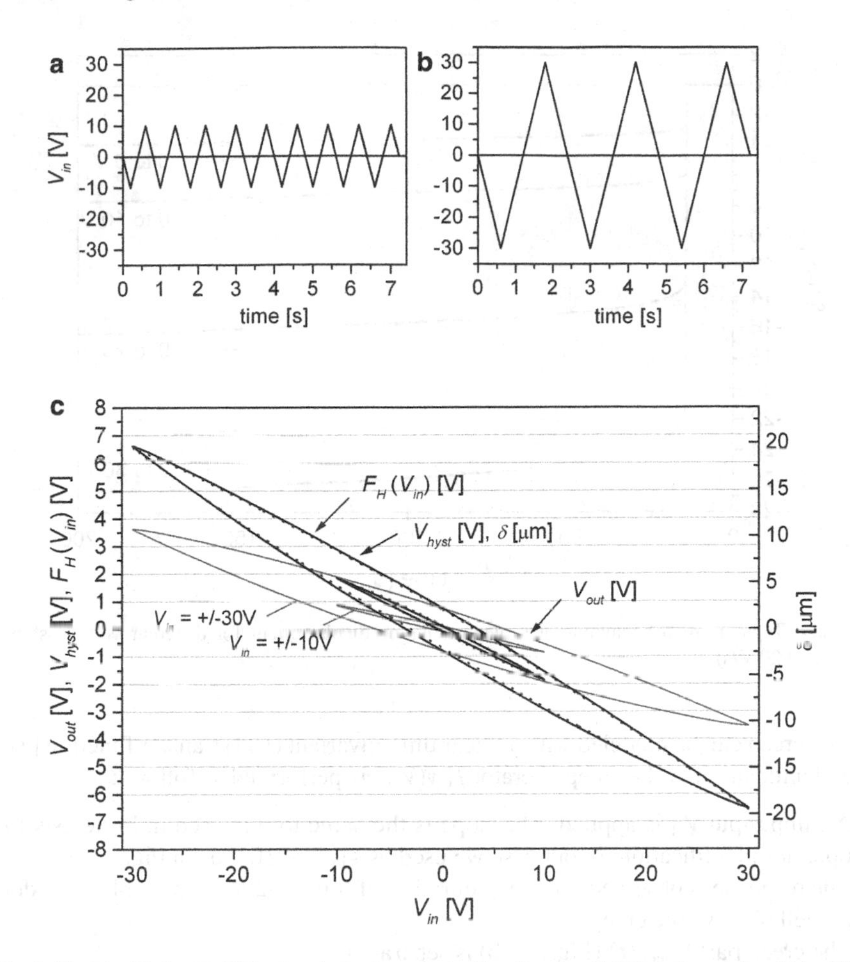

Fig. 2.28 (**a**)–(**b**) Applied Vin signals for hysteresis identification. (**c**) Experimental and identified curves

In Fig. 2.28c, experimental and identified hysteresis curves superpose in detail. Identified F_C operator consisted of $n = 40$ elements. For space-saving reasons, individual values are not represented; however, bw_i varied from 1.40 to 57.90 and w_i from -0.1179 to 0.2938.

2.4.5.6 Creep Transfer Function Identification

Creep effect is, like hysteresis, related to piezoelectric coefficients nonlinearity. Upon an applied V_{in} step signal, it is recorded that displacement will continue to drift for several hundred seconds (Fig. 2.29).

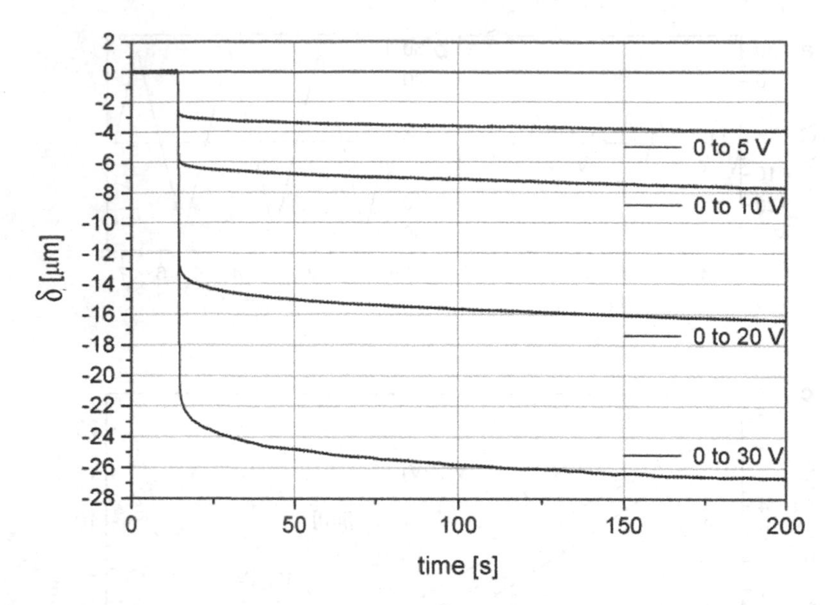

Fig. 2.29 Typical creep behaviour of a piezo actuator displacement for different voltage steps (ramped at 20 V/s)

The creep can be modeled with a linear time invariant (LTI) transfer function [3]. The identification of the creep operator $F_{Cy}(V_{\text{in}})$ is performed as follows:

- A ramp input V_{in} is applied. The slope is the same to that used in hysteresis PI operator identification. In our case we used as $V_{\text{in}} = -10\,\text{V}$ as in Fig. 2.30a.
- The response is observed for a long duration of time (>200 s if possible) in order to well observe the creep.
- The creep part $V_{\text{creep}}(t)$ (Fig. 2.30b) is separated:

$$V_{\text{creep}} = \left(V_{\text{out}} - \frac{C_{\text{R}}}{C} V_{\text{in}} + \frac{1}{C} \int \left(\frac{V_{\text{in}}(t)}{R_{\text{FP}}} + i_{\text{BIAS}} \right) \mathrm{d}t \right) - F_{\text{H}}(V_{\text{in}}). \tag{2.37}$$

- Having the input $V_{\text{in}}(t)$ and the output $V_{\text{creep}}(t)$, the model $F_{\text{C}}(V_{\text{in}})$ remains to be identified.

Several models with different orders were identified using the ARMAX method and Matlab. It seems that from the 3rd order, the error between the identified model and the experimental curve stops decreasing exponentially. We chose a model of 4rd order:

$$F_{\text{C}}(s) = \frac{V_{\text{creep}}(s)}{V_{\text{in}}(s)} = \frac{a_0 s^4 + a_1 s^3 + a_2 s^2 + a_3 s^1 + a_4}{s^4 + b_1 s^3 + b_2 s^2 + b_3 s^1 + b_4} \tag{2.38}$$

where $a_0 \ldots a_4$ and $b_1 \ldots b_4$ are numerator and denominator polynomial coefficients of creep transfer function.

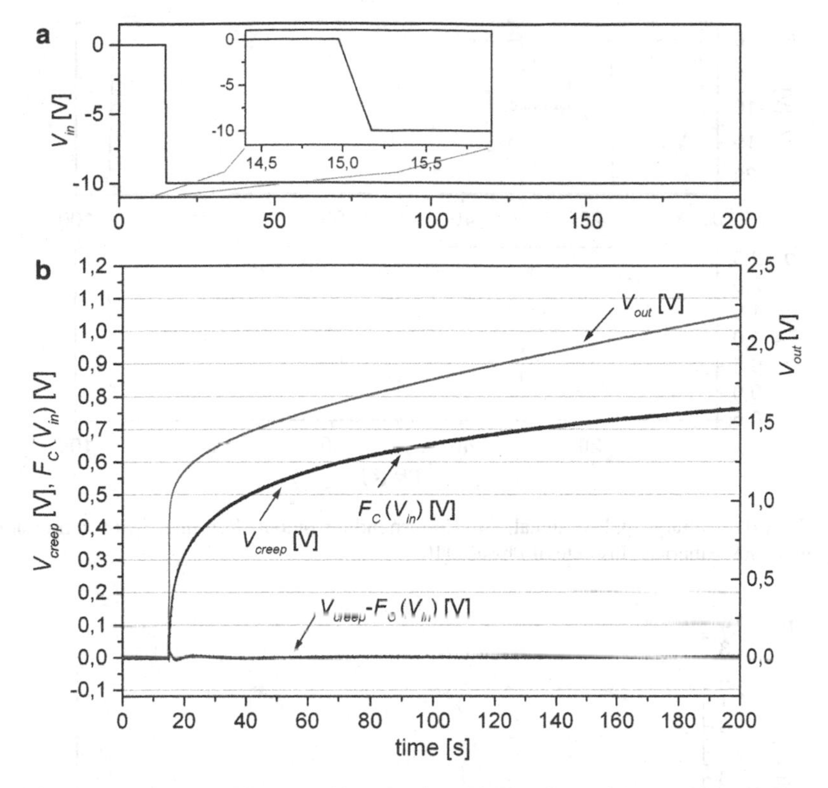

Fig. 2.30 (**a**) Applied signal for creep identification. (**b**) Experimental data and identified transfer function. Reprinted with permission, Copyright 2009 American Institute of Physics [10]

For instance, data plotted above were identified with the following transfer function:

$$F_C(s) = \frac{0.1601s^4 - 39.03s^3 - 767.1s^2 - 488.6s - 11.57}{s^4 + 130.3s^3 + 6.241e4s^2 + 1.229e4s + 145.5}. \quad (2.39)$$

2.4.6 Results and Discussion

In this section, some self-sensing results are presented issuing the explained schematic from Fig. 2.21 and subsequent identified parameters. A unimoph PZT on *Ni* cantilevered actuator ($15 \times 1 \times 0.28$ mm) is brought near the force sensor (close to contact) then a series of ramped negative periodic steps (Fig. 2.31) ranging from 0 to -10 and -20 V is applied, making the actuator to enter in contact and push the sensor. Maximum-recorded force was of 1.15 mN. As seen from Fig. 2.32, error between estimated and measured force sums 0.16 mN, meaning 14% (from +8% to -6%).

From Fig. 2.33 it results that total displacement is of 3 μm. Tip entered into contact with force sensor at 1.5 μm. Error between self-sensing method and optical sensor readings ranges from -0.65 to $+0.45$ μm (36% in total).

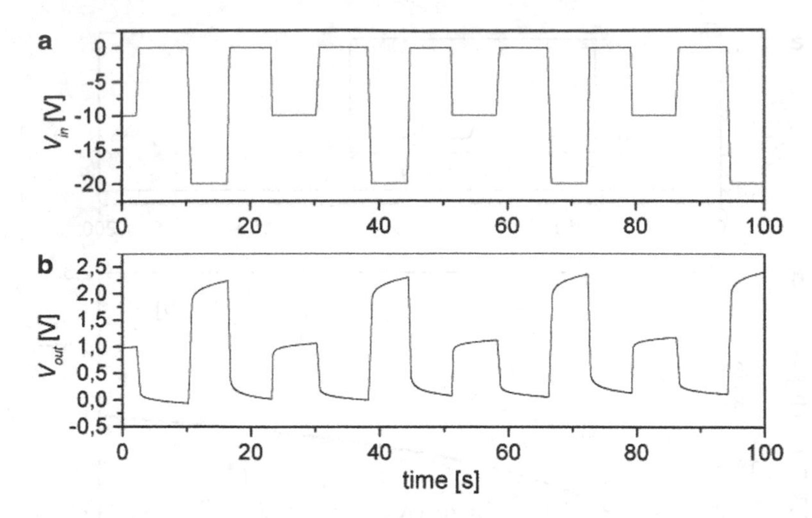

Fig. 2.31 (**a**) Arbitrary applied signal. (**b**) Corresponding output. Reprinted with permission, Copyright 2009 American Institute of Physics [10]

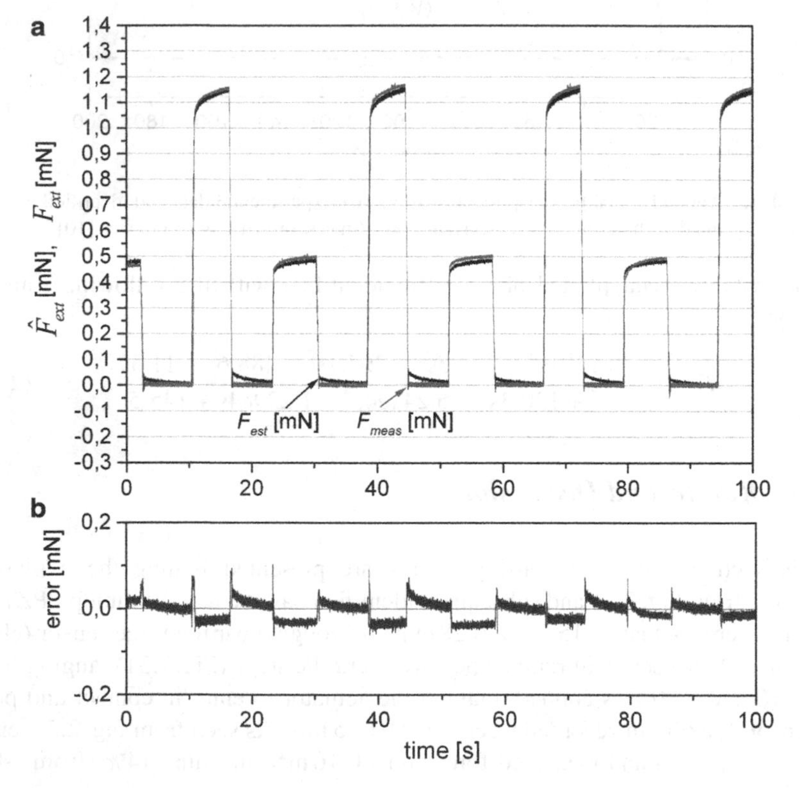

Fig. 2.32 (**a**) Measured and estimated force. (**b**) Absolute error. Reprinted with permission, Copyright 2009 American Institute of Physics [10]

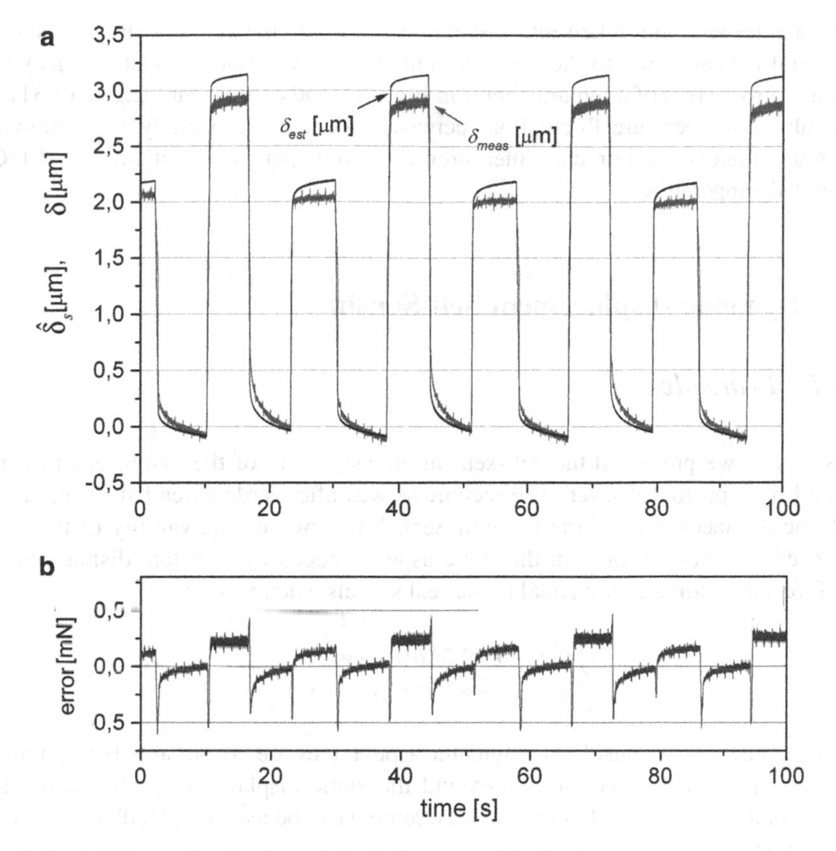

Fig. 2.33 (**a**) Measured and estimated tip displacement. (**b**) Absolute error

There are some limitations of the method. Displacement results are, as expected, slightly inferior in term of precision relative to those reported in only displacement self-sensing (see Sect. 2.3) but, in exchange, supplementary information about force was able to be provided. Also, as expected, force estimation is slightly less accurate in self-sensing actuator mode than in sensor-only mode (cantilever submitted only to external force) but the advantage of self-sensing consist in its double role. For instance self-sensing method would provide useful displacement and/or force information from the arms of a micro-gripper.

The displacement range or dynamics of the actuator are not influenced by the self sensing circuit, but as applied voltage increases, nonlinearties cause relative error to grow significantly, exceeding 50% for ±40 V operation. To overcome that, more complex hysteresis operator compensation should be employed and extra close contact information would also help in ameliorating precision. Indeed, a very accurate hysteresis and creep modeling is required because 1% of error in charge estimation due to applied voltage may cause nearly 20–30% of error in force detection.

Dynamics were not taken into account as we tested the actuator in quasi-static at several Hz compared to the first resonant frequency at 630 Hz, faster tasks will require introduction of a dynamic term in Fig. 2.21 and expressions (2.29)–(2.31).

Ambient temperature fluctuation increase uncertainty especially for unimoph layered actuators; in our case measurements were performed at around 0.1°C stabilized temperature.

2.5 Dynamic Displacement Self-Sensing

2.5.1 *Principle*

In Sect. 2.3, we presented the self-sensing measurement of the displacement performed by a piezocantilever. The technique was afterwards extended to measure both the displacement and the force in Sect. 2.4. However, the validity of the two proposed schemes are only in the static aspect. Indeed for constant displacement and force, the estimates are equal to the real signals when $t \longrightarrow \infty$:

$$\begin{cases} \hat{\delta}_s(t \longrightarrow \infty) \cong \delta(t \longrightarrow \infty) \\ \hat{F}_{\text{ext}}(t \longrightarrow \infty) \cong F_{\text{ext}}(t \longrightarrow \infty) \end{cases}. \tag{2.40}$$

Since they do not consider the dynamic aspect, they are not suitable for a control point of view. In this section, we extend the static displacement self-sensing in Sect. 2.3 into dynamic such that the measurement can be used in a feedback control application.

To extend the static displacement self-sensing into dynamic, we consider the initial scheme in Fig. 2.8 and add a dynamic part in the observer. We, therefore, obtain a dynamic estimator as depicted in Fig. 2.34 such that δ_{d} is the estimate displacement. The objective is to compute the gains of the dynamic part such that:

$$\hat{\delta}_{\text{d}}(t) \cong \delta(t);\ \forall t\,. \tag{2.41}$$

2.5.2 *Computation of the Gains of the Dynamic Estimator*

To compute the dynamic estimator, we first provide the transfer function between the static estimate $\hat{\delta}_s$ and the input voltage V_{in}. For that, denote $H(s)$ the transfer between V_{out} and V_{in}:

$$V_{\text{out}}(s) = H(s)V_{\text{in}}(s). \tag{2.42}$$

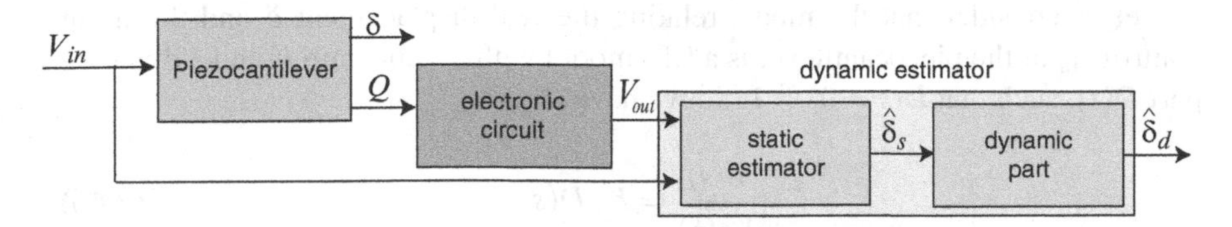

Fig. 2.34 Dynamic displacement self-sensing

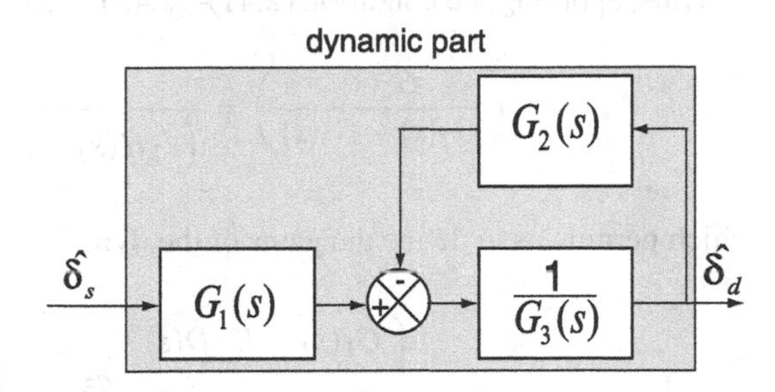

Fig. 2.35 Block-scheme for the dynamic estimator

Applying the Laplace transformation to (2.8), and using (2.9) and (2.42), we derive the following transfer function:

$$\frac{\hat{\delta}_s}{V_{\text{in}}} = \left[c_1 H(s) + c_2 - \frac{c_3}{s} - \frac{k_s}{\tau s + 1}\right]. \tag{2.43}$$

Consider the structure in Fig. 2.35, called inverse multiplicative structure, as the block diagram of the dynamic part of the estimator [11]. In the structure, $G_1(s)$, $G_2(s)$ and G_3 are gains of the estimator to be computed. The main advantage of this scheme is that there is no direct inversion of transfer function which prevents us therefore from some conditions (bi-causality[1] and bi-stability) on the different models, except for $G_3(s)$ which is, as we will see, a non-null real number (and therefore invertible).

From Fig. 2.35, we have:

$$\frac{\hat{\delta}_d(s)}{\hat{\delta}_s(s)} = \frac{G_1(s)}{G_2(s) + G_3(s)}. \tag{2.44}$$

Employing (2.43) and (2.44), we obtain the full transfer function relating the estimate displacement $\hat{\delta}_{\text{d}}$ and the input control V_{in}:

$$\frac{\hat{\delta}_d(s)}{V_{\text{in}}(s)} = \left(\frac{G_1(s)}{G_2(s) + G_3(s)}\right)\left(c_1 H(s) + c_2 - \frac{c_3}{s} - \frac{k_s}{\tau s + 1}\right). \tag{2.45}$$

[1]Bi-causality means causality of the model and of its inverse.

Let us consider that the model relating the real displacement δ and the input control V_{in} of the piezocantilever is a LTI-model with a static gain K and a dynamic part $D(s)$, such that $D(s=0)=1$. So we have:

$$\frac{\delta(s)}{V_{\text{in}}(s)} = K \cdot D(s). \tag{2.46}$$

Thus, applying the condition (2.41)–(2.45) and (2.46) yields:

$$\left(\frac{G_1(s)}{G_2(s)+G_3(s)}\right) = \frac{K \cdot D(s)}{\left(c_1 H(s) + c_2 - \frac{c_3}{s} - \frac{k_s}{\tau s+1}\right)} \tag{2.47}$$

which permits us to define the gains of the dynamic part of the estimator as follows:

$$\begin{cases} G_1(s) = K \cdot D(s) \\ G_2(s) = c_1 H(s) - \dfrac{c_3}{s} - \dfrac{k_s^*}{\tau s+1} \\ G_3(s) = c_2 \end{cases}. \tag{2.48}$$

From this result, it is confirmed that there is indeed no inversion of the transfer functions $D(s)$ and $H(s)$ as expected.

2.5.3 *Parameters Identification*

The experimental setup already used before is used here. Thus, the different parameters related to the static self-sensing are known and only K, $D(s)$ and $H(s)$ have to be identified. For that, we apply a step voltage $V_{\text{in}} = 20\,\text{V}$ to the cantilever. The output deflection δ of the cantilever is captured using the optical sensor. It allows the identification of K and $D(s)$ accordingly to (2.46). At the same time, the output voltage V_{out} is also captured which, according to (2.42), allows us identify $H(s)$. Figure 2.36 pictures the experimental results.

Applying the ARMAX systems identification to these experimental results, we obtain:

$$\begin{cases} K = 0.69 \left[\dfrac{\mu\text{m}}{\text{V}}\right] \\ D(s) = \dfrac{5.752\times10^{-3}(s+3.062\times10^4)(s^2-1.95\times10^4 s+3.076\times10^8)}{(s+3976)(s^2+54.37s+1.362\times10^7)} \end{cases} \tag{2.49}$$

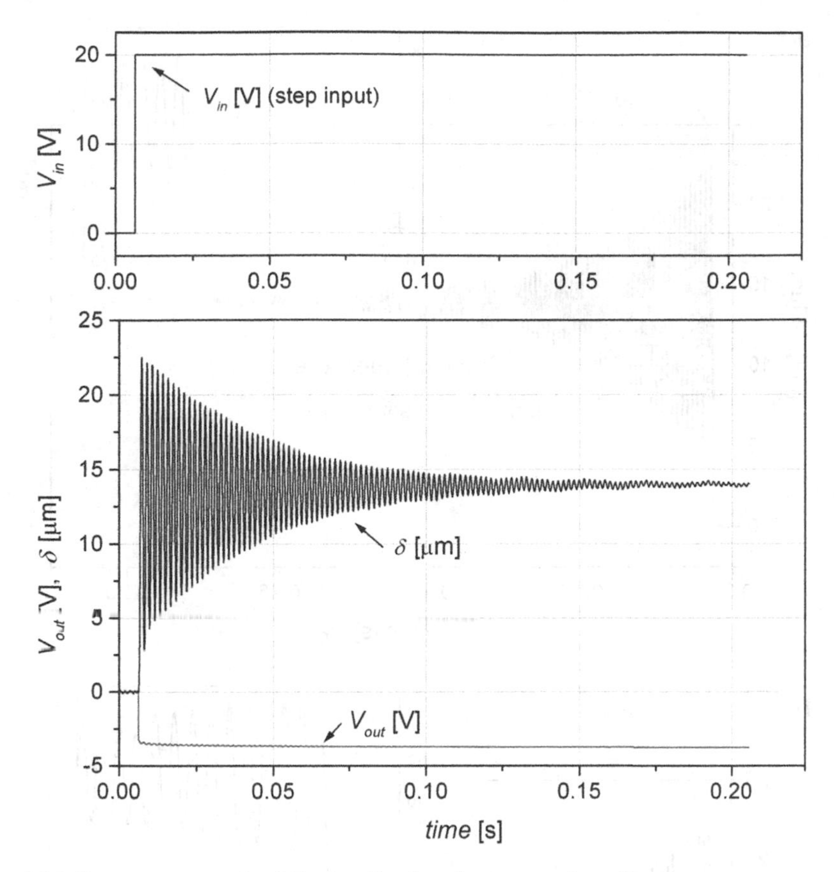

Fig. 2.36 Step response on the deflection δ and on the output voltage V_{out}

and

$$H(s) = \frac{-0.1584(s+5.911\times 10^4)(s+236)(s+13.74)}{(s+5.541\times 10^4)(s+224)(s+12.99)}. \tag{2.50}$$

2.5.4 Dynamic Displacement Self-Sensing Experimental Results

In order to prove the efficiency of the dynamic displacement self-sensing, it is implemented and the result is compared with that of the static displacement self-sensing. For that, a step input voltage $V_{\text{in}} = 20\,\text{V}$ is applied to the piezocantilever. The estimate $\hat{\delta}_{\text{s}}$ corresponding to the static self-sensing is captured. We also captured the estimate $\hat{\delta}_{\text{d}}$ of the dynamic self-sensing. Figure 2.37 shows the results. They clearly show that the measurement $\hat{\delta}_{\text{d}}$ from the dynamic self-sensing well captures both the transient part and the final value of the deflection δ while $\hat{\delta}_{\text{s}}$ from the static self-sensing only captures the final value.

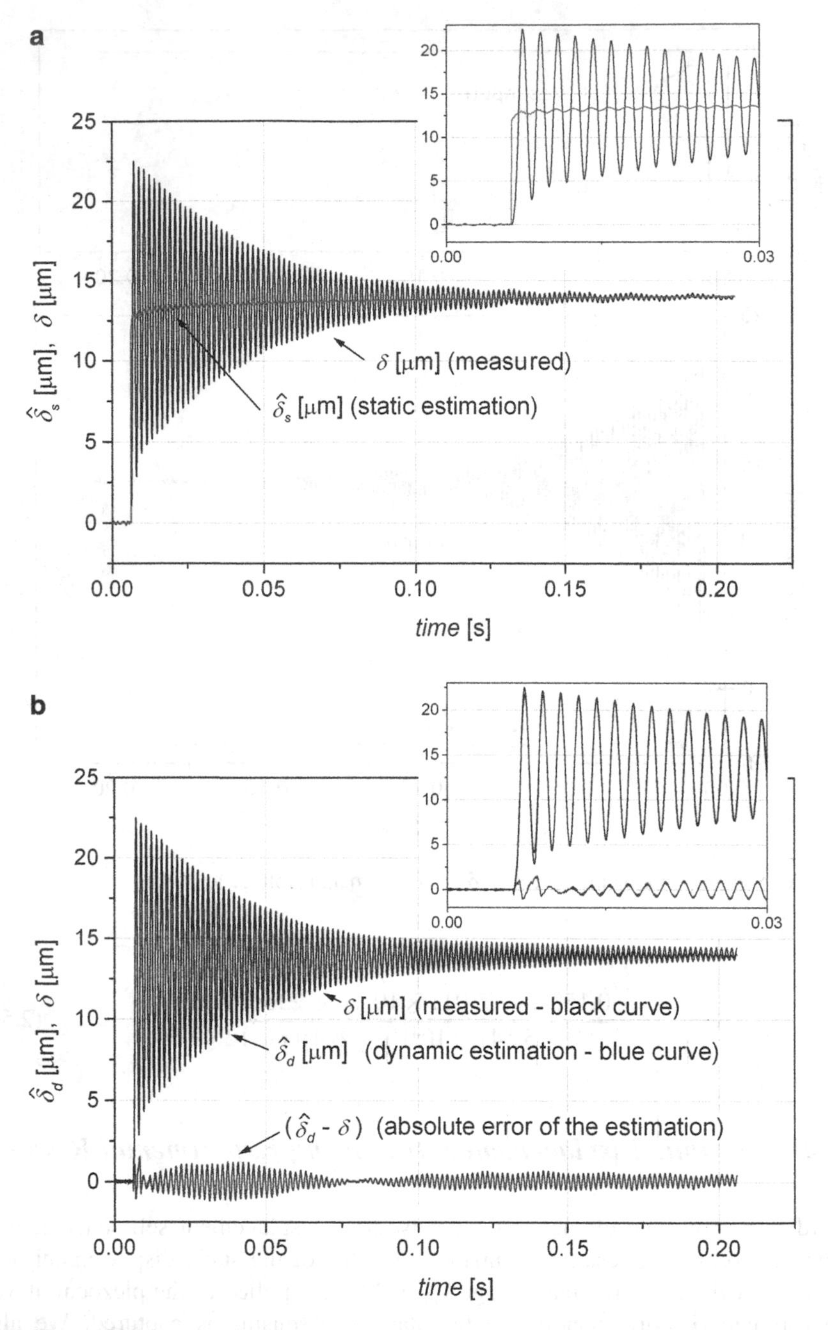

Fig. 2.37 (**a**) Result with the static displacement self-sensing. (**b**) Result with the dynamic displacement self-sensing

2.6 Conclusion

This chapter documented a self-sensing method to observe either displacement-only or both displacement/force observation at the tip of cantilevered piezoelectric actuators, generally when the latter is in contact with objects. Both the static and dynamic aspectes are accounted. The nonlinearities of piezoelectric coefficients causing static hysteresis and creep unwanted behaviour were taken into account and compensated allowing improving the proposed techniques in term of the precision. While the static hysteresis was identified with a Prandtl-Ishlinskii model, the creep was approximated by a 4th order LTI transfer function whose parameters were identified using ARMAX method. Furthermore, leakage resistance was also compensated. The proposed method is convenient for quasi-static applications as well as for feedback control utilization, such as in micromanipulation/microassembly tasks where several seconds of periods and control context are required.

References

1. Micky Rakotondrabe, Cédric Clévy and Philippe Lutz, "Modelling and robust position/force control of a piezoelectric microgripper", IEEE International Conference on Automation Science and Engineering (CASE), pp. 39–44, Scottsdale AZ USA, Sept 2007.
2. Micky Rakotondrabe and Alexandru Ivan, "Development and force/position control of a new hybrid thermo-piezoelectric microGripper dedicated to micromanipulation tasks", IEEE - Transactions on Automation Science and Engineering (T-ASE), DOI.10.1109/TASE.2011.2157683, 2011.
3. M. Rakotondrabe, C. Clέy and P. Lutz, "Complete open loop control of hysteretic, creeped and oscillating piezoelectric cantilevers", IEEE Transactions on Automation Science and Engineering (TASE), Vol.7(3), pp. 440–450, July 2010.
4. Jeffrey J. Dosch, Daniel J. Inman and Ephrahim Garcia, "A self-sensing piezoelectric actuator for collocated control", Journal of Intelligent Material Systems and Structures, Vol.3(1), pp. 166–185, January 1992.
5. W.W. Law, W-H. Liao and J. Huang, "Vibration control of structures with self-sensing piezoelectric actuators incorporating adaptive mechanisms", Smart Materials and Structures, Vol. 12, pp. 720–730, 2003.
6. Chee Khiang Pang, Guoxiao Guo, B. M. Chen, Tong Heng Lee "Self-sensing actuation for nanopositioning and active-mode damping in dual-stage HDDs", IEEE/ASME Transactions on Mechatronics (T-mech), Vol.11(3), pp. 328–338, 2006.
7. S. O. R. Moheimani and Y. K. Yong, "Simultaneous sensing and actuation with a piezoelectric tube scanner", Review of Scintific Instruments, 79, 073702, 2008
8. S. O. R. Moheimani, "Invited Review Article: Accurate and fast nanopositioning with piezoelectric tube scanners: Emerging trends and future challenges", RReview of Scintific Instruments, 79, 071101, 2008.
9. Alexandru Ivan, Micky Rakotondrabe, Philippe Lutz and Nicolas Chaillet, "Quasi-static displacement self-sensing method for cantilevered piezoelectric actuators", Review of Scientific Instruments (RSI), Vol.80(6), 065102, June 2009.
10. Alexandru Ivan, Micky Rakotondrabe, Philippe Lutz and Nicolas Chaillet, "Current integration force and displacement self-sensing method for cantilevered piezoelectric actuators", Review of Scientific Instruments (RSI), Vol.80(12), 2126103, December 2009.

11. Micky Rakotondrabe, Ioan Alexandru Ivan, Sofiane Khadraoui, Cédric Clévy, Philippe Lutz and Nicolas Chaillet, "Dynamic displacement self-sensing and robust control of cantilevered piezoelectric actuators dedicated to microassembly tasks", IEEE/ASME - AIM, (International Conference on Intelligent Materials) 2010, pp:557–562, Montreal Canada, July 2010.
12. J.G. Smits and S.I. Dalke, "The constituent equations of piezoelectric bimorphs", IEEE Ultras. Symp. pp. 781–784, 1989.
13. S. Muensit et al., "Piezoelectric coefficients of multilayer Pb(Zr,Ti)O3 thin films", Appl. Phys. A, 92, pp. 659–663, 2008.
14. T.S. Low and W. Guo, "Modelling of a three-layer piezoelectric bimorph beam with hysteresis", J. Microelectro mech. Syst., vol. 4, no. 4, pp. 230–236, 1995.
15. H. J. M. T. A. Adriaens, W. L. de Koning, and R. Banning, "Modelling piezoelectric actuators", IEEE/ASME Trans. on Mechatronics, vol. 5, no. 4, pp. 331–341. 104–110, 2008.
16. Y-G. Zhou et al., "Modeling of sensor function for piezoelectric bender elements", J. Zhejiang Univ. Sci. A. 9(1), 1–7, 2008.
17. R. J. Wood, E. Steltz, and R. S. Fearing, "Nonlinear performance limits for high energy density piezoelectric bending actuators", Proc. to IEEE ICRA, Spain, 2005.
18. Takigami et al., "Application of self-sensing actuator to control of a soft-handling gripper", Proc. to IEEE ICCA, pp. 902–906, Italy, 1998.
19. H. Jung, J.Y. Shim and D. Gweon, "New open-loop actuating method of piezoelectric actuators for removing hysteresis and creep" Review of Scientific Instruments, 71 (9), pp. 3436–3440, 2000.
20. K. Kuhnen, H. Janocha, "Compensation of the creep and hysteresis effects of piezoelectric actuators with inverse systems", Actuator 98, Proceedings 6th International Conference on New Actuators (Bremen 17.-19.6.1998), Bremen: ASCO-Druck, pp. 426–429, 1998.
21. R. Changhai and S. Lining, "Hysteresis and creep compensation for piezoelectric actuator in open-loop operation", Sensors and Actuators, 2005, vol. 122, no1, pp. 124–130.
22. J. Minase, T-F. Lu and B.Cazzolato, "Adaptive identification of hysteresis and creep in piezoelectric stack actuators", Proc. to 8th Int. Symp. for Meas. Tech. and Intell. Instr. (ISMTII), Japan, 2007.
23. Y. Cui et al. "Self-sensing compounding control of piezoceramic micro-motion worktable based on integrator", Proc. to 6th World Cong. On Intell. Con. and Autom., China, 2006.
24. A. Badel et al. "Self sensing force control of a piezoelectric actuator", IEEE Trans. on UFFC, vol. 55, no. 12 pp. 2571–2581, 2008.
25. A. Badel et al., "Self-sensing high speed controller for piezoelectric actuator", J. of Intell. Mater. Syst. and Struct., vol 19, no. 3, pp. 395–405, 2008.
26. M. Goldfarb and N. Celanovic, "Modeling piezoelectric stack actuators for control of micromanipulation", IEEE Control Systems Magazine, vol. 3, no. 3, pp. 69–79, 1997.
27. A. S. Putra et al., "With Adaptive Control in Applications With Switching Trajectory", IEEE/ASME Trans. on Mechatronics, vol. 13, no. 1, pp. 104–110, 2008.
28. D. Campolo, R. Sahai, R.S. Fearing, "Development of piezoelectric bending actuators with embedded piezoelectric sensors for microelectromechanical flapping mechanisms", Proc. to IEEE Int. Conf. on Robot. Autom. Vol. 3, pp. 3339–3346, Taipei, 2003.
29. J. G. Smits and W-S. Choi., "The constituent equations of piezoelectric heterogeneous bimorphs", IEEE Ultrasonic Symposium, pp. 1275–1278, 1990.
30. A. Dubra and J. Massa and C.l Paterson, "Preisach classical and nonlinear modeling of hysteresis in piezoceramic deformable mirrors", Optics Express, pp. 9062–9070, 2005.
31. K. Kuhnen and H. Janocha, "Inverse feedforwrad controller forcomplex hysteretic nonlinearities in smart-materials systems". Control and Intelligent Systems, vol. 29, no. 3, pp. 74–83, 2001.
32. B. Mokaberi and A. A. G. Requicha, "Compensation of scanner creep and hysteresis for AFM nanomanipulation", IEEE Trans. on Autom. Sci. and Eng., pp. 197–208, 2008.
33. Micky Rakotondrabe, "Bouc-Wen modeling and inverse multiplicative structure to compensate hysteresis nonlinearity in piezoelectric actuators", IEEE - Transactions on Automation Science and Engineering (T-ASE), Vol.7, Issue.4, DOI.10.1109/TASE.2010.2081979, December 2010.

34. A. Badel, J. Qiu and T. Nakano, "A New Simple Asymmetric Hysteresis Operator and its Application to Inverse Control of Piezoelectric Actuators", IEEE Trans. on UFFC, Vol. 55–5, pp. 1086–1094, 2008.
35. M.A. Krasnosel'skii and A.V.Pokrovskii, "Systems with Hysteresis", Springer-Verlag, Berlin, New York, 1989 (410p).

Chapter 3
Kalman Filtering Applied to Weak Force Measurement and Control in the Microworld

Mokrane Boudaoud, Yassine Haddab, and Yann Le Gorrec

Abstract Dexterous manipulation of small components and assembly of microsystems require measurement and control of gripping forces. In the microworld,[1] the main limitation for force sensing is the low signal to noise ratio making very difficult to achieve efficient force measurements when useful signals magnitudes are close to noise level. Thus, optimal filters allowing both filtering the noise without loss of dynamic measurements and an easy real time implementation for force control are required. This chapter deals with gripping force measurement and control in the microworld describing successful uses of the optimal Kalman filtering to overcome the limitations due to noise. Two applications are then presented: the first one focuses on the improvement of strain gauges micro-forces measurement using an optimal Kalman filter, and the second one describes a successful implementation of a LQG (Linear–quadratic–Gaussian) gripping force controller on an electrostatic microgripper for the dexterous manipulation of 80 µm glass balls.

Keywords Micro-force measurement • Measurement noise • Kalman filtering • Gripping force control • Piezoelectric cantilevers • Strain gauges • Electrostatic microgripper • Micro-glass balls manipulation

3.1 Introduction

Force measurement is the key issue to perform efficient and dexterous manipulation tasks in the microworld. Indeed, sensing and controlling gripping forces allow

[1]The microworld defines the world of very small objects whose typical size is ranging from several micrometers up to some millimeters.

M. Boudaoud (✉)
FEMTO-ST Institute, 32, avenue de l'Observatoire 25044 Besançon, France
e-mail: mokrane.boudaoud@femto-st.fr

C. Clévy et al. (eds.), *Signal Measurement and Estimation Techniques for Micro and Nanotechnology*, DOI 10.1007/978-1-4419-9946-7_3,

the manipulation of fragile samples without destroying or damaging them. This is often needed in microassembly [1], minimally invasive surgery and cell mechanical characterization [2–4]. In that sense, various methods can be used to measure the force applied on manipulated objects. Generally, force information is evaluated from position measurement. Then, the deformation of small devices (cantilevers, membranes, etc.) whose stiffness is known allows force sensing. Moreover, according to the desired resolution, various physical effects and materials can be used [5]. Small strain gauges glued on deformable devices such as cantilevers lead to the *mN* resolution when using centimeter sized cantilevers and the μN resolution with microfabricated cantilevers, piezoelectric materials, in particular PVDF (polyvinylidene fluoride) allow measurements below 0.1 µN and the electrostatic principle used in microfabricated sensors offers the possibility of measuring forces with resolutions reaching the *nN*. Moreover, *pN* resolutions can be obtained when using optical sensors such as often performed on AFM (Atomic Force Microscope) cantilevers.

Many papers reported the successful integration of force sensors into micromanipulation systems such as microgrippers [6–8]. In many cases, due to the weak amplitude of measurements, signal to noise ratio is often very low leading to a significant loss of resolution. This undesired physical phenomenon is a severe limitation for an efficient gripping force measurement making it impossible to be used in feedback control. The range of forces that can be controlled is then considerably limited in the microworld. Noise is partly due to electronic devices (Johnson-Nyquist noise, 1/f noise, shot noise,etc.) and also to the vibrations of mechanical structures[2] due to external perturbations such as the environmental noise.[3] In order to reach wanted resolutions, filtering the noise without loosing the dynamics of useful signals is of great interest for the improvement of achievable closed loop performances in control point of view. For this purpose, to be used in feedback control, the filter must allow an easy real time implementation. Kalman filtering can be advantageously used to deal with such requirements, for which traditional filters are often limited. For instance, low-pass filtering allows improving the resolution of static and slowly varying forces but fast forces are filtered. On the other hand, filtering noise signals without loosing dynamic measurements is possible with algorithms such as the spectral subtraction [9] but a real time implementation for feedback control is not possible.

Kalman filter is a powerful tool which addresses the general problem of estimating states of a noisy process in order to enhance the sensing resolution. This filter is theoretically attractive because among many existing filters, it is one of the few for which working principle is based on the minimization of the variance error on the state estimation. For this reason, the filter is considered as an optimal observer and is generally used in feedback control. Moreover, thanks to the

[2]When end effectors of a microgripper are used as sensors, vibrations are reflected in the sensor output as a measurement noise.

[3]Environmental noise can include effects of external temperature, humidity, pressure, etc. and also relates about ground motion and acoustic noises at a given location.

separation principle, the Kalman filter can be computed and adjusted independently of a controller. Kalman filters can deal with linear time invariant systems and also with nonlinear systems using the extended form (Extended Kalman Filter – EKF). For force estimation, several authors have reported successful uses of this filter, mainly in macroscopic systems. In [10], a Kalman filter has been used for the identification of contact and grasping uncertainties and for monitoring the force control in assembly operations, in [11, 12] a Kalman filter has been used to estimate forces applied on beam structures and in [13] a Kalman filter is applied for force-controlled robot tasks and especially for estimating the contact state based on force sensing between a grasped object and its environment. Otherwise, the filter is widely applied in various fields of scientific research for instance dynamic positioning, radar tracking, speech enhancement, navigation systems, and finds interesting applications in microrobotics such as described in this paper.

The chapter is organized as follows. In Sect. 3.2, a very brief summary of the standard Kalman filter applied to linear time invariant systems is presented. To illustrate the efficiency of the optimal filter, results on the improvement of strain gauges micro-forces measurement using Kalman filtering are presented in Sect. 3.3. In Sect. 3.4, the use of the filter in feedback control through the LQG (Linear–quadratic–Gaussian) algorithm is described. This controller is implemented on an electrostatic microgripper for the dexterous manipulation of 80 μm glass balls. Discussion and remarks are given for both applications. Conclusions and perspectives for further work are presented in the last section.

3.2 The Standard Kalman Filter

In this section, the original formulation of the Kalman filter [14] is presented. The process is described into a discrete state space representation and noise signals are considered through a statistical description.

Consider a discrete time invariant process governed by the following linear state space representation:

$$X(k+1) = A.X(k) + B.u(k). \tag{3.1}$$

$$y(k) = C.X(k). \tag{3.2}$$

Where $A \in \Re^{n \times n}$, $B \in \Re^{n \times p}$, and $C \in \Re^{q \times n}$ relate to the state, input and output matrices, respectively, and $X(k) \in \Re^{n}$ denotes the state vector at a time step k.

In the Kalman theory, two random variables, namely the process noise $w(k) \in \Re^{n \times p}$ and the measurement noise $v(k) \in \Re^{q \times n}$ are considered, leading to a statistical description of the noisy process, such as:

$$X(k+1) = A.X(k) + B.u(k) + M.w(k). \tag{3.3}$$

$$y(k) = C.X(k) + v(k). \tag{3.4}$$

The M matrix models the relationship between the state and the process noise.

The process and measurement noises are characterized by their W and V variances respectively, and are assumed to be independent of each other, white and with a normal probability distribution.

The filter provides a mathematical algorithm which recursively estimates the process state vector taking into account process and measurement noises in a way of minimizing the estimation error variance. This is conducted following the repetition of two steps: time update and measurement update.

During time update step, the a priori state $X_e(k/k-1)$ and error variance $P_e(k/k-1)$ are predicted at time k according to the model at time $k-1$:

$$X_e(k/k-1) = A.X_e(k-1/k-1) + B.u(k-1). \tag{3.5}$$

$$P_e(k/k-1) = A.P_e(k-1/k-1).A^T + M.W.M^T. \tag{3.6}$$

With: $P_e(k/k-1) = E[(X(k) - X_e(k/k-1)).(X(k) - X_e(k/k-1))^T]$

Then, a gain matrix $K_e \in \Re^{n\times q}$ is computed such as:

$$K_e(k) = P_e(k/k-1).C^T.[C.P_e(k/k-1).C^T + V]^{-1}. \tag{3.7}$$

Thereafter, during measurement update step, the measurement $y(k)$ and the gain matrix are used in order to correct the state and the error variance to obtain their a posteriori form:

$$X_e(k/k) = X_e(k/k-1) + K_e(k).[y(k) - C.X_e(k/k-1)]. \tag{3.8}$$

$$P_e(k/k) = [\text{eye}(n) - K_e(k).C].P(k/k-1), \tag{3.9}$$

With: $P_e(k/k) = E[(X(k) - X_e(k/k)).(X(k) - X_e(k/k))^T]$

From (3.8), when the gain matrix is high, the state correction is mainly based on the measurement $y(k)$ and little on the model. This is the case when the a priori error variance $P_e(k/k-1)$ is high (see (3.9). Otherwise when the a priori state is correctly predicted from the process model, the measurement is 'trusted' less and less.

Moreover, as explained above, the prediction of the a priori state at a time k requires only information about the model at the previous step time. This is an important advantage of the filter which allows an easy real time implementation.

Applying the Kalman filter on a noisy process requires (Fig. 3.1) first an accurate modeling of the process leading to the description of the state space representation free from undesired noises (see (3.1) and (3.2)). For this purpose, the process can be identified using measurements from an external high resolution sensor or described by a knowledge based model. Afterwards, the characterization of the noise has to be performed. Process and measurement noises have to be in accordance with the specifications of the Kalman theory (Gaussian, centered and white) and their variance have to be evaluated. With the assumption of stationary noises, the gain matrix can be computed offline, and its steady state can be thereafter

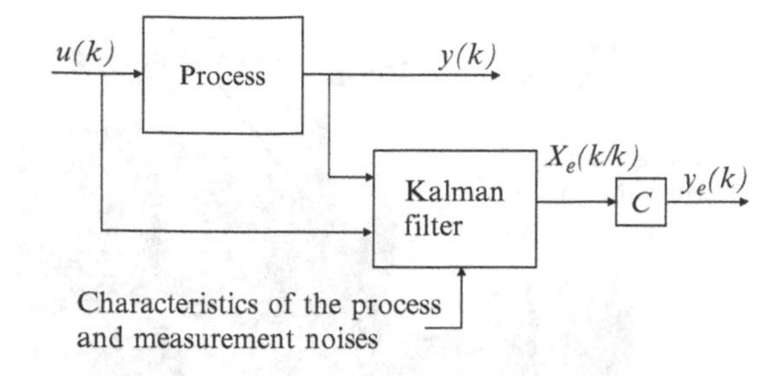

Fig. 3.1 Estimation of the process state and output using a Kalman filter

implemented in real time into the process. However, when the environmental noise has a significant effect on the process, the gain matrix has to be computed online and noise characteristics must be periodically updated. For more explanations about the Kalman filter, the reader can refer to [14] and [15].

3.3 Kalman Filtering Applied to Micro-Forces Measurements Using Strain Gauges

Integration of small sensors into microgrippers allows performing efficient manipulation tasks in confined environments and leads to more flexibility and mobility than the use of external sensors. In this application, strain gauges are glued on a micromanipulation system. The sensor provides a low signal to noise ratio reducing significantly the sensing resolution. A Kalman filter is used for the improvement of micro-forces measurements. Measurement and process noises are characterized in time and frequency domains and the standard Kalman filter is implemented in real time for weak force measurements.

3.3.1 Force Measurement Using the Strain Gauges

A micromanipulation system integrating strain gauges for gripping force measurement is designed. The system is made up of two parts (Fig. 3.2): a linear motor from PiezoMotor and a piezoelectric bimorph cantilever equipped with electrodes. The linear motor is used to perform displacements of the cantilever with a 10-nm positioning resolution and a stroke of 35 mm. The linear motor is not studied in this section. The dimensions of the cantilever are: $16 \times 2 \times 0.5$ mm. When a voltage V_{in} is applied across the electrodes of the piezoelectric cantilever, the latter bends. Two small strain gauges from ENTRAN are glued on the faces of the cantilever close to the fixed end. The cantilever and the strain gauges constitute a half of a gripper and the tip of the cantilever is used to handle micro-objects. The dimensions of the stain gauges are: 1.27×0.38 mm.

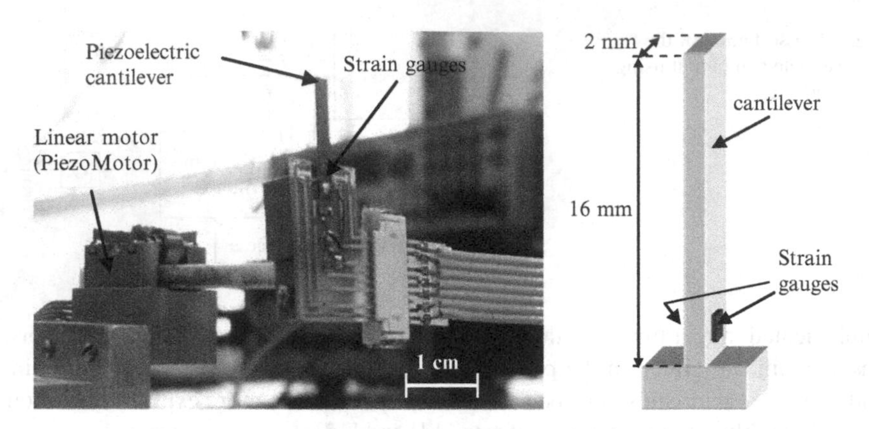

Fig. 3.2 Constitution of the designed system

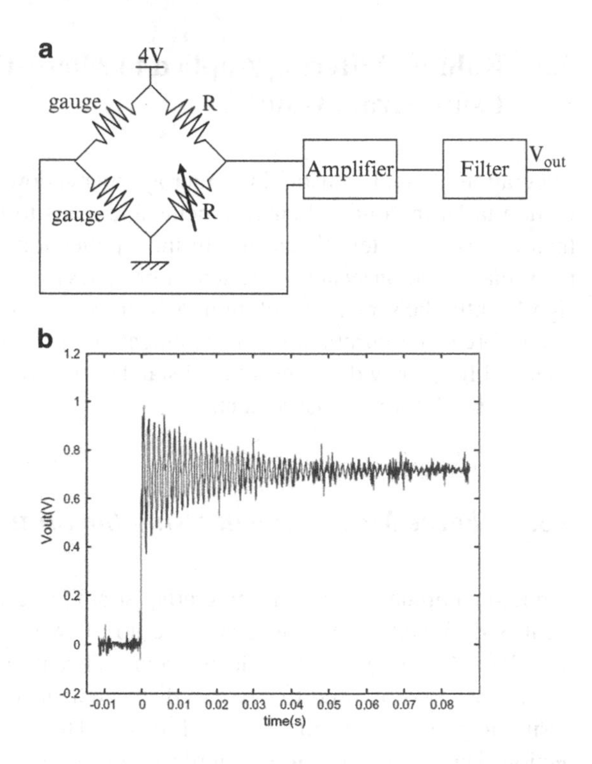

Fig. 3.3 Setup for force measurement using strain gauges and measured voltage V_{out} in response to a step voltage of 15 V

In order to achieve efficient force measurements from the strain gauges, a thermal compensation is needed. Then, the two strain gauges are included in a Wheatstone bridge and two other resistors are used to complete the bridge. An amplifier with a gain of 51.3 is used to increase the voltage level, and the signal is filtered using a low pass filter (Fig. 3.3a). The cutoff frequency of the filter is more than four times greater than the resonant frequency of the cantilever. The filter is used to eliminate high frequency noise which is outside the system's frequency bandwidth.

The measurement system has been calibrated following the technique proposed in reference [16] and derives a sensor gain of 52344.6 V/m. On the other hand, the compliance of the cantilever is about 1.23 μm/mN. The gripping force applied at the tip of the cantilever can be then derived from the output voltage V_{out} (Fig. 3.3a) taking into account the calibration parameters.

In response to a step voltage (voltage applied to the piezoelectric cantilever) of 15 V, the output measured signal V_{out} (Fig. 3.3b) shows that the passband of the measurement system is sufficient to reproduce the dynamic behavior of the cantilever. However, the level of the measured noise is high and it limits the measurement's accuracy. This noise can reach 1.98 mN in the worst case leading to measurements errors. Reducing the cutoff frequency of the filter can attenuate the noise but it will result in a loss of dynamic measurements.

3.3.2 Noise Characterization and Kalman Filter Implementation

From [16], a discrete state space model of the measurement system is obtained. The model has been identified using a high resolution (10 nm) laser sensor from KEYENCE which allows the measurement of the cantilever tip displacement when applying a 15 V step voltage. The input of the model is the supply voltage V_{in} while the output is the deflexion δ at the tip of the cantilever:

$$\begin{aligned} X(k+1) &= A.X(k) + B.V_{in}(k) \\ \delta(k) &= C.X(k) \end{aligned} \tag{3.10}$$

$$A = \begin{pmatrix} 0 & 1 & 0 & 0 \\ 0 & 0 & 1 & 0 \\ 0 & 0 & 0 & 1 \\ 0.04258 & 0.70475 & -2.50860 & 2.75595 \end{pmatrix} B = \begin{pmatrix} 0 \\ 0 \\ 0 \\ 1 \end{pmatrix}$$

$$C = \begin{pmatrix} 0 & 3.6139 \times 10^{-8} & -6.60332 \times 10^{-8} & 3.47543 \times 10^{-8} \end{pmatrix}.$$

The model is then presented into its statistical description:

$$X(k+1) = A.X(k) + B.V_{in}(k) + M.w(k). \tag{3.11}$$

$$\delta(k) = C.X(k) + v(k). \tag{3.12}$$

Process and measurement noises have to be characterized experimentally and noise variances must be evaluated for the Kalman filter implementation.

The noise characterization is performed under the following assumptions:

- The process noise is mainly due to the input (noise produced by the voltage generator).
- The process and measurement noises are independent from each other because they are generated by different devices.

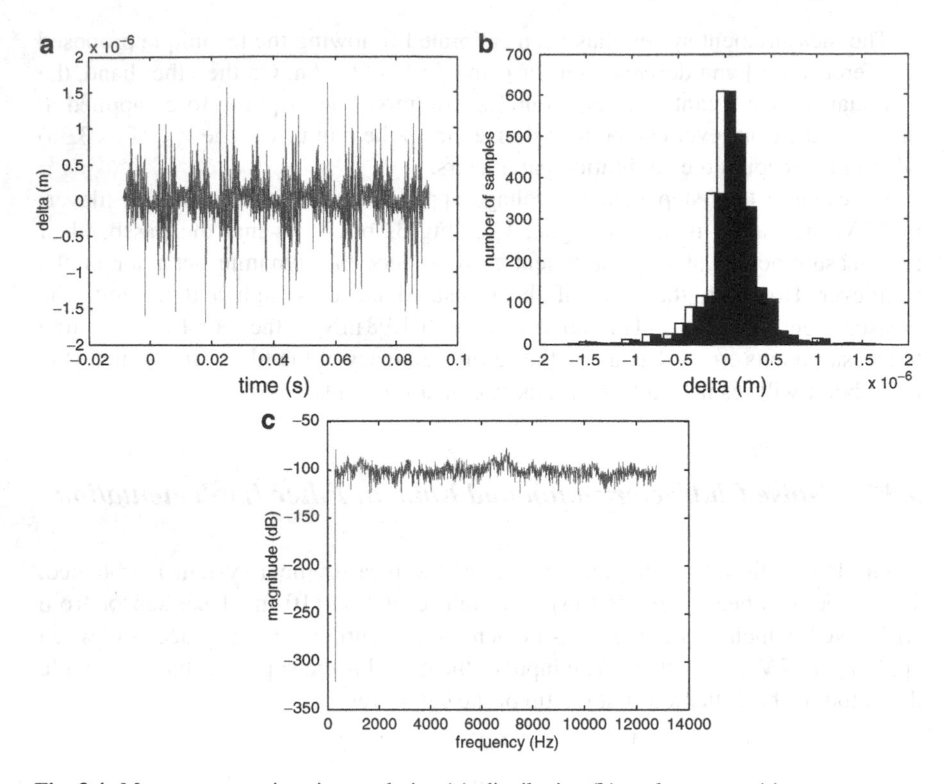

Fig. 3.4 Measurement noise: time evolution (**a**), distribution (**b**), and spectrum (**c**)

The process noise $v_n(k)$ is then defined such as: $w(k) = B.v_n(k)$ and is measured at the input (supply voltage) when 'zero input is applied'. The measurement noise (Fig. 3.4a) is recorded using the strain gauges measurement system when no input is applied (disconnected from the supply voltage).

From the measurement noise, the statistical distribution is evaluated (Fig. 3.4b) and the spectrum is computed (Fig. 3.4c). According to this analysis, the recorded signal is a white noise characterized by a Gaussian distribution with a mean of 0. The same analysis has been performed on the process noise, it has shown similar characteristics to those of the measurement noise.

Both measurement and process noises are in accordance with the Kalman theory hypothesis. The variances of the process and measurement noises have been extracted from recorded data for the computation of the filter.

To assess the effectiveness of the filter, the deflection of the cantilever in response to a 15 V step voltage is analyzed. Then, the Kalman filter is implemented in real time on the system. The deflection δ given from the output of the measurement system (Fig. 3.5a) and the one estimated by the filter (Fig. 3.5b) are recorded. As we can see in experimental results, the Kalman filter allows filtering the measurement noise without loss of the dynamic measurement. The measurement noise amplitude

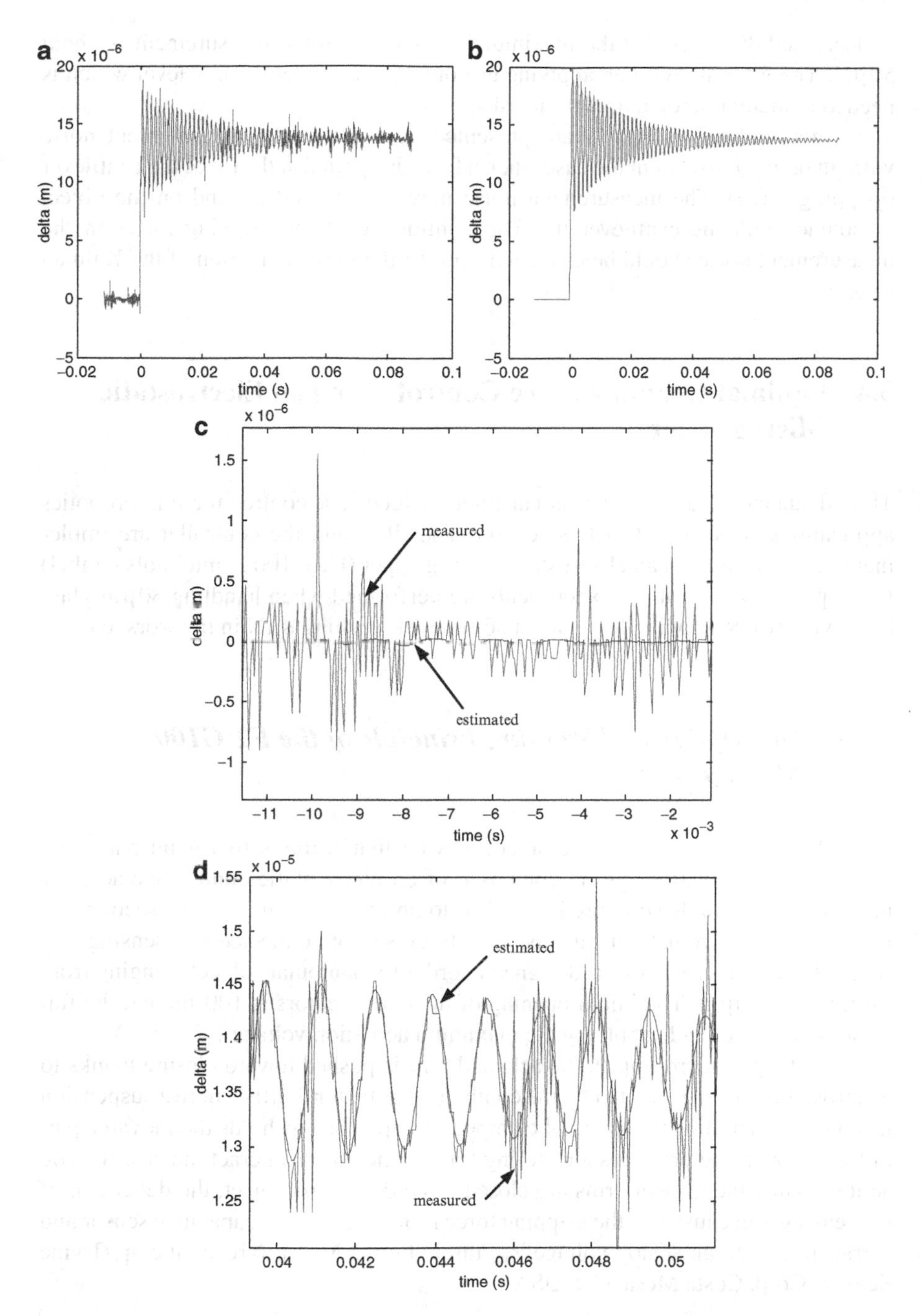

Fig. 3.5 Deflection δ in response to a step voltage of 15 V: measurement (**a**) and Kalman estimation (**b**). Comparison between the measured and estimated deflection δ in static (**c**) and dynamic (**d**) mode

is decreased 40 times and the maximum error in the force measurement is about 50 μN. The filter allows then applying gripping forces under the *mN* level which is needed in many micromanipulation tasks.

In this application, results are presented considering the measurement noise without taking into acount the case when a force is applied at the tip of the cantilever (gripping force). The measurement noise may change and depend on the object in contact with the cantilever tip. Then, influence of the gripping force on the measurement noise should be evaluated prior to the implementation of the Kalman filter.

3.4 Optimal Gripping Force Control Using an Electrostatic Microgripper

The advantage of using the Kalman filter in feedback control for microrobotics applications is presented in this section. The filter and the controller are implemented in real time on an electrostatic microgripper (FT-G100 FemtoTools GmbH) for gripping force control. Experiments are performed when handling 80 μm glass balls with 10 μN gripping force and a noise level reaching 8 μN in the worst case.

3.4.1 Description and Working Principle of the FT-G100 Microgripper

The FT-G100 microgripper is a silicon device that features two main parts: the first one is an actuation system composed of an electrostatic comb-drive actuator in which the movable structure is attached to an actuated arm, and the second one is a sensing system including a capacitive force sensor connected to a sensing arm (Fig. 3.6). The microgripper is designed in order to manipulate objects ranging from 1 μm up to 100 μm. The initial opening of the end effectors is 100 μm and the full close is achieved while applying the maximum actuation voltage $V_{in} = 200\,V$.

To pick up a micro-object, the actuated arm is pushed toward closure thanks to electrostatic forces generated by the comb drive actuator. Furthermore a suspension mechanism including two pairs of clamped-clamped beams holds the movable part of the actuator and produces a restoring force aimed to counteract the electrostatic action. While the gripper arms are closed around a micro-object, the deflection of the sensing arm caused by the gripping force is detected by the capacitive sensor and is translated into an analog voltage V_{out} throughout a MS3110 readout chip (Irvine Sensors Corp, Costa Mesa, CA, USA).

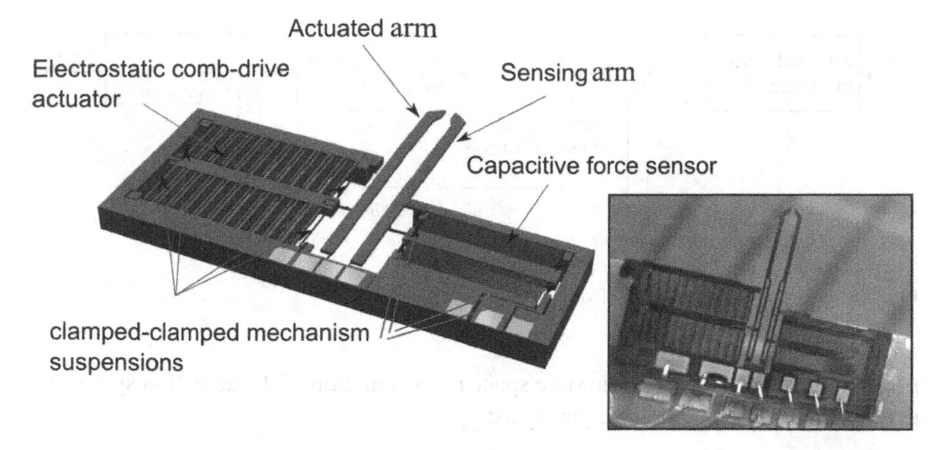

Fig. 3.6 Structure of the FT-G100 microgripper (FemtoTools GmbH)

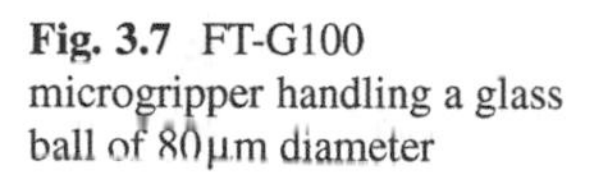

Fig. 3.7 FT-G100 microgripper handling a glass ball of 80 μm diameter

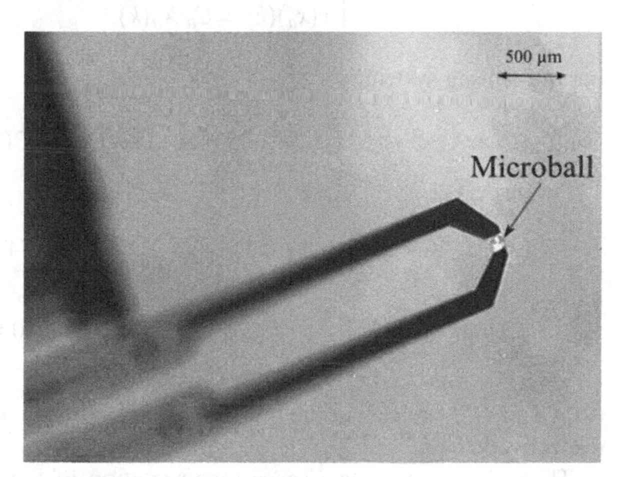

3.4.2 *Microgripper Modeling*

In order to achieve gripping force control, a knowledge based model of the microgripper has been developed in [17] when the end effectors are closed around a glass ball of 80 μm diameter (Fig. 3.7). Then, both the actuation and sensing systems have been modeled separately. After that, the obtained models have been coupled[4] taking into account the stiffness of the manipulated object (Fig. 3.8). The mass, the damping of the microgripper and the stiffness of the manipulated object have been identified using experimental data and least squares identification method. Using

[4]The state vector $X(k)$ of the coupled model includes the states $X_a(k)$ and $X_b(k)$ of the actuation and sensing systems respectively.

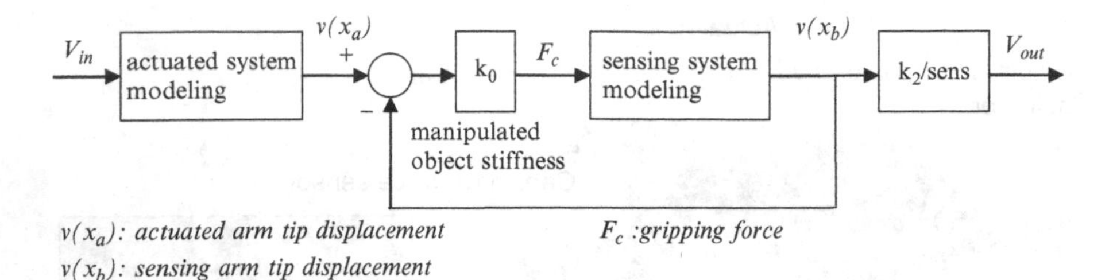

Fig. 3.8 Bloc diagram of the coupled model

linear assumptions, the discrete state space representation of the actuation system is given for 20 KHz sampling frequency as:

$$\begin{cases} X_a(k+1) = A_a.X_a(k) + B_a.V_{\text{in}}(k) \\ v(x_a)(k) = C_a.X_a(k). \end{cases} \tag{3.13}$$

For the sensing system:

$$\begin{cases} X_b(k+1) = A_c.X_b(k) + B_c.F_c(k) \\ v(x_b)(k) = C_c.X_b(k). \end{cases} \tag{3.14}$$

$$A_a = \begin{pmatrix} 1.6621 & -0.9536 \\ 1 & 0 \end{pmatrix} B_a = \begin{pmatrix} 0.25 \\ 0 \end{pmatrix}, \quad A_c = \begin{pmatrix} 1.142174 & -0.947527 \\ 1 & 0 \end{pmatrix} B_c = \begin{pmatrix} 0.5 \\ 0 \end{pmatrix}$$

$$C_a = \begin{pmatrix} 0.1104 \; 0.1086 \end{pmatrix} \qquad C_c = \begin{pmatrix} 0.132080 \; 0.129656 \end{pmatrix}$$

$v(x_a)$ and $v(x_b)$ relates to the tip displacements of the actuated and sensing arms, respectively.

Then, the general state space representation of the coupled system has been defined as follow:

$$\begin{cases} \begin{bmatrix} X_a(k+1) \\ X_b(k+1) \end{bmatrix} = \begin{bmatrix} A_a & 0 \\ B_c.k_k.C_a & A_c \end{bmatrix} \cdot \begin{bmatrix} X_a(k) \\ X_b(k) \end{bmatrix} + \begin{bmatrix} B_a \\ 0 \end{bmatrix} .V_{\text{in}}(k) \\ Vout(k) = \begin{bmatrix} 0 \; \frac{k_s}{\text{sens}} .C_c \end{bmatrix} \cdot \begin{bmatrix} X_a(k) \\ X_b(k) \end{bmatrix} \end{cases} . \tag{3.15}$$

$$k_k = \frac{k_0.k_s}{k_0 + k_s},$$

k_s and k_0 are the stiffness of the sensing system and the one of the manipulated object respectivelly. Moreover, sens $= 50\,\mu\text{N/V}$ is the sensitivity of the capacitive sensor.

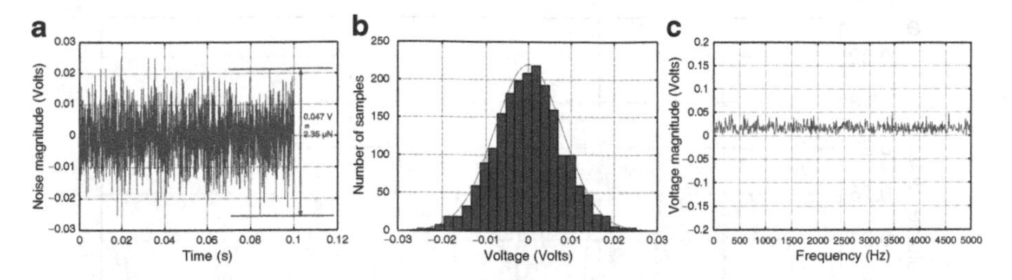

Fig. 3.9 Measurement noise: time evolution (**a**), distribution (**b**), and spectrum (**c**)

As explained in [17], the actuation system is highly nonlinear due to the suspension mechanism integrated in the actuator. For this reason, the coupled model presented above is valid only for small gripping forces around the operating point ($V_{in} = 60\,\text{V}$). Moreover, only the stiffness of the manipulated object is taken into account. A sufficient approach for modeling the manipulated micro-object is to also consider its effective mass and viscous damper.

In Sect. 3.4.4, the coupled model will be presented as:

$$\begin{cases} X(k+1) = A.X(k) + B.V_{in}(k) \\ V_{out}(k) = C.X(k) \end{cases} . \tag{3.16}$$

The gripping force is deduced from the output voltage V_{out} using the sensitivity of the capacitive sensor.

To deal with the noise at the output of the capacitive sensor, a noise characterization is performed and a Kalman filter is implemented in real time.

3.4.3 Noise Analysis and Kalman Filter Implementation

Process and measurement noises characterization is performed with the assumptions given is Sect. 3.3.2. Measurement noise is recorded at the output of the capacitive sensor when the tip of the sensing arm is free (no gripping force). Then, the statistical distribution and the spectrum of the noise are computed (see Fig. 3.9). The process noise is recorded at the output of the voltage generator providing a zero input. It shows similar characteristics to those of measurement noise. Process and measurement noises are then white in the frequency bandwidth of interest, centered with normal probability distributions. The variance of the measurement noise is $V = 6.33 \times 10^{-5}\,\text{V}^2$, and the one of the process noise $W = 9.62 \times 10^{-5}\,\text{V}^2$.

The matrix gain of the filter is obtained in the steady state as $K_e = \left[-0.11\ 0.06\ 0\ -0.008\right]^T$. The filter is then implemented in real time with 20 KHz sampling frequency using the Simulink software and the dSPACE controller board. The implementation is performed when the microgripper is handling the glass ball. Then, a 10 V step voltage is applied to the comb drive actuator. The measurement

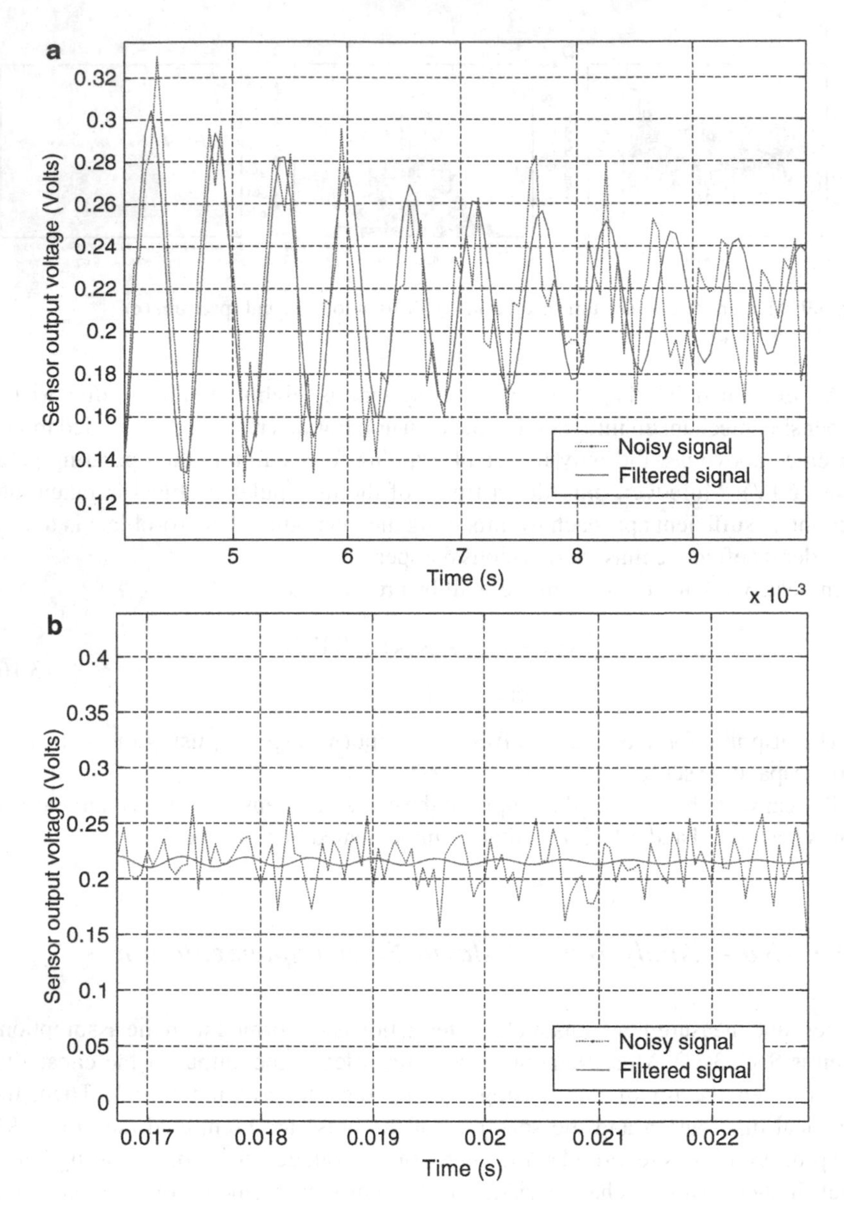

Fig. 3.10 Comparison between the noisy measurement and the filtered one: dynamic (**a**) and static (**b**) modes

at the output of the force sensor and the one given by the Kalman filter are recorded and compared (Fig. 3.10). The filter allows a reduction of about 97% of the measurement noise while keeping the dynamic measurement. This can be necessary when performing the manipulation of very fragile samples for which exceeding a

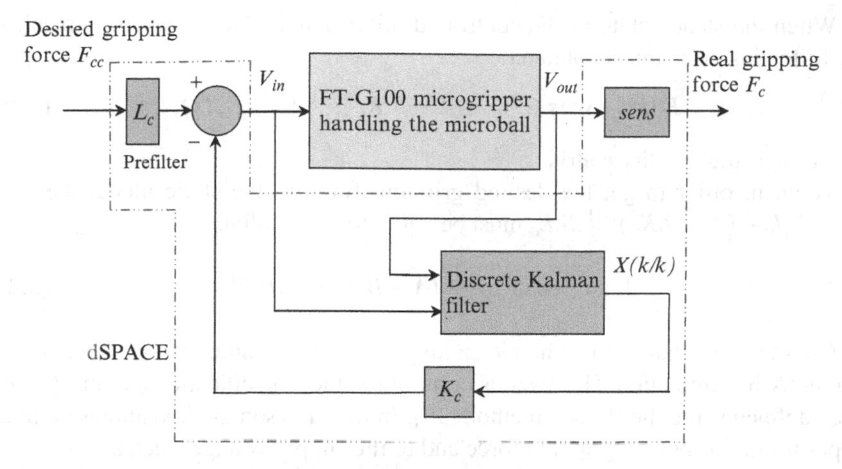

Fig. 3.11 Block diagram of force-controlled microgripping

threshold of force leads to the sample destruction. Then, if this threshold is near to noise level, performing dexterous micromanipulation tasks would be impossible without additional treatments such as the use of optimal filters.

3.4.4 Gripping Force Control

A Linear Quadratic Gaussian (LQG) controller is used for the gripping force control. The control law is composed of a Kalman filter and a state optimal feedback controller. Both algorithms are computed independently thanks to the separation principle. As shown in Fig. 3.11, the Kalman filter provides the posteriori state estimation of the process at each time step k, then a feedback matrix K_c corrects the dynamic performances and the prefilter L_c corrects the static one. The desired gripping force is set into the reference input F_{cc}.

Consider the following energetic criteria for the input and the states of the model (3.16):

$$J(k) = \sum_{i=k}^{\infty} X^T(i).Q.X(i) + V_{\text{in}}^T(i).R.V_{\text{in}}(i). \tag{3.17}$$

Where Q is a positive definite matrix and R is a positive scalar.

Parameters Q and R define the weighting given to the state and the control voltage within the optimal criteria. It is possible to adjust the dynamics of the closed loop system with these parameters.

The linear quadratic problem focuses on computing the optimal feedback vector K_c which minimizes the energetic criteria with respect to the Lyapunov stability. For a discrete computation, this vector is given by [18]:

$$K_c = (R + B^T.P_c.B)^{-1}.B^T.P_C.A. \tag{3.18}$$

P_c is the solution of Riccati's equation.

When the steady state of the controlled micro-manipulation system is reached, the following equation is obtained:

$$F_c(k) = sens.C.[I_n - (A - B.K_C)]^{-1}.B.L_c.F_{cc}(k), \tag{3.19}$$

Where I_n is the identity matrix.

Then, in order to get the desired gripping force in the static mode, the term $sens.C.[I_n - (A - B.K_C)]^{-1}.B.L_c$ must be equal to one leading to:

$$L_c = [sens.C.[I_n - (A - B.K_C)]^{-1}.B]^{-1}. \tag{3.20}$$

On the other hand, in dynamic mode, R and Q parameters can be adjusted manually by simulation. However, according to study specifications, R and Q have been defined using the Bryson method [19]. In fact, Bryson method allows defining upper limitations to the gripping force and to the supply voltage, such as:

$$R = \frac{1}{\sup(V_{\text{in}})^2} \quad Q = \begin{bmatrix} \frac{1}{\sup(x_1)^2} & 0 & 0 & 0 \\ 0 & \frac{1}{\sup(x_2)^2} & 0 & 0 \\ 0 & 0 & \frac{1}{\sup(x_3)^2} & 0 \\ 0 & 0 & 0 & \frac{1}{\sup(x_4)^2} \end{bmatrix}.$$

With: $\left[\sup(x_1)\ \sup(x_2)\ \sup(x_3)\ \sup(x_4)\right]^T = \sup(X(k))$

The performances needed in this study are given as follow: $\sup(V_{\text{in}}) = 10\,\text{V}$ and $\sup(F_c) = 10\,\mu\text{N}$ (no-overshoot).

Translating into canonical form, the output matrix of the coupled model (16) is given as: $C_{\text{canon}} = [0.0119\ 0.0002\ 0.0092\ {-0.0026}]$.

Then: $\sup(F_c) = sens.C_{\text{canon}}.sup(X)$.

Many set parameters of $sup(X)$ are possible to achieve study requirements, one of them is: $sup(X) = [9.5\ 3\ 9.5\ 0.3]^T$. The weighting parameters are then given as:

$$R = 1\,,\, Q = \begin{bmatrix} 1 & 0 & 0 & 0 \\ 0 & 10 & 0 & 0 \\ 0 & 0 & 1 & 0 \\ 0 & 0 & 0 & 1000 \end{bmatrix}.$$

This leads to: $K_c = \left[-0.0005\ -\ 0.0506\ {-0.9675}\ 1.6038\right]$

The LQG controller has been implemented in real time on the microgripper when handling the glass ball. As shown in (Fig. 3.12), the whole micro-manipulation station is made up of a 3-DOF manual micro-positioning table (M-UMR 5.16, Newport) where the FT-G100 micro-gripper is attached and tilted with an angle of 45°

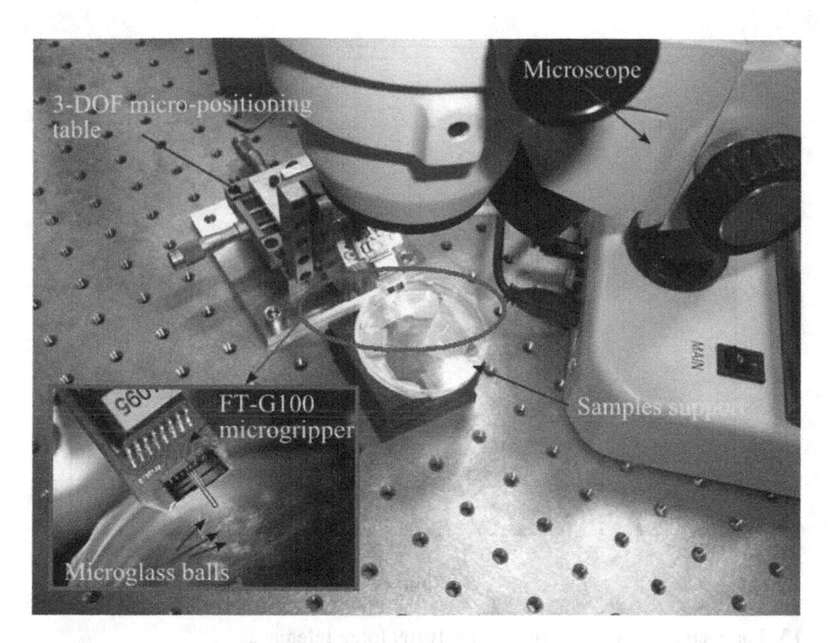

Fig. 3.12 Force-controlled micromanipulation setup

to allow gripping the samples. Moreover, due to the small size of the manipulator, a microscope is used and all the components are mounted on a vibration isolation table. The control algorithm is developed using the Matlab/Simulink software (r2007b) and implemented in real time using a dSPACE control board with a sampling frequency of 20 Khz.

A supply voltage of 60 V has been applied to the comb drive actuator allowing the displacement of the actuated arm tip of 20 μm to come in contact with the glass ball. Then, from the dSPACE control board, a 10 μN gripping force reference has been set at the in input of the feedback system.

The 80-μm glass ball has been handled successfully with 10 μN gripping force according to the desired dynamic force. Figures 3.13 and 3.14 show that the desired performances are obtained, i.e. the response time of the gripping force reaches 6 ms without overshoot. The voltage does not exceed 10 V around the operating point (60 V). A small peak is also observed on the actuation voltage dynamic at approximately 0.4 ms (Fig. 3.14) giving the impulse to the gripping force and inducing a fast response time. Moreover, the high quality of the griping force estimated by the Kalman filter can be observed in Fig. 3.13.

In this application, the variance of the process noise is close to the one of the measurement noise. Then, the a posteriori state estimation depends on both the measurements and the model. Uncertainties in the model can lead to an erroneous state estimation. For this reason, it is important to know whether the filtered signal reflects the real gripping force. Thus the micro-motions on the tip of the sensing arm have been recorded while applying the force feedback control. This has been conducted thanks to a high resolution (0.01 nm) laser interferometer from

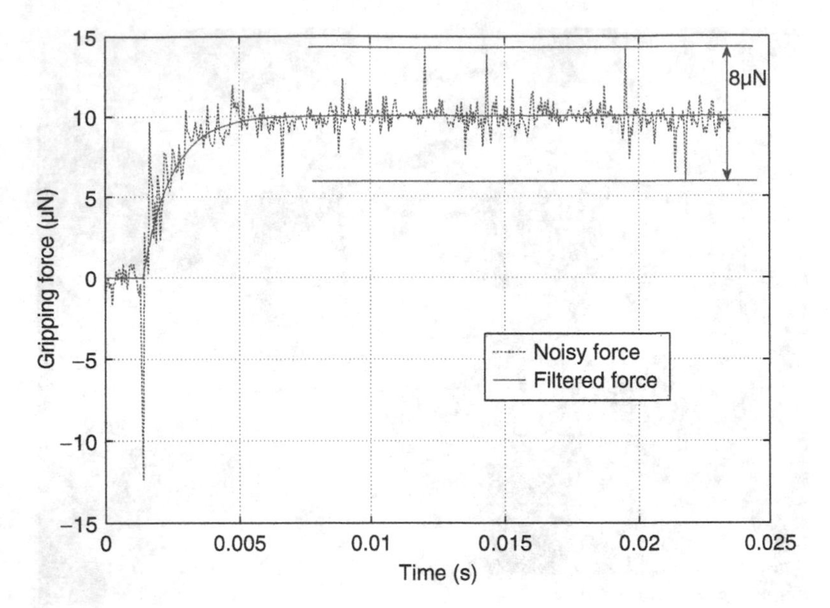

Fig. 3.13 Force-controlled microgripping for 10 μN force reference

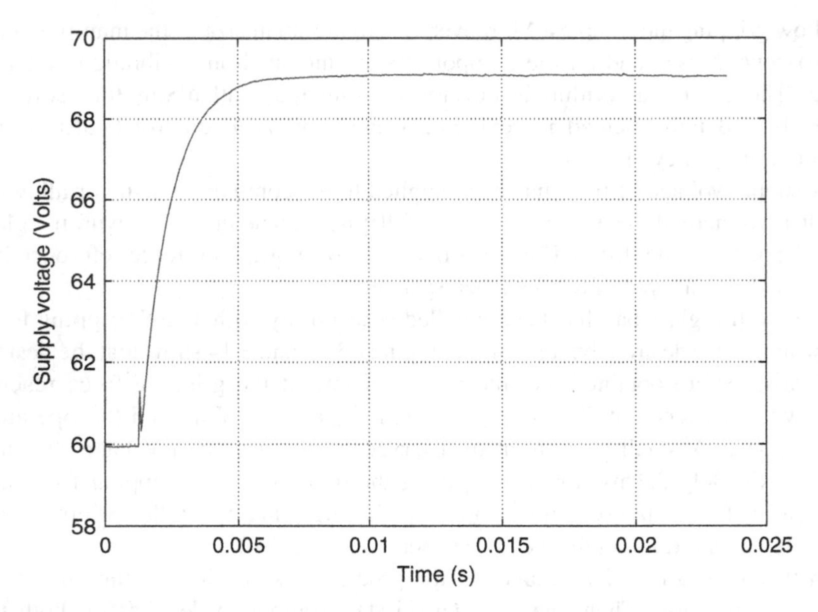

Fig. 3.14 Evolution of the control voltage for 10 μN force reference

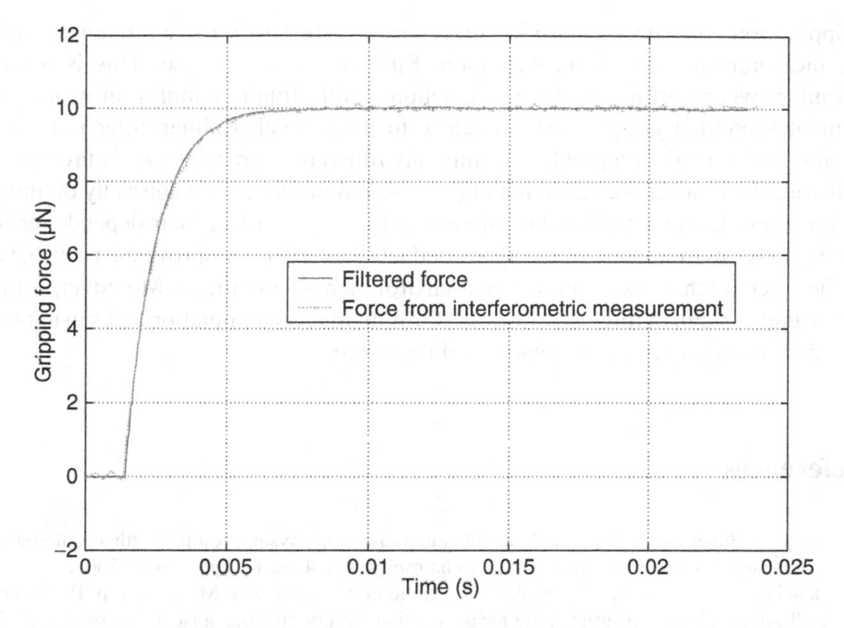

Fig. 3.15 Comparison between the filtered force and the one given through the interferometric measurement

Metechnik GmbH. The real gripping force has been deduced from the knowledge of the sensing system stiffness (6.45 N/m) identified in [17]. Figure 3.15 shows both the force estimated by the Kalman filter and the one deduced from the interferometric measurement, the error between the two signals is less than 0.1 µN. The effectiveness of the Kalman filter is then entirely proven.

In both applications, process and measurement noises characteristics are considered constant. However, noises can vary if the working conditions are different (influence of the environmental noise). In this case, the noise characterization must be periodically updated in order to obtain better performances. The model can also include the effects of the environmental noise leading to the improvement of the controller robustness. Physical origins of process and measurement noises can be considered differently.

3.5 Conclusion

In this chapter, we have presented advantages of using Kalman filtering for force measurement and control in the microworld. The effectiveness of the filter has been observed experimentally through two applications. In the first one, the filter allowed increasing the signal to noise ratio by a factor of 40 making possible to perform measurements below the *mN* force level using strain gauges. In the second application, the filter has been used in a feedback control and implemented on an electrostatic microgripper for the dexterous manipulation of 80 µm glass balls.

Gripping force has been controlled successfully with 10 μN force reference while the measurement noise level was about 8 μN in the worst case. This is a real advantage when performing the manipulation of very fragile samples for which the limit of supported gripping force is close to noise level. Kalman filter is then a reliable tool to deal with problems commonly found in microrobotics. Future works will concern a fine characterization of process and measurement noises by defining their origins, the manner in which they affect the system and their dependence to the working environment. This will be conducted in order to improve the robustness of the filter when working in different environmental conditions. Moreover, using the extended Kalman filter will allow performing micromanipulation tasks using the electrostatic microgripper in a wide working range.

References

1. Clevy C., Hubert A., Chaillet N., "Flexible micro-assembly system equiped with an automated tool changer", Journal of Micro-Nano Mechatronics, vol .4, no.1–2, pp. 59–72, 2008.
2. Menciassi A., Eisinberg A., Scalari G., Anticoli C., Carrozza M C., Dario P., "Force feedback-based microinstrument for measuring tissue properties and pulse in microsurgery", In Proceedings of the IEEE International Conference on Robotics and Automation, Seoul, Korea, 2001.
3. Sun Y., Wan K T., Roberts K P., Bischof J. C., Nelson B J., "Mechanical property characterization of mouse zona pellucida",IEEE Trans Nanobioscience,vol 2, pp. 279–286, 2003.
4. Liu X Y., Kim K Y., Zhang Y., Sun Y., "Nanonewton force sensing and control in microrobotic cell manipulation", International Journal of Robotics Research, vol. 28, no. 8, pp. 1065–1076, 2009.
5. Lu Z, Chen P. C. Y., Lin W., "Force sensing and control in micromanipulation". IEEE Trans Syst Man Cybern Part C 2006;36(6):713–24.
6. Kim D. H., Lee M. G., Kim B., Sun Y., "A superelastic alloy microgripper with embedded electromagnetic actuators and piezoelectric force sensors: a numerical and experimental study", Smart Materials and Structures, vol. 14, no. 6, pp. 1265–1272, 2005.
7. Park J., Moon W., "A hybrid-type micro-gripper with an integrated force sensor", Microsystem technologies, vol. 9, no. 8, pp. 511–519 , 2003.
8. Greitmann G., Buser R. A., "Tactile microgripper for automated handling of microparts", Sensors and actuators. A, Physical, vol. 53, no. 1–3, pp. 410–415, 1996.
9. He D. F., Yoshizawa M., "A method of background noise cancellation of SQUID applications". Supercond. Sci. Technol, vol 16, pp. 1422–1425. 2003.
10. Katupitiya J., Dutre S., Demay S De Geeter J., "Estimating contact and grasping uncertainties using kalman filters in force controlled assembly". In: Proceedings of the 1996 IEEE/RSJ international conference on intelligent robots and systems, Osaka, Japan; 1996. p. 696–703.
11. Ma CK., Chang JM., Lin DC. "Input forces estimation of beam structures by an inverse method". J Sound Vib 2003;259(2):387–407.
12. Liu JJ., Ma CK., Kung IC., Lin DC. "Input force estimation of a cantilever plate by using a system identification technique computer methods in applied mechanics and engineering", vol. 190. Elsevier; 2000. p. 1309–22.
13. Lefebvre T., Bruyninckx H., de Schutter J. "Nonlinear Kalman filtering for force-controlled robot tasks". Springer tracts in advanced robotics, vol. 19, 2005.
14. Kalman R. E., "A new approach to linear filtering and prediction problems". Transactions of the ASME. J Basic Eng 1960:35–45.
15. Maybeck P. S., "Stochastic models, estimation and control", vol. 1. Academic Press, Inc.; 1979.

16. Haddab Y., Chen Q., Lutz, P., "Improvement of strain gauges micro-forces measurement using Kalman optimal filtering", Mechatronics, vol 19, issue 4, pp. 457–462, 2009.
17. Boudaoud M., Haddab Y., Le Gorrec Y., "Modelling of a MEMS-based microgripper: application to dexterous micromanipulation", IEEE/RSJ IROS 2010, Taipei, Taiwan.
18. Anderson B. D. O., Moore, J. B., "Optimal control: linear quadratic methods", Prentice Hall, 1990.
19. Bryson A. E., Ho Y. C., "Applied optimal control", Hemisphere, 1975.

Chapter 4
Microforce-Sensing Tools and Methodologies for Micromechanical Metrology

Simon D. Muntwyler, Felix Beyeler, and Bradley J. Nelson

Abstract The increasing interest in investigating ever smaller samples and the industrial trend toward miniaturization has created a need for new metrological tools and methodologies for micromechanical measurements. Within this chapter, MEMS-based single-axis as well as multi-axis capacitive microforce-sensing tools enabling the measurement of forces in the micronewton to nanonewton range are presented. By combining sensing as well as actuation elements on a single chip, a monolithically integrated multi-axis microtensile-tester chip is shown, allowing the direct measurement of the mechanical as well as electrical properties of a microscale sample along multiple directions. The design, microfabrication, and characterization are discussed. Motivated by the unavailability of reference standards in the nanonewton range, a methodology for the calibration of microforce sensors is developed. In combination with the implementation of the latest advancement in the field of multivariate uncertainty analysis using a Monte Carlo method, this allows for SI-traceable microforce measurements in the nanonewton to micronewton range. At the end of this chapter, the application of the microtensile tester chip is demonstrated for the quantitative micromechanical investigation of individual plant cells in their living state.

Keywords Capacitive microforce-sensing • Electrostatic actuation • Microelectromechanical systems • Capacity-to-voltage conversion • Calibration • Primary standard • Guide to the expression of uncertainties in measurements • Monte Carlo method • SI-traceability • Sensor tuning

---

S.D. Muntwyler (✉)
Institute of Robotics and Intelligent Systems, Tannenstrasse 3, 8092 Zurich, Switzerland
e-mail: msimon@ethz.ch

C. Clévy et al. (eds.), *Signal Measurement and Estimation Techniques for Micro and Nanotechnology*, DOI 10.1007/978-1-4419-9946-7_4,

4.1 Introduction

Advances in miniaturization technologies have had great impacts on our lives. Radios, computers, and telephones that once occupied large spaces now fit in the palm of a hand [1]. Using specially developed microfabrication processes, large electronic circuits and components are integrated into millimeter-sized integrated circuit (IC) chips. In addition to the benefit of enabling smaller components, microfabrication offers other advantages, such as reduced production costs due to a highly parallelized batch fabrication and better performance. Microelectromechanical systems (MEMS) uses these fabrication processes originally developed for the IC industry to produce mechanical structures and components as well as integrate mechanical systems such as sensors and actuators with electronics on a common silicon substrate. "Small is beautiful [2]" because small is better: With increasing miniaturization, devices often yield better performance, such as faster response time or higher sensitivity. But due to the microscopic dimensions of these devices or components, conventional, macroscopic tools for the testing and especially mechanical characterization have reached their limitations. Besides this industrial trend toward ever smaller components, the focus in various fields of research such as plant biology has shifted from studying the organization of the whole body or individual organs toward the behavior of the smallest units of the organism, the individual cell [3]. Measuring the mechanical characteristics of a cell by applying a force can give insights into its structures or functionalities since the mechanical properties of biological materials are tightly coupled to their physiological functions.

The most commonly used system for mechanical investigation of microscopic samples and devices is the atomic force microscope (AFM). Primarily developed for dimensional metrology, the AFM also allows for the measurement of small forces. It consists of a bending beam with a sharp tip at its end, attached to a micropositioner, allowing it to be scratched over a surface or pushed against a sample. The deflection of the beam, which is proportional to the force acting on its tip, is calculated by measuring the displacement of an optical beam reflected from its surface. This approach offers high spatial as well as force resolution but involves a number of limitations, which makes it a poor choice for a variety of applications.

For the micromechanical testing of small samples such as the measurement of the stiffness of biological cells or microfabricated structures, the AFM-based technique is limited in its versatility. Due to the large size of the optical beam deflection measurement system, the AFM is difficult to integrate into a complete testing system with a high-resolution microscope, such as an inverted optical microcopy or a scanning electron microscope (SEM). Although much effort has been made in the development of fully integrated AFMs, where the optical beam deflection principle is replaced by, e.g., a piezoresistive sensing element [4, 5], these sensors are usually limited to single-axis measurements and are based on the deflection of a cantilever. Cantilever-based sensors are sensitive to off-axis loads and induce lateral motions when they are deflected, inducing slippage in the worst case. Given the metrological limitations of the cantilever as a force sensor, which are acknowledged and appreciated by most in the AFM community, it is interesting that viable alternatives are still scarce [6].

For the continuous, successful industrial miniaturization and ongoing investigation of smaller samples in research, a new set of metrological tools for testing and characterization are required.

With the advancement of MEMS technology, fabrication techniques have become available for the development of a new generation of microforce-sensing tools with the potential to overcome the limitations of the cantilever-based techniques. Driven by the increasing number of applications requiring the measurement of small forces in different fields such as mechanobiology, material sciences, microrobotics, and life sciences, microforce sensors based on various different approaches have been developed.

In [7, 8], the design of a single-axis capacitive force sensor and its application to study insect flight control, the mechanical characteristics of mouse embryo cells, and the threshold for touch sensation in *Caenorhabditis elegans* [9] are shown. In [10], a sensor based on an optical, diffractive micrograting demonstrates the measurement of the injection forces into drosophila embryos, and in [11] the use of an AFM, based on optical beam deflection, shows the measurement of molecular interaction forces. A sensor based on the trapping of a magnetic particle in a magnetic field is shown in [12] and is used to measure the force-extension curves of DNA molecules. More recently, many of these principles have been used to develop multi-axis sensors. For many applications, additional force components offer a great advantage, e.g., in the case of automated cell injection [13] since a misalignment of the cell and the injection pipette can be detected and compensated. In [14, 15], two three-axis force sensors based on piezoresistive materials are presented and used for the manipulation of embryo cells. A three-axis capacitive force/torque sensor for the measurement of forces and torque acting on a magnetic microrobot has been developed in [16], and in [17] a six-axis force/torque sensor is described.

Despite the need for high-resolution multi-axis microforce-sensing tools, currently neither a standardized calibration methodology nor an SI-traceable reference standard exists for the calibration of forces in the nanonewton range. Typically, microforce sensors, such as AFMs, are calibrated based on a model of the sensor, with the disadvantage of having an unknown accuracy, since the results are not traceable back to SI units. Thus, within this chapter, a methodology is presented, enabling traceable microforce sensor calibration down to the nanonewton. And by implementing the most recent advancements in multivariate uncertainty analysis, the uncertainties in the characteristic parameters of the sensors could be calculated. This methodology should create the basis for the utilization of microforce-sensing tools to perform SI-traceable force measurements in the nanonewton to micronewton force range and the calculation of their measurement uncertainty.

This chapter starts with an overview and comparison of different microforce-sensing technologies and the selection of the most appropriate approach for micromechanical measurements in the micronewton to nanonewton range followed by the presentation of the design and fabrication of multiple microforce-sensing tools. Furthermore, the methodology for traceable calibration of single as well as multi-axis microforce sensors and the calculation of their measurement uncertainties is derived. To finalize, the application of one of the microforce-sensing tools is shown for the combined, quantitative micromechanical and dimensional measurement of individual plant cells in their living state while still attached to the plant.

4.2 Microforce-Sensing Tools

The limiting factor for the development of standardized micromechanical testing systems is the unavailability of microforce-sensing probes that can measure forces down to the nanonewton level along multiple axes. In this chapter, after an overview of the most relevant sensing principles, the design and fabrication of novel microforce-sensing probes as well as a complete measurement system on a chip are described.

4.2.1 Microforce-Sensing Technologies

A microforce-sensing probe is a transducer that converts a small applied force into an electrical signal. Different approaches and sensing technologies for the development of microforce sensors, measuring forces in the micronewton to nanonewton range, have been pursued. These range from large-scale machines such as the electrostatic force balance (EFB) developed by the National Institute of Standards and Technology (NIST) [18] to microscopic sensors developed by Physikalisch-Technische Bundesanstalt (PTB) [19], which was fabricated using the MEMS technology.

Depending on the requirements of the application for which the sensor is developed, such as the force range or whether static or dynamic forces need to be measured, the appropriate approach and sensing technology needs to be chosen. For the micromechanical testing of samples with micrometer characteristic dimensions, such as individual cells, MEMS-based mechanisms, or biological fibers, the measurement of quasi-static forces (<1kHz) in the micronewton to nanonewton range is required. The sensor needs to be compatible with different environments, such as air, liquid, and vacuum. And to enable the integration into a measurement system, the sensor needs to be compact in size and light in weight.

For the development of miniature sensors to measure forces in the micronewton to nanonewton range, four commonly used technologies exist: piezoresistive, piezoelectric, capacitive, and optical-based force sensing. More exotic principles such as magnetic-based sensing [20] or tunneling microforce sensing [21] are presented in literature, but are currently not mature enough for the development of reliable sensors. A short overview of the four main sensing technologies, as well as the selection of the most suitable approach for micromechanical measurements, is given.

(a) ***Piezoresistive Force Sensing***: A force applied to a semiconducting material such as silicon (SI), germanium (Ge), or gallium arsenide (GaAs) will result in a modulation of the electrical resistance. The modulation is a result of two effects: the geometry effect and the piezoresistive effect. The geometry effect refers to the change in resistance due to the change in the material's geometry as it gets deformed by the applied force. The piezoresistive effect refers to the variation of the specific resistance of a material induced by the applied stress, which for semiconducting material is the dominating cause of resistance change [22]. Piezoresistive sensing allows for the highest level of miniaturization (among the four discussed sensing technologies) but is limited to a relatively low sensitivity.

(b) ***Piezoelectric Force Sensing***: A piezoelectric material generates a voltage across the material when it is mechanically deformed. Piezoelectric materials are mostly crystalline materials such as gallium arsenide (GaAs) or zinc oxide (ZnO) in which the positive and negative charges are separated but symmetrically distributed. Therefore, the material is overall electrically neutral. When a stress is applied, this symmetry is destroyed and the charge asymmetry generates a voltage. The typical deformations are in the nanometer range resulting in very stiff sensors with a high resonant frequency, enabling the measurement of forces at high frequencies. The disadvantage of piezoelectric sensing is the rapid decay of the electrical signal after the force is applied, rendering this technology unsuitable for the measurement of static forces.

(c) ***Capacitive Force Sensing***: Capacitive force sensors measure a force-induced displacement of an elastic element, such as a flexure, as a variation of the capacity between attached capacitive electrodes. Compared with piezoresistive-based sensors, capacitive sensors have the advantage of no hysteresis, better long-term stability, and higher sensitivity [22].

(d) ***Optical Force Sensing***: The very well-known optical beam deflection technique to measure the deformation of a cantilever is the principle used by AFM. This is the most commonly used microforce sensor and has demonstrated its potential in a large variety of applications. It offers very high sensitivity and resolutions. Other optical-based microforce sensors that are, e.g., based on optical encoders or optical grids exist; however, all these sensors have the disadvantage of being relatively large, making a system integration in most cases impossible.

Each of the four force-sensing technologies has advantages and disadvantages. For the mechanical characterization of microscopic samples, forces at low frequencies (<1 kHz) need to be measured. The sensor must be compact to enable integration into an automated measurement system. Additionally, it should be stable, highly sensitive to its primary input, and insensitive to variations in environmental conditions. This leaves either piezoresistive or capacitive force sensing as a possible option. The superior characteristics in terms of stability and hysteresis [22] make capacitive force sensing the most suited technology for micromechanical material testing systems.

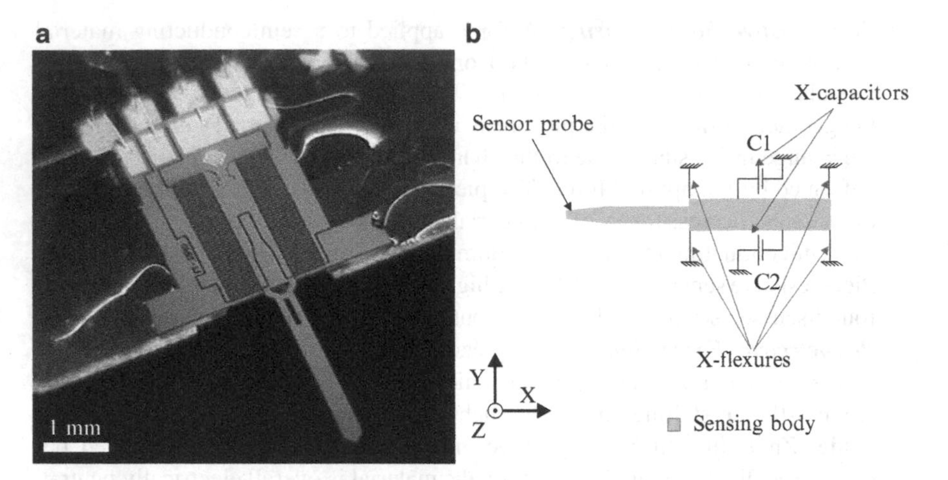

Fig. 4.1 (**a**) Photograph and (**b**) schematic of a single-axis MEMS-based capacitive microforce-sensing probe from Femtotools GmbH (the immovable outer frame is not shown)

4.2.2 *Single-Axis Capacitive Microforce-Sensing Probes*

The working principle of single-axis MEMS-based capacitive microforce-sensing probes is schematically shown in Fig. 4.1. The sensor consists of a movable body with an attached probe suspended by four flexures within an outer frame. A force applied to the probe in the *x*-direction results in a relative motion of the body and the outer frame, which can be measured by attached capacitive electrodes as a change in capacitance.

The benefits of using this four-flexure configuration lie in its parallel motion as it is deflected, thereby eliminating lateral tip motion such as observed in cantilever-based sensors.

The transversal capacitive displacement-sensing method benefits from a great sensitivity to displacements, but lacks in linearity when measuring only one of the capacitance (C_1 or C_2). However, by measuring the two capacitive changes of C_1 and C_2 (with opposite signs) differentially, the relationship is linearized as shown in Fig. 4.2 and equations (4.1)–(4.3), where C is the capacitance, $\varepsilon = 8.85 \times 10^{-12} C^2/(\mathrm{Nm}^2)$ is the permittivity of air, d is the initial gap between the capacitor electrodes, Δd is the change of this gap (induced by the applied force), and A is the overlapping total area of the all capacitor electrodes. For $x = \Delta d/d$ and $|x| < 1$ the Maclaurin series of $(1-x)^{-1}$ (infinite geometric series) can be used to simplify (4.1)–(4.3).

$$C_1 - C_2 = \varepsilon \cdot A \left(\frac{1}{d - \Delta d} - \frac{1}{d + \Delta d} \right) \tag{4.1}$$

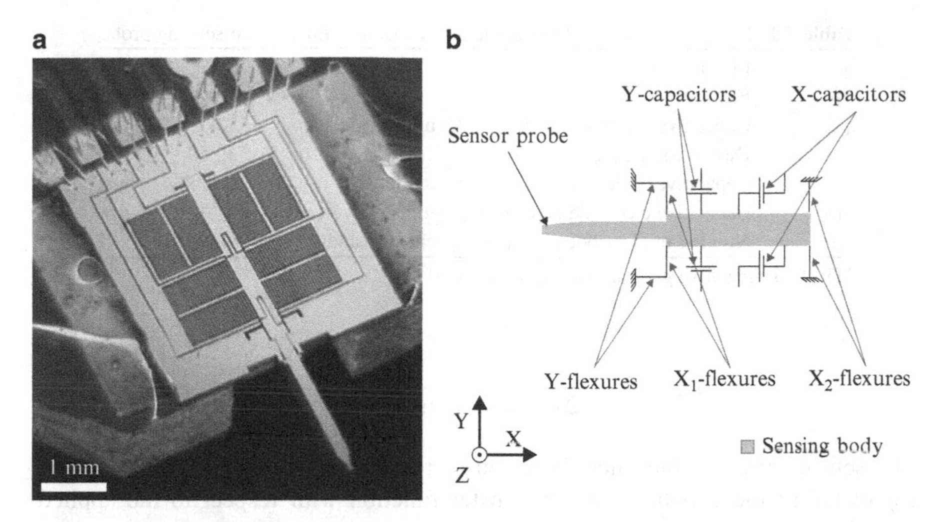

Fig. 4.2 (**a**) Photograph and (**b**) schematic of a two-axis MEMS-based capacitive microforce-sensing probe (the immovable outer frame is not shown)

$$\frac{1}{1-x} = \sum_{n=0}^{\infty} x^n \approx 1 + x + x^2 \tag{4.2}$$

$$C_1 - C_2 = \frac{\varepsilon \cdot A}{d}\left(\frac{1}{1-\frac{\Delta d}{d}} - \frac{1}{1+\frac{\Delta d}{d}}\right) \approx \frac{\varepsilon \cdot A}{d}\left(2\frac{\Delta d}{d}\right) = \varepsilon \cdot A\frac{2 \cdot \Delta d}{d^2} \tag{4.3}$$

The differential capacity is converted to a voltage using a commercial capacity-to-voltage converter (MS3110, Irvine Sensors Inc.). For simplicity this relationship is described by (4.4) using the constant C_{CVC} and offset $\mathrm{Offset}_{\mathrm{CVC}}$. Details about the capacitance-to-voltage (CVC) converters used for the different force-sensing tools are presented in Sect. 4.2.6.

$$V_{\mathrm{Out}} = C_{\mathrm{CVC}}(C_1 - C_2) + \mathrm{Offset}_{\mathrm{CVC}} \tag{4.4}$$

The theoretical transfer function (4.6) of the single-axis microforce sensor, describing the relationship between the output voltage of the sensor V_{Out} and the applied force F (in x), can be found using (4.3), C_{CVC} and $\mathrm{Offset}_{\mathrm{CVC}}$ and the stiffness of the sensor in the x-direction given by (4.5), where E_{F} is the Young's modulus, t_{F} the thickness, w_{F} the width, and l_{F} the length of each of the four flexures. Additionally, d_{c}, the gap between the capacitive electrodes, A_{c}, the overlapping area of the capacitive electrodes, and ε_{c}, the permittivity of air, are needed.

$$k_x = 4\frac{E_{\mathrm{F}} \cdot t_{\mathrm{F}} \cdot w_{\mathrm{F}}^3}{l_{\mathrm{F}}^3} \tag{4.5}$$

Table 4.1 Design parameters of the single-axis capacitive microforce-sensing probe

l_F	Flexure length	540	μm
w_F	Flexure width	10	μm
t_C,t_F	Capacitive electrode thickness, flexure thickness	50	μm
F_S	Design force range	±250	μN
l_C	Capacitive electrode overlapping plate length	500	μm
d_1,d_2	Capacitive electrode gap spacing widths[a]	5, 20	μm
n_C	Number of electrode pairs per capacitor C_1 and C_2	50	

[a]The two capacitive gaps are shown in Fig. 4.10

$$V_{\text{Out}} = C_{\text{CVC}} \frac{\varepsilon_c \cdot A_c \cdot l_{\text{F}}^3}{2 \cdot d_c^2 \cdot E_{\text{F}} \cdot t_{\text{F}} \cdot w_{\text{F}}^3} F_x + \text{Offset}_{\text{CVC}} \tag{4.6}$$

The sensitivity S_x of the microforce sensor to a force in the x-direction can be calculated as the derivative of the transfer function with respect to the applied force in x.

$$S_x = \frac{\partial V_{\text{Out}}}{\partial F_x} = C_{\text{CVC}} \frac{\varepsilon_c \cdot A_c \cdot l_{\text{F}}^3}{2 \cdot d_c^2 \cdot E_{\text{F}} \cdot t_{\text{F}} \cdot w_{\text{F}}^3} \tag{4.7}$$

The sensor is microfabricated out of silicon-on-insulator (SOI) wafer substrates, which refers to the use of a layered silicon–insulator–silicon wafer. The insulating oxide layer between two silicon layers makes it possible to use one of the two silicon layers, called the devices layer, to form the active elements, such as the movable body, the flexures, and the capacitive electrodes, and the other layer, called the handle layer, is used to form the outer frame for the sensor mechanically holding all these components together. Due to the isolating silicon oxide layer, all the active components in the device layer are held together, but are electrically isolated from each other. More details about the microfabrication process used for the different sensing tools can be found in Sect. 4.2.5.

A large number of different single-axis sensing probes with flexure lengths ranging from 200 to 3,000 μm have been realized. But due to the unavailability of traceable reference standards and methodologies for the calculation of their uncertainties, none of these sensors have been fully characterized, making the estimation of their quality and their measurement accuracy impossible. Therefore, within the course of this chapter, a methodology for the calibration of these types of microforce sensors is presented and demonstrated for a sensor design with the parameters presented in Table 4.1.

4.2.3 Multi-Axis Capacitive Microforce-Sensing Probes

A force is a vector in three-dimensional space; therefore, using a single-axis microforce-sensing probe is only justified to measure its magnitude when its sensitive direction is perfectly aligned with the direction of the force vector under

Table 4.2 Design parameters of the two-axis capacitive microforce-sensing probe. Copyright 2010 *IEEE Journal of Microelectromechanical Systems* [24]

l_{Fx1}	X_1-flexure length	300	μm
l_{Fx2}	X_2-flexure length	200	μm
l_{Fy}	Y-flexure length	130	μm
w_F	Flexure width	5	μm
t_C, t_F	Capacitive electrode thickness, flexure thickness	50	μm
F_S	Designed force range (x and y)	±70	μN
l_C	Capacitive electrode overlapping plate length	470	μm
d_1, d_2	Capacitive electrode gap spacing widths[a]	5, 20	μm
n_C	Number of electrode pairs per capacitor C_1 and C_2	50	

[a]The two capacitive gaps are shown in Fig. 4.10

investigation. In the case of microscopic samples, this alignment is often challenging and can lead to faulty measurements. Different approaches for the simultaneous measurement of the complete force vector with its three components have been made based on either capacitive [17] or piezoresistive [23] sensing. However, none of these approaches resulted in the ability to measure submicronewton forces and involve either complex fabrication processes or the need for microassembly, both rendering large-scale fabrication of these devices impossible.

In a first step, the development of a two-axis (in-plane) microforce-sensing probe is presented, before introducing the approach for the simultaneous in- and out-of-plane full three-axis microforce sensing. The basic working principle of a multi-axis capacitive microforce sensor is similar to the single-axis case. The sensor consists of a movable body suspended by flexures within an outer frame. A force applied to the probe results in a relative motion of the body and the frame, which can be measured by capacitive electrodes as a change in capacitance. By designing the flexures in such a way that they allow the body to move in multiple directions, and by using several of these capacitive displacement sensors, forces/displacements in multiple directions can be measured.

Figure 4.2 shows the design of a two-axis microforce-sensing probe. The stiffness of the flexures in the x- and y-directions is used to design the force-sensing range along its two sensitive directions. The force range has been designed to be up to ±70 μN along both axes. To ensure equal sensitivity of the sensor along both axes, it has been modeled in a finite element model (FEM) environment (ANSYS Inc.). Using this model in an optimization loop, the dimensions of the flexure can be found that will result in the desired sensitivity along both directions. Table 4.2 presents the design parameters of the two-axis microforce-sensing probe.

In the case of the single-axis and two-axis sensing probes, only forces in the sensor plane are measured using a transversal capacitive electrode configuration. Measuring forces/displacements out of the sensor plane is more challenging. In [17], out-of-plane displacements are measured by bonding an additional silicon layer onto the substrate, thereby enabling an out-of-plane parallel-plate capacitor configuration. Although the transversal out-of-plane displacement sensing achieved a high sensitivity, the high fabricational complexity resulted in a low yield and expensive

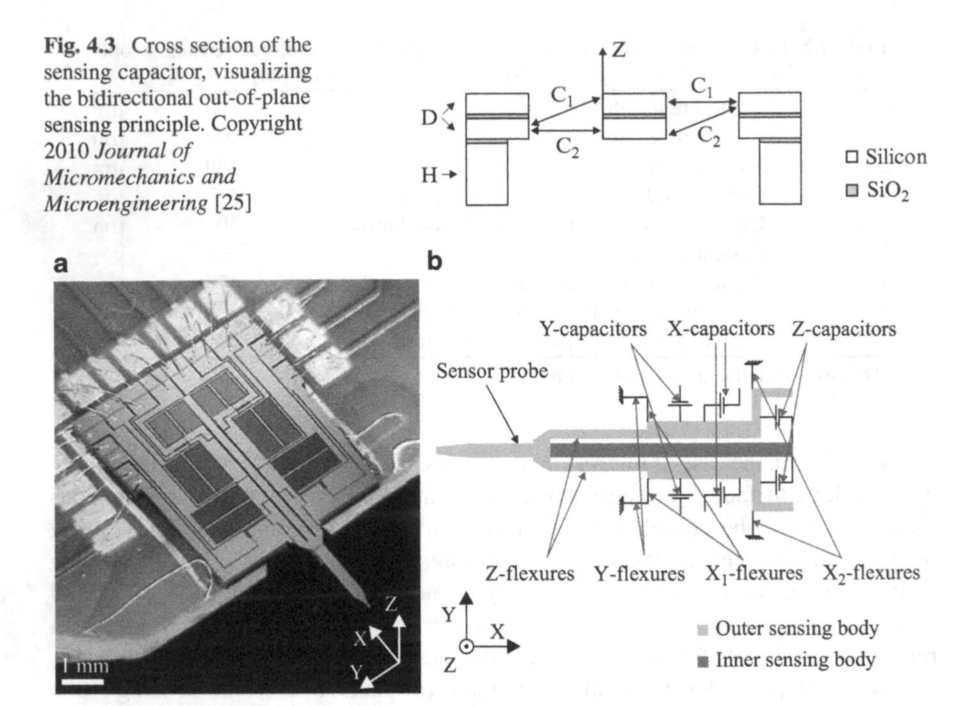

Fig. 4.3 Cross section of the sensing capacitor, visualizing the bidirectional out-of-plane sensing principle. Copyright 2010 *Journal of Micromechanics and Microengineering* [25]

Fig. 4.4 (**a**) Photograph and (**b**) schematic of a three-axis MEMS-based capacitive microforce-sensing probe (the immovable outer frame is not shown). Copyright 2010 *Journal of Micromechanics and Microengineering* [25]

fabrication. Therefore, a new approach based on lateral sensing was chosen to measure the out-of-plane forces. In this case, the variation of the capacitance is induced by a change in the vertical overlapping area of the parallel plates. The disadvantage of this approach is that it cannot be distinguished between positive and negative forces in the z-direction, since in both cases the overlapping area of the capacitive electrodes gets smaller, and, therefore, the capacitance gets smaller as well.

To overcome this problem, two layered capacitive electrodes are used; these are schematically shown in Fig. 4.3. This electrode configuration can be achieved by fabricating the sensor from a double SOI wafer. The relatively thick silicon handle layer (H) forms the outer frame of the sensor, and the two thin silicon device layers (D) form the active elements and the movable body. All the layers are electrically isolated by an SiO_2 layer. The additional device layer enables the distinction between positive and negative forces in the z-direction. The differential capacitance $(C_1 - C_2)$ is negative when the inner sensing body moves up $(z > 0)$ and positive when it moves down $(z < 0)$.

Figure 4.4 shows the schematic of the three-axis sensing probe using this multilayer electrode configuration. Due to the unequal sensitivity of transversal in-plane sensing compared with lateral out-of-plane sensing, these degrees of freedom

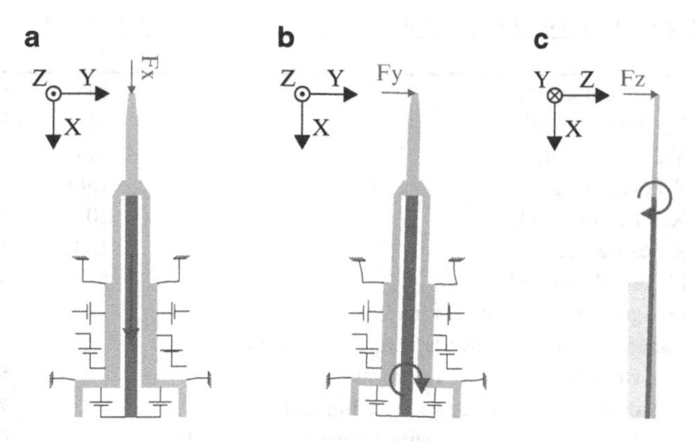

Fig. 4.5 Mechanical response of the three-axis MEMS-based capacitive microforce-sensing probe to an applied force in: (**a**) the x-direction, (**b**) the y-direction, and (**c**) the z-direction. Copyright 2010 *Journal of Micromechanics and Microengineering* [25]

are divided into two sensing bodies suspended within each other. The outer body measures displacements in the x- and y-directions (and, therefore, forces relative to them), while the inner sensing body measures forces/displacements relative to the outer sensing body in the z-direction. The stiffness of the flexures is used to design the force-sensing range. The sensor is designed to measure forces of up to $\pm 200\,\mu\text{N}$ in the x-, y-, and z-directions. Multiple sensor configurations (position and geometry of flexures, capacitors, and movable bodies) have been analytically compared. Besides the sensitivity criterion for each axis, the most important factor in multi-axis sensor design is the decomposability of the force components. To ensure a minimum cross coupling between the different axes, each capacitor pair is dedicated to a single force component and placed such that the main contribution to an output signal can be directly related to the force in the corresponding direction. Therefore, the x-capacitor is mainly sensitive to forces in the x-direction, the y-capacitor to forces in the y-direction, and the z-capacitor to forces in the z-direction. Similar considerations have been made for the flexures, such that by changing the dimensions of the flexures the mechanical stiffness of the sensor can be independently adjusted for each axis.

Figure 4.5 shows the mechanical response of the sensor to each force component. A force applied in the x-direction will result in a parallel movement of the sensing bodies and, therefore, only produces a change in the x-capacitance. Since one of the parallel-plate capacitor electrodes is always longer than the other, a parallel relative movement (as is the case for the y-capacitor) will not result in a change of the overlapping area and thus will not affect their capacitance, as shown in Fig. 4.5a.

Forces in the y-direction result in a rotation of the two sensing bodies relative to the outer frame. To ensure high sensitivity to these rotations, the y-capacitor pair is placed as far from the point of rotation as possible, as shown in Fig. 4.5b. The absolute capacitances in the x-capacitors will change due to a force in y, but both of them will have equal signs, resulting in no change in the differential capacitance.

Table 4.3 Design parameters of the three-axis capacitive microforce-sensing probe

		A	B	
l_{FX1}	X_1-flexure length	835	865	μm
l_{FX2}	X_2-flexure length	537	537	μm
l_{FY}	Y-flexure length	314	344	μm
l_{FZ}	Z-flexure length	1,900	1,900	μm
w_{FXY}	X,Y-flexure width	10	10	μm
w_{FZ}	Z-flexure width	183	183	μm
t_F	Flexure thickness[a]	52	52	μm
F_S	Designed force range (*x*, *y*, and *z*)		200	μN
l_C	Capacitive electrode overlapping plate length		470	μm
t_{CXY}	X,Y-capacitive electrode thickness[a]		50	μm
d_{1XY}, d_{2XY}	X,Y-capacitive electrode gap spacing widths		7, 20	μm
n_{CXY}	Number of *x*, *y*-electrode pairs per capacitor C_1 and C_2		60	
t_{CXY}	Z-capacitive electrode thickness		25	μm
d_{1Z}, d_{2Z}	Z-capacitive electrode gap spacing widths		7, 7	μm
n_{CZ}	Number of *z*-electrode pairs per capacitor C_1 and C_2		100	

[a]The active elements consist of two 25-μm thick silicon layers and one 2-μm thick SiO_2 layer

Due to the lower sensitivity of lateral sensing, forces in the *z*-direction need to produce a much larger deflection of the capacitor electrodes than forces in the *x*- or *y*-directions. Therefore, an amplification lever is integrated, as shown in Fig. 4.5c, and the *z*-capacitor is placed as far from the sensor tip as possible to maximize its leverage effect. Using this method, the *z*-flexure stiffness does not need to be significantly reduced. Forces in the *z*-direction will only result in a signal in the *z*-capacitors, since due to the aspect ratio of the flexures in the *x*- and *y*-directions (which are much thicker than they are wide), the out-of-plane motion can be neglected. These design considerations for separating the three force components into the three sensing capacitor pairs are experimentally validated and presented in Sect. 4.3.6.

An FEM (ANSYS Inc.) has been created to calculate the quantitative mechanical response of the sensor to an applied force at its probe. This enables the optimization of the flexure geometry for a certain target sensitivity along the three axes. The deflections at the position of each capacitor and the corresponding differential signal change were used as the design criteria and the flexure dimensions as the design parameters. Using the FEM analysis in an optimization loop and starting with an initial estimate of the flexure dimensions, the difference from the target deformations in each capacitor and for each force direction was found. By scaling the flexure dimensions with these errors, the ideal flexure geometry could be found, which not only ensures the desired signal change at the target force in the corresponding capacitor pair, but also minimizes the signals in all the other capacitor pairs. The resulting flexures dimensions are shown in Table 4.3. Since all the capacitors need to have electrical contact to the outer frame, two flexures (A and B) at each point must be fabricated instead of one to produce the required electrical connections.

4.2.4 Monolithically Integrated Microtensile Tester on a Chip

The microforce-sensing probes presented in the previous sections are only one component of a measurement system. Only the combination with micropositioners and position feedback sensors enables micromechanical measurements.

In this section, the combination of all the necessary components for a multi-axis microtensile-tester on a single chip is demonstrated, which enables the batch fabrication of such a complete measurement system.

Tensile testers have been used by material scientists for decades to gather quantitative information on the mechanical properties of materials by stretching them while measuring the applied force. Increasing effort is being made to develop novel microtensile testers for the ever smaller sample sizes being studied. With the advancement of MEMS technology, the development of novel electromechanical tools has become possible, enabling the integration of such systems into single measurement chips. The main components of a tensile tester chip are the displacement actuator, the force sensor, and the position feedback sensor. These components enable the stretching of a sample while measuring its deformation and the applied force. For MEMS-based actuation, thermal actuators are most commonly used where high forces and small displacements are needed [26, 27]. Alternatively, electrostatic comb drive-based actuators are used where relatively low forces and large displacements are needed [28, 29]. For most microtensile testers, the deformation of the sample is visually observed through a high-resolution microscope, as shown in [26–28]. A different approach is demonstrated in [29], where the elongation of the sample is measured using a capacitive displacement sensor integrated into the actuator. For the measurement of the applied force, different approaches have also been used. In [26] a capacitive-based force sensor has been developed, and in [27] and [29] the force has been calculated from the discrepancy between the expected and the measured displacement.

In this section, the fully integrated two-axis microtensile tester shown in Fig. 4.6 is presented, allowing for compression, tensile testing, and shear testing of microscopic samples. It offers dedicated sensors for force and displacement measurements for each of its two axes as well as two independent actuators. This enables an optimal separation of the actuation and sensing elements, resulting in minimal cross sensitivity. Due to the complete integration of the tensile tester, visual feedback is not needed for the measurement of the sample deformation or the applied force. This full integration additionally allows for the tensile tester to be mounted on a micromanipulator as an end effector, greatly facilitating the alignment to the sample under investigation.

The microtensile tester can be subdivided as shown schematically in Fig. 4.6b. The sample is measured between the tips of the two end effector arms. The right arm is connected to a two-axis capacitive microforce sensor as described in Sect. 4.2.3 with its characteristic design parameters presented in Table 4.2 allowing it to simultaneously measure forces and positions in the x- and y-directions. The left arm is connected to a platform suspended by flexures within two orthogonally

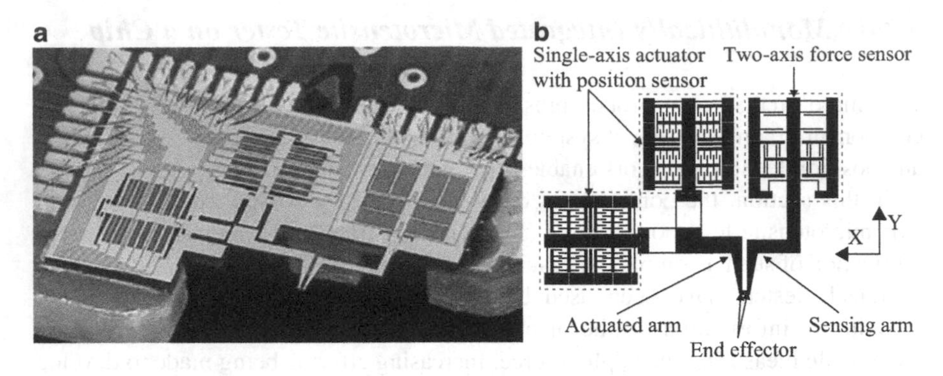

Fig. 4.6 (**a**) Photograph and (**b**) schematic of the monolithically integrated MEMS-based two-axis microtensile tester (the chip size is 7 mm by 10.8 mm). Copyright 2010 *IEEE Journal of Microelectromechanical Systems* [24]

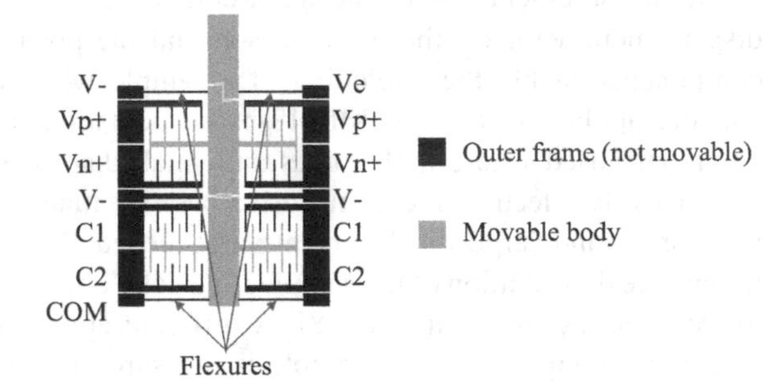

Fig. 4.7 Schematic of one single-axis electrostatic actuator with integrated capacitive position feedback sensor. Copyright 2010 *IEEE Journal of Microelectromechanical Systems* [24]

attached electrostatic actuators. Both actuators can move along one axis and offer capacitive position feedback. Therefore, the platform can be actuated along two axes, and assuming a rigid body, the position of the end effector can be measured in both the x- and y-directions. To ensure a parallel rather than rotational movement, the platform is attached to the actuators by two parallel beams [30]. This is important to ensure linearity of the position feedback sensors. The schematic of the single-axis microactuator with position feedback sensor is shown in Fig. 4.7. Like the forces sensor, it is made up of a movable body suspended by flexures within an outer frame. In this case, an electrostatic actuator and a capacitive displacement sensor are attached to the body. The actuating element is an array of parallel-plate electrodes (comb drive). For the relatively large displacements of $\pm 16\,\mu\text{m}$, a lateral movement (rather than transversal) of the comb drive plates relative to each other is chosen. By applying a voltage difference between the central movable body ($V-$) and one of the outer comb drive electrodes ($Vp+$ or $Vn+$), the movable comb drive part gets pulled into the static one, resulting in a linear motion. Table 4.4 lists the design parameters of these linear actuators. For the position feedback sensors, two additional comb drives are used to measure the displacement of the actuators as a change in capacitance. Due to the large displacements of the actuators, lateral sensing, rather than transverse sensing, had to be chosen.

Table 4.4 Design parameters of the microtensile-tester's actuator and position feedback sensors. Copyright 2010 *IEEE Journal of Microelectromechanical Systems* [24]

Actuator			
V_{AC}	Actuation voltage	0–120	V
d_{AC}	Actuator electrode gap spacing width	5	μm
n_{AC}	Number of electrode gaps per actuator	340	
w_A	Flexure width	6	μm
L_{AC}	Actuator flexure length	1,200	μm
L_{PF}	Actuator platform flexure length	1,000	μm
t	Material thickness	50	μm
Position feedback sensor			
l_{PS}	Initial capacitive electrode overlapping plate length	70	μm
d_{PS}	Capacitive electrode gap spacing widths	5	μm
n_{PS}	Number of capacitive electrode gap spacings	170	μm

A special feature of the tensile tester is the ability to electrically contact each of the end effector arms. This allows for the simultaneous electromechanical characterization of the samples under investigation.

4.2.5 Microfabrication Process

For the fabrication of the MEMS-based microforce-sensing tools with features down to a few micrometers, traditional fabrication techniques such as milling, drilling, and sawing cannot be used. Microfabrication based on lithographic processes enables the fabrication of structures with features down to the micrometer level as well as the batch fabrication of a large number of devices in parallel, greatly reducing the cost of such devices.

Two different process sequences are used. One sequence for the fabrication of devices with only in-plane sensing and actuating and another sequence for devices that incorporate in- and out-of-plane sensing. Both of these sequences involve only a combination of photolithography and dry etching steps, greatly facilitating the development of the microforce-sensing tools.

The microforce-sensing tools involving only in-plane sensing or actuation are fabricated using a bulk silicon microfabrication process described in [31]. It is based on an SOI wafer, of which the devices are etched by deep reactive-ion etching (DRIE) only incorporating two photolithographic masks. The detailed microfabrication process flow is depicted in Fig. 4.8 and described hereafter.

(a) A 100-mm diameter SOI wafer is used as the substrate material, with a handle layer thickness of 400 μm, device layers thickness of 50 μm, and an intermediate SiO_2 layer with a thickness of 2 μm. Both silicon layers have a <100> orientation and are highly p-doped.

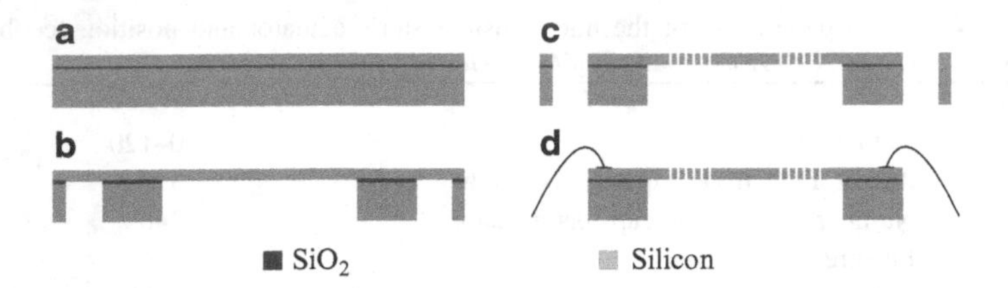

Fig. 4.8 Schematic of the SOI-based microfabrication process for in-plane sensing and actuation

(b) The handle layer is etched with DRIE where the SiO_2 acts as an etch stop and is removed subsequently with reactive-ion etching (RIE). This will form the outer frame of the sensors. Since the sensor features a probe overhanging the rest of the device, dicing cannot be performed to separate the individual dies. Therefore, a cavity surrounding the sensor is etched in the wafer in this step.
(c) The wafer is mounted onto a support wafer using heat conductive grease (Cool grease, AI Technology Inc.), and the device layer is etched with DRIE to form the active parts of the sensor such as comb drives, flexures, and the movable body.
(d) In the last step, the remaining photoresist is stripped in oxygen plasma, and the sensors are glued onto a printed circuit board and wire bonded.

For the microfabrication of MEMS-based devices incorporating three-axis (in- and out-of-plane) sensing and actuation, only relative complex process flows are published, such as those presented in [23] and [17] involving a large number of masks, wafer bonding, or microassembly steps. Therefore, here, a different microfabrication process, similar in complexity to the SOI process, is presented, greatly facilitating the fabrication of three-axis sensors and actuators. It is based on a double SOI substrate with sequential etching of the two device layers using dry etching. Although commercially available double SOI substrates are more expensive, the reduction of photomasks (only three mask necessary) and fabrication steps results in a higher yield rate and, therefore, in a more economical fabrication. The fabrication process is depicted in Fig. 4.9. The photoresist layers are only shown in the steps involved in the sequential etching of the two device layers (D–F).

(a) A 100-mm diameter double SOI wafer is used as a substrate, with a handle layer thickness of 400 μm, along with two device layers with a thickness of 25 μm and three SiO_2 layers with a thickness of 2 μm. All silicon layers have a <100> orientation and are highly p-doped.
(b) The SiO_2 on the top device layer is structured with RIE to form an etch mask in the regions, where in the last step, wires are bonded to the lower device layer.
(c) The handle layer is etched with DRIE where the SiO_2 acts as an etch stop and is removed subsequently with RIE. This will form the outer frame of the sensor. Since the sensor features a probe overhanging the rest of the device, dicing cannot be performed to separate the individual dies. Therefore, a cavity surrounding the sensor is etched in the wafer.

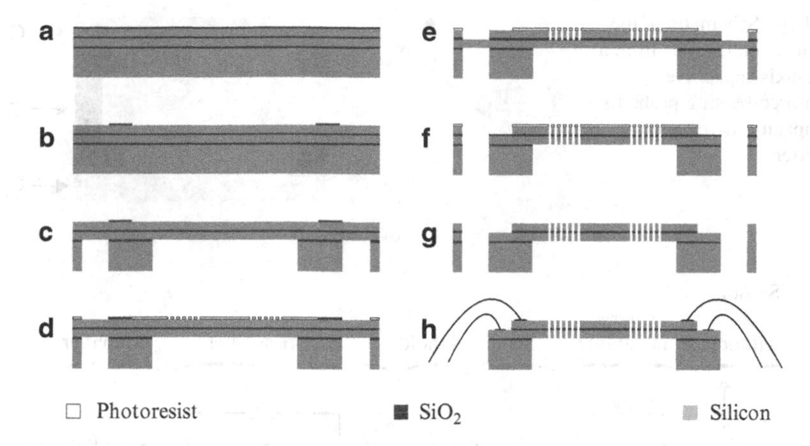

Fig. 4.9 Schematic of the double SOI-based microfabrication process for in- and out-of-plane sensing. Copyright 2010 *Journal of Micromechanics and Microengineering* [25]

(d) The photoresist (AZ 4562) is applied with a thickness of 5 μm and structured to form a second etch mask on the top device layer. This defines the active parts of the sensor such as comb drives, flexures, and the movable bodies.
(e) The wafer is mounted onto a support wafer using heat conductive grease (Cool grease, AI Technology Inc.), and the top device layer is etched with DRIE. Subsequently the underlying SiO_2 is etched with RIE. The SiO_2 etch mask, formed in A, is removed during this step as well.
(f) The lower silicon device layer is structured with DRIE with the same photoresist mask completely releasing the devices. At the same time, the top device layer is etched in the regions where the SiO_2 mask was removed in step E.
(g) The SiO_2 on the lower bonding regions is removed with RIE, and the remaining photoresist is stripped in oxygen plasma.
(h) In the last step, the sensors are glued onto a printed circuit board and wire bonded.

This three-mask fabrication process is not limited to the development of three-axis microforce sensors. The process could be used for the development of any kind of three-axis sensor or actuator with a major reduction in fabrication complexity.

4.2.6 Capacity-to-Voltage Conversion

The microforce-sensing tools convert an applied force into a displacement, which causes the capacitance of the attached electrodes to change. To measure this change in capacitance, it is converted into an analog voltage using commercially available CVC converter integrated circuits (ICs).

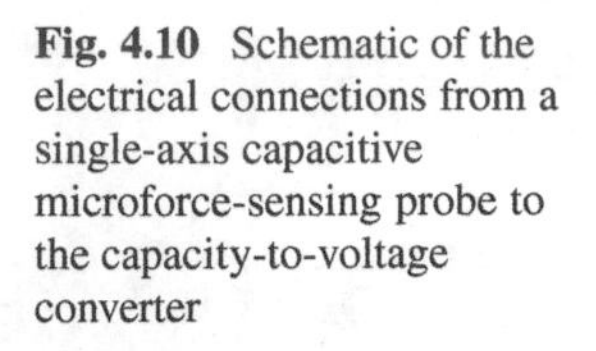

Fig. 4.10 Schematic of the electrical connections from a single-axis capacitive microforce-sensing probe to the capacity-to-voltage converter

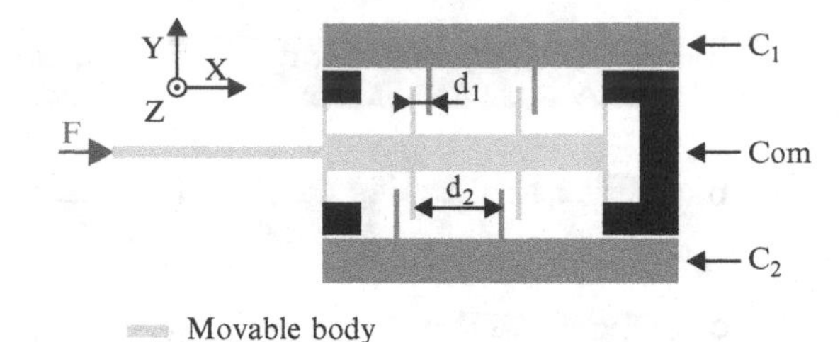

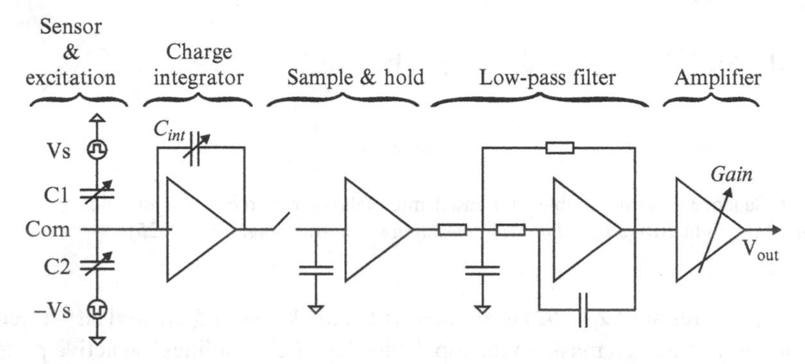

Fig. 4.11 Block diagram of the capacity-to-voltage converter IC (MS3110 and CVC 1.1). Copyright 2010 *Journal of Micromechanics and Microengineering* [25]

Two different CVC ICs are used: The universal capacitive readout MS3110 (Irvine Sensors Inc.) and the capacity-to-voltage-converter CVC1.1 (GEMAC Chemnitz GmbH). The working principle of both readout ICs is based on the impedance relation measurement, where two periodic 180° phase-shifted excitation signals are applied to a capacitor pair (C_1 and C_2) as shown in Fig. 4.10. The demodulated response of the common (COM) electrode is then proportional to the ratio of the two capacities. The analog part of the block diagram of these mixed-signal ICs is shown in Fig. 4.11. They consist of a charge integrator with integration capacitance C_{int}, a sample hold cell, a low-pass filter where the cutoff frequency has been set to 5 kHz, and an amplifier stage with an additional gain (Gain) that can be set by a serial interface. The transfer function of the MS3110 CVC is shown in (4.8), where $V2P25$ is the excitation voltage and V_{REF} is the offset voltage (both set to 2.25 V). Due to fabricational imperfections or parasitic capacitances, the C_1 and C_2 capacitors on the sensor will not be exactly equal. They can, however, be trimmed to an equal value using the serial interface of the IC, so that the output signal of the CVC is equal to V_{REF} when no load is applied. All the settings are stored in an integrated electrically erasable programmable read-only memory (EEPROM) cell. The serial interface and the analog voltage readout have been realized using Labview (National Instruments Corp.) and a data acquisition card (NI PCI-6259, National Instruments Corp.).

$$V_{\mathrm{Out}} = \mathrm{Gain} \cdot V2P25 \cdot 1.14 \frac{C_{1T} - C_{2T}}{C_{\mathrm{int}}} + V_{\mathrm{REF}} \tag{4.8}$$

$$C_{1T} = C_1 + C_{\mathrm{Trim1}} \tag{4.9}$$

$$C_{2T} = C_2 + C_{\mathrm{Trim2}} \tag{4.10}$$

A special feature of this readout is the ability to tune the electrical sensitivity of each capacitive sensor by changing C_{int} and Gain, as shown in (4.8), using the serial interface. This feature is used in the calibration step for tuning the sensitivity of the sensors to the desired value. More details about the tuning of these sensing tools are presented in Sect. 4.3.6.

The MS3110 CVC is used for single-axis sensing probes, where only a single output channel is required. For multi-axis sensing tools, multiple output channels are needed. Multiple MS3110 CVCs cannot be used to read out multiple capacitor pairs on a single sensing chip, since the excitation voltages cannot be synchronized, leading to cross coupling of the excitation signals form the different MS3110 ICs onto the different common electrodes. Therefore, for multi-axis sensing tools the CVC1.1 is used, which allows for synchronization of multiple ICs for multichannel capacity-to-voltage conversion. The exact transfer function of the CVC is not known, but the output voltage is again proportional to the capacitive change. This IC also allows for capacitance trimming and sensor tuning using a serial interface. With synchronization capability, a dedicated capacitive-to-voltage converter IC is used for each of the capacitive sensors of the force-sensing tools (up to four on the microtensile tester).

4.3 Sensor Calibration and Uncertainty Analysis

For the measurement of a force or a displacement, the sensing tools provide a related output voltage. But, since the exact relationship of the output voltage and the force/displacement is not precisely known, the only way to achieve accurate measurements is to initially calibrate the sensing tools. Analytical and FEM models of the sensors have been used for their design, but cannot be used to predict the forces/displacements from their outputs, since their accuracy is not known. The main reason for this limitation is the lack of knowledge of the exact geometry of the active elements, such as the flexures, due to imperfections in the microfabrication processes.

In this section, the calibration of the force as well as the displacement sensors used for the different sensing tools is presented, along with the methodology to estimate the accuracy of their outputs.

4.3.1 SI-Traceable Small-Force Reference Standard

Precise calibration of microforce sensors is difficult for several reasons, such as the lack of an accurate reference-force standard, the lack of standardized calibration procedures, and the need to apply known force vectors at precise positions and orientations on these small and fragile microdevices. The most commonly used microforce sensor, the AFM, has led to the development of a large number of methods for calibrating forces in the micronewton and nanonewton range [32]. However, the accuracy of these methods is unknown since most of them are based on a model of the sensor and are therefore not SI-traceable, resulting in nonquantitative measurement results.

SI-traceable calibration requires that each calibration step is a comparison back to the primary standard, and that the uncertainty associated with its propagation from one device to the next is evaluated at each step to place bounds on the actual value of the unit after its propagation through this calibration chain [18].

Force is a derived unit. The 11th General Conference on Weights and Measures (1960) has adopted the Newton as the unit of force in the International System of Units (SI) derived from the basic units of mass, length, and time as its primary standards. One Newton is the force required to accelerate a mass of 1 kg to 1 m/s^2. And with the kilogram remaining the only SI base unit defined by a material artifact, it is constantly in danger of being damaged or destroyed.

A number of scientific laboratories and National Metrology Institutes (NMIs), such as the National Institute of Standards and Technology (NIST), Physikalisch-Technische Bundesanstalt (PTB), and National Physical Laboratory (NPL), are currently investigating different approaches for a novel primary realization of the kilogram. The most promising solutions are the "Watt balance project" and the "Avogadro project." To date, there are still discrepancies between these two approaches. Until they have been resolved the kilogram remains the only SI base unit defined by a material artifact. Therefore, for the SI-traceable calibration of a force sensor, it must be referenced back to its primary standard, the International Prototype Kilogram, and is currently subjected to all its variations over time demonstrating an additional challenge in the calibration of small forces.

Macroscopic force sensors are calibrated using force standard machines, which apply a variety of SI-traceable mass artifacts combined with a suitably accurate estimate of the local gravitational acceleration while recording the resulting output signal of the sensor under investigation.

The lowest SI-traceable mass artifact available on the market is 1 mg (corresponding to approximately 10 μN) [6]. Thus, calibrating sensors down to approximately 10 μN is relatively simple. Currently efforts are being made to fabricate smaller SI-traceable calibration weights. The Korea Research Institute of Standards and Science (KRISS) has developed and calibrated a microweight set consisting of 0.05, 0.1, 0.2, and 0.5 mg artifacts for the calibration of AFM cantilevers [33]. However, there are limitations with this approach, since with smaller weights, their dimensions reach the resolving power of the human eye, making their handling increasingly challenging. Due to the mostly microscopic dimensions of the microforce sensor's probes, the application of deadweights is often not possible. For these reasons, a new method for the calibration of micronewton and nanonewton forces needs to be found. Multiple NMIs have working groups focusing on the development of such systems; first and foremost NIST, NPL, KRISS, and PTB.

Since none of the NMI's approaches to realizing a primary small-force reference standard has matured enough to be used, a different approach for the calibration of these sensors (or transfer standards) has been developed. A combination of the highly accurate, compensated SI-traceable semi-microbalance (XS205DU, Mettler-Toledo International Inc.), deadweights and a macroscale, custom build reference-force sensor is used. The goal of this combination is to profit from the advantages of the mature microbalance technology while eliminating their disadvantages. Deadweights are used as a primary transfer artifact calibrated on a high-precision microbalance before each force sensor calibration. This step takes advantage of the precision balance's high accuracy, but avoids the issue of its slow reaction time and its influence on the calibration uncertainty due to drifting of the sensor's signal. At the same time, it eliminates the effects of deadweight degradation and contamination, since they are recalibrated before each sensor calibration. In the second step, the calibrated deadweights are used to calibrate the microforce sensors. Due to the small geometry of the MEMS-based sensing probes, the additional macroscale reference-force sensor is introduced into the calibration chain, which can be calibrated by applying these deadweights. Finally, the MEMS-based sensors are pushed against this reference-force sensor using a micropositioner, while recording both signals. Since the reference-force sensor does not incorporate active force compensation and offers a fast readout (up to 10 kHz), the limitations of the microbalance are successfully avoided while still taking advantage of its high accuracy.

4.3.2 Uncertainty Analysis

The result of a measurement or calibration is only an approximation of the value of the measurand and, thus, is complete only when accompanied by a statement of the uncertainty of that estimate [34]. The measurement uncertainty is a parameter associated with the results of a measurement that characterizes the dispersions of the values that could reasonably be attributed to the measurand [35]. Therefore, for SI-traceability, besides the measurement result, its uncertainty also needs to be measured and propagated throughout the entire calibration chain, starting with the primary reference standard and its uncertainty.

The joint committee for guides in metrology (JCGM) of the BIPM has a working group responsible for the expression of uncertainty in measurements. They have published the International Organization for Standardization's (ISO) guide to the expression of uncertainties in measurements (GUM) [34], which has become the internationally accepted master document for the evaluation and combination of these uncertainties. In the GUM, a deterministic method based on the law

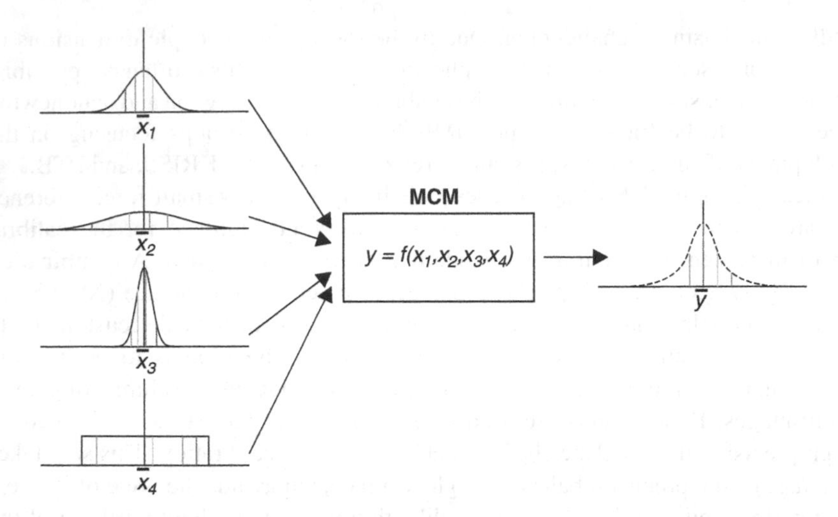

Fig. 4.12 Schematic of the Monte Carlo method (MCM)

of propagation of uncertainties is described. It is based on the characterization of the uncertainties of all the measured input quantities by either normal or Student's t-distribution, allowing the calculation of coverage intervals for the output quantities.

To deal with problems that are not linear or involve probability density functions (PDFs) other than the normal- or the Student's t-distribution, the supplement one has been added to the GUM describing the Monte Carlo method (MCM) [36]. The uncertainty in a measurement is a result of several uncertainties. They can be categorized into two groups depending on how their values are estimated. The type A uncertainties are those that are evaluated by statistical methods and can often be described using a normal- or Student's t-distribution. The type B uncertainties are those that are evaluated by other means, such as a prior knowledge of the system or specifications of the equipment, e.g., given by an upper and lower bound, defining a rectangular PDF. Both types of uncertainties can be incorporated into the analysis using the MCM.

The MCM evaluates the propagation of distributions by performing random sampling from the PDFs of all input quantities to predict the PDF of the output. And in the latest supplement two, these methods have been extended to multivariate problems with any number of output quantities. For the calculation of the uncertainty in the calibration coefficients as well as in force predications of microforce-sensing probes, the multivariate adaptive MCM is used. For comprehensive reasons, a short introduction to the method depicted schematically in Fig. 4.12 is given here, and the complete description can be found in [37]. In the following sections, the MCM is applied for the calculation of the uncertainties in the calibration results of the various sensing tools.

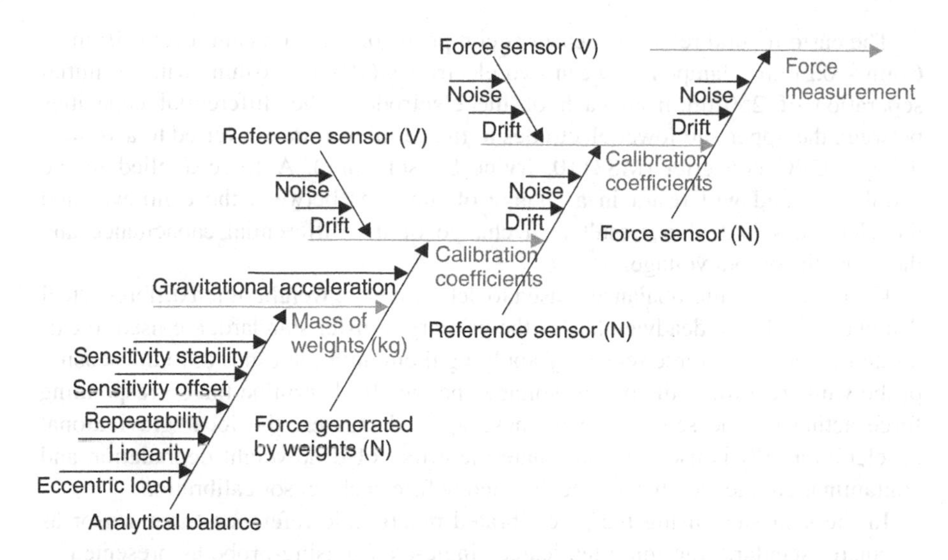

Fig. 4.13 Cause and effect diagram for the propagation of the diverse sources of uncertainty in the calibration chain

A number of M random samples are drawn from the PDFs of all input quantities of the transfer function f under investigation (e.g., ordinary least squares). Using the transfer function f, M multivariate model results can be calculated giving a discreet representation of the distribution function of the results. These M model results can then be used to calculate the best estimate of the results (e.g., the calibration coefficients), as well as their covariance and expansion coefficients. The effectiveness of the MCM depends on an adequately large value for the number of M Monte Carlo trials. Therefore, in the adaptive MCM an increasing number of Monte Carlo trials are carried out until the standard deviation of the results has stabilized in a statistical sense.

In the next sections, the MCM is applied in each step of the calibration chain for the calibration of the microforce-sensing tools, shown with the cause and effect diagram in Fig. 4.13 in which the most relevant sources of uncertainty are listed.

4.3.3 Calibration of Single-Axis Microforce-Sensing Probes

Due to the lack of an accurate reference standard for the calibration of forces below 10 μN, an SI-traceable compensated semi-microbalance (XS205DU, Mettler-Toledo International Inc.) is used as a primary reference, which is precalibrated by the manufacturer with a given uncertainty. To transfer this reference standard to the different MEMS-based microforce-sensing tools, a combination of deadweights and the macroscale, custom build reference-force sensor is used.

The custom build reference sensor is composed of a silicon cantilever (40 mm × 6 mm × 0.2 mm) clamped between two electrodes (10 mm × 6 mm) with an initial separation of 250 μm from each of the electrodes. The differential capacities between the upper and lower electrodes to the cantilever are converted to a voltage using a CVC converter (MS3110, Irvine Sensors Inc.). A force applied to the cantilever's end will result in a change of the gaps between the cantilever and the electrodes and, thus, result in a change of the differential capacitance and therefore the output voltage.

First, the semi-microbalance is used to determine the weight of ten different steel deadweights. These deadweights, as the primary transfer standard, are used to calibrate the reference-force sensor by applying them multiple times onto the sensing probe while recording the output voltage change. To determine the corresponding force acting on the sensor due to these applied masses, the local gravitational acceleration [38] is used. To eliminate the effect of deadweight degradation and contamination, the weights are recalibrated before each sensor calibration.

In the next step, using the precalibrated macroscale reference-force sensor as a transfer standard, the microfabricated single-axis sensing-probe as presented in Sect. 4.2.2 can be calibrated. By pushing this reference sensor stepwise against the MEMS-based sensor using a motorized linear stage (MT1-Z6, Thorlabs Inc.) and recording the voltage change from both sensors, the calibration curve of these microfabricated-sensing probes can be recorded as shown in Fig. 4.14a. For accurate positional and rotational alignment of the two sensors, the MEMS-based sensor is mounted on a three-axis micropositioner (MP-285, Sutter Instrument Co.), and the entire setup is fixed under a high-resolution microscope (A-ZOOM, Signatone Corp.).

For the macroscale reference sensor as well as for the microfabricated single-axis sensing-probe, all the different sources of uncertainty shown in the cause and effect diagram in Fig. 4.13 with their corresponding PDFs are used to create one joint PDF for each applied force and for each output voltage change in the calibration data. By randomly sampling from these joint PDFs and using the method of ordinary least squares, a third-order polynomial function as shown in (4.11) is fit into the calibration data for each of the $M = 10^4$ Monte Carlo trials, minimizing the residual $\boldsymbol{r}_i$. These M sets of the calibration coefficients (k_1, k_2, k_3) give a discrete representation of the multivariate PDF of the result. From this PDF, the best estimate, its standard uncertainties, and the correlation coefficient of the calibration coefficients can be calculated.

$$\boldsymbol{F}_i = \begin{pmatrix} \boldsymbol{V}_i & \boldsymbol{V}_i^2 & \boldsymbol{V}_i^3 \end{pmatrix} \begin{pmatrix} k_{1,i} & k_{2,i} & k_{3,i} \end{pmatrix}^{\mathrm{T}} + \boldsymbol{r}_i \quad \text{for } i = 1 - M \tag{4.11}$$

The MCM is carried out multiple times until the results have stabilized. This occurs when twice the standard deviation of all the calibration characteristics is smaller than the numerical tolerance, as defined by the number of relevant significant digits [37]. The final multivariate PDF of the calibration coefficients, necessary to make force predictions, is calculated taking the entire $h \times M$ (with h the number of Monte Carlo iterations) realization of the calibration coefficients and is shown for the microfabricated-sensing probe in Table 4.5. The corresponding calibration data are

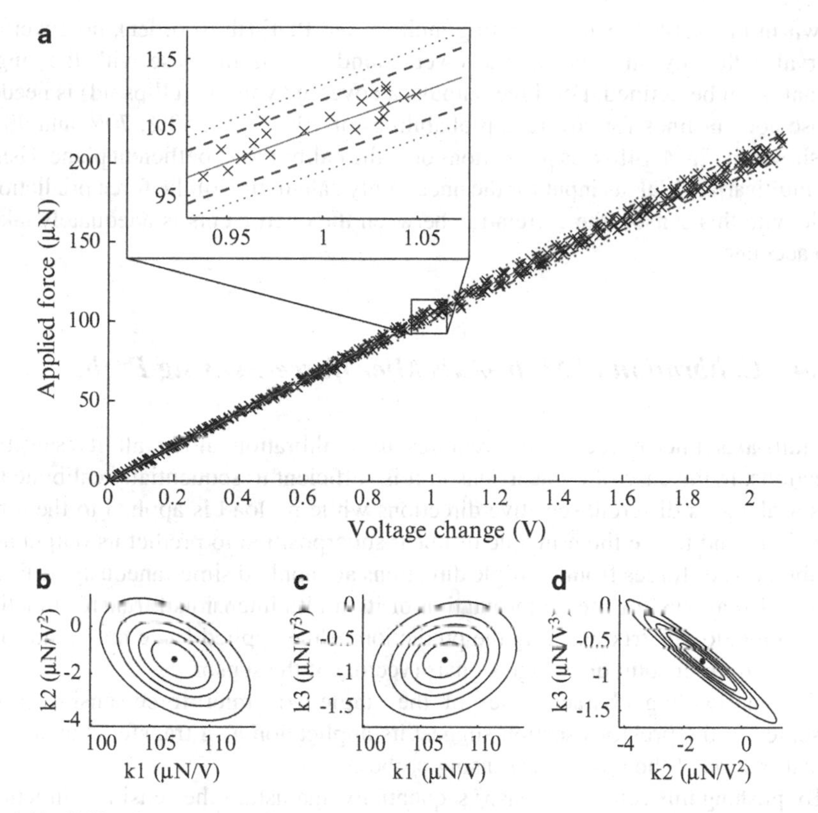

Fig. 4.14 Calibration results of the single-axis reference-force sensor consisting of (**a**) the calibration data (*cross*) as well as the best estimate (*solid line*), the 68% (*dashed line*), and the 95% (*dotted line*) coverage intervals of the calibration curve, (**b–d**) contour lines of the multivariate PDF of the calibration coefficients for coverage probabilities of 10%, 30%, 50%, 70%, and 90%, projected onto the calibration coefficient plane of (**b**) k_1 and k_2, (**c**) k_1 and k_3, (**d**) k_2 and k_3.

Table 4.5 Calibration results: single-axis MEMS-based microforce-sensing probe

Input range F_x (μN)	±230
Output range (V)	0–4.5
Calibration coefficients (μN/V μN/V^2 μN/V^3)^T	(107.07 −1.38 −0.81)^T
Standard uncertainties (μN/V μN/V^2 μN/V^3)^T	(2.52 1.02 0.34)^T
Correlation coefficients	
r_{12}	−0.468
r_{13}	0.247
r_{23}	−0.960
u_{Noise} at 10 Hz (μN)	0.02
u_{Drift} ($t = 30$s) (μN)	0.07

shown in Fig. 4.14. For this resulting multivariate PDF (third order), no coverage interval with only an upper and a lower bound – as is the case with the single variant – can be defined. For three outputs, a coverage volume (ellipsoid) is needed whose contour lines for coverage probabilities of 10%, 30%, 50%, 70% and 90%, are shown in Fig. 4.14b–d as projections onto the calibration coefficient plane. Using this multivariate PDF as input for the uncertainty calculation of the force predictions made with this sensor, the correlation between the coefficients is adequately taken into account.

4.3.4 Calibration of Multi-Axis Microforce-Sensing Probes

A multi-axis microforce sensor requires the calibration along all its sensitive directions. In the case of a linear sensor, it is sufficient to sequentially calibrate the sensor along its different sensitive directions while no load is applied to the other directions, and to use the principle of linear superposition to predict its output as a combination of forces from multiple directions are applied simultaneously. A linear sensor also allows for the representation of its multidimensional transfer function by a calibration matrix enabling the prediction of the applied force magnitude and direction from the output voltage signals (vector) of the sensor.

The outstanding characteristics of the single-axis microforce-sensing probe presented in the previous section suggest its application as a transfer standard for the calibration of the multi-axis sensing probes.

By pushing this reference sensor sequentially against all the sensitive directions of the multi-axis sensor and measuring all the output voltages of the two sensors, the calibration curves can be found. These data can then be used (in the case of a linear sensor) to extract the calibration matrix $\boldsymbol{A}$, describing the relationship between any combination of output voltages of the sensor and the corresponding applied force. The calibration curves of the two-axis microforce sensor presented in Sect. 4.2.3, integrated into the microtensile tester (Sect. 4.2.4), are shown in Fig. 4.15. Due to the linear relationship, the system can be represented by a calibration matrix $\boldsymbol{A}^{2\times 2}$, describing the relationship between the two output voltages of the two-axis force sensor and the applied forces in the x- and y-directions acting on its end effector.

The uncertainties in the calibration matrix are calculated using the adaptive multivariate MCM. All the PDFs defined by the different sources of uncertainty are combined into a joint PDF for each of the calibration data points. By taking $M = 10^4$ random samples from all the joint PDFs, M calibration data sets (4.12) and (4.13) are created, where $\mathbf{V}c_i$ is the voltage from the two-axis force sensor of the tensile tester and $\mathbf{F}c_i$ is the applied force, given by the reference-force sensor with i from 1 to M. Using the ordinary least squares method (4.14), M calibration matrices can be calculated. This MCM is carried out multiple times (h) until all the results have stabilized as described in Sect. 4.3.3.

The best estimate of the calibration matrix (from all the $h \times M$ realizations), the standard uncertainties, and the correlation coefficients (no correlation between $\hat{\mathbf{A}}_{\mathbf{x}}$

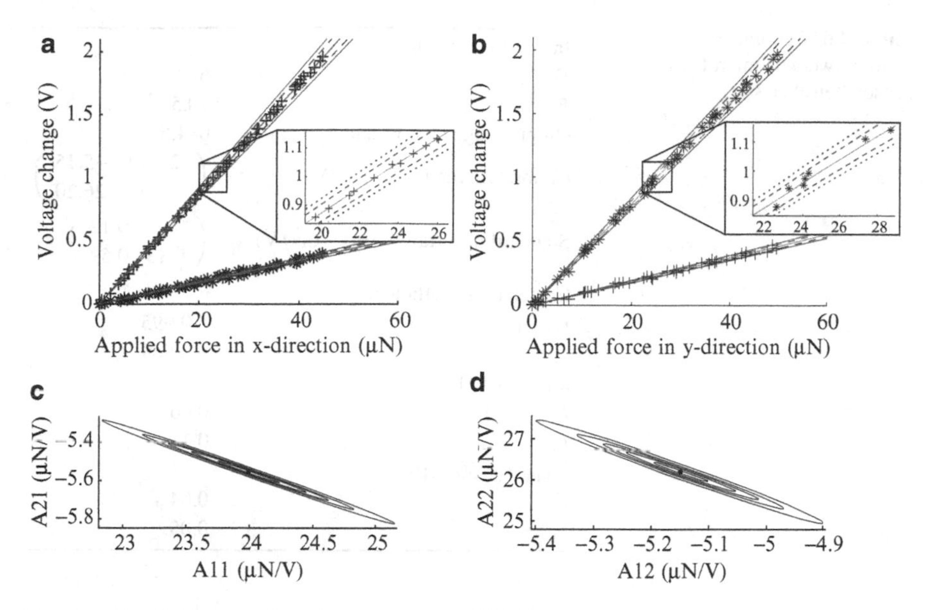

Fig. 4.15 Calibration results of the two-axis MEMS-based microforce sensor (tensile tester) consisting of the calibration data (*plus, asterisk*) as well as the best fit (*solid line*), 68% (*dashed line*), and 95% (*dotted line*) coverage intervals of the calibration curve in (**a**) the *x*-direction and (**b**) the *y*-direction, and contour lines of the multivariate PDF of the calibration coefficients in (**c**) *Ax* and (**d**) *Ay* for coverage probabilities of 10%, 30%, 50%, 70%, and 90%. Copyright 2010 *Journal of Microelectromechanical Systems* [24]

and $\hat{\mathbf{A}}_\mathbf{y}$) are shown in Table 4.6, and the contour lines of the multivariate PDF of the calibration coefficients are shown in Fig. 4.15c, d.

For the calibration of the full three-axis microforce sensor presented in Sect. 4.2.3, the same approach is used, where the reference force is applied sequentially along all the three sensitive directions while recording the output voltages. The result of the three-axis sensor calibration is presented together with its tuning capabilities in Sect. 4.3.6.

$$\mathbf{F}c_i = \begin{bmatrix} Fx_{1,i} & \cdots & Fx_{N,i} & 0 & \cdots & 0 \\ 0 & \cdots & 0 & Fy_{N+1,i} & \cdots & Fy_{2\cdot N,i} \end{bmatrix}^\mathrm{T} \tag{4.12}$$

$$\mathbf{V}c_i = \begin{bmatrix} Vx_{1,i} & \cdots & Vx_{N,i} & Vx_{N+1,i} & \cdots & Vx_{2\cdot N,i} \\ Vy_{1,i} & \cdots & Vy_{N,i} & Vy_{N+1,i} & \cdots & Vy_{2\cdot N,i} \end{bmatrix}^\mathrm{T} \tag{4.13}$$

$$\mathbf{F}c_i = \mathbf{V}c_i \cdot \hat{\mathbf{A}}^{2\times 2} + \boldsymbol{r}_{i,\min} = \mathbf{V}c_i \left[\hat{\mathbf{A}}_x^{2\times 1}\ \hat{\mathbf{A}}_y^{2\times 1}\right] + \boldsymbol{r}_{i,\min} \tag{4.14}$$

$$\hat{\mathbf{A}}_i^{2\times 2} = \left(\mathbf{V}c_i^\mathrm{T}\mathbf{V}c_i\right)^{-1}\mathbf{V}c_i^\mathrm{T} \cdot \mathbf{F}c_i \tag{4.15}$$

Table 4.6 Calibration results: two-axis microforce sensor (tensile tester)

Input range (μN)	
F_x	66.5
F_y	70.5
Output range (V): V_x and V_y	0–4.5
Calibration matrix $\hat{\mathbf{A}}$ (μN/V)	$\begin{pmatrix} 23.99 & -5.15 \\ -5.55 & 26.20 \end{pmatrix}$
Standard uncertainty $\mathbf{u}(\hat{\mathbf{A}})$ (μN/V)	$\begin{pmatrix} 0.54 & 0.12 \\ 0.13 & 0.59 \end{pmatrix}$
Correlation coefficients	
r_{11-21}	−0.995
r_{21-22}	−0.989
u_{Noise} at 10 Hz (μN)	
F_x	0.06
F_y	0.1
u_{Drift} ($t = 30$s) (μN)	
F_x	0.04
F_y	0.05

4.3.5 Calibration of the Multi-Axis Position Feedback Sensors

The relationship between the output voltages and the position of the end effector arms of the tensile tester is also unknown, and must be found by calibration. For the actuated arm, a position feedback sensor is integrated into each of the two orthogonally aligned actuators, whereas in the force-sensing arm, the output of the force sensor can be correlated with the movements of its end effector.

In order to calibrate the position feedback sensors of the actuated as well as the force-sensing end effector arms, their positions need to be measured by other means and compared with their output voltage signals. A microscope (A-ZOOM, Signatone Corp.) with an attached camera (A622f, Basler AG) which has been precalibrated using an optical target (1951 USAF Resolution Targets, Edmund Optics Inc.) is used to visually measure the position changes of the end effectors. High magnification pictures are taken at each calibration step and are analyzed using a visual rigid body tracker to extract the relative movement of the end effector arms.

The rigid body tracker is based on the fitting of a geometrical model of each of the end effector arms onto the images taken during the calibration using the least squares method, as shown in Fig. 4.16. The position of each of the end effector tips can be extracted from each image. The results are correlated with the output voltages measured at the instant in which the corresponding picture was taken. More details about the visual rigid body tracker can be found in [39].

For the calibration of the actuated end effector arm, the relationship between the output voltages of the actuator feedback sensors and the position of the end effector is calibrated by subsequent actuation in the x- and y-direction. Voltage ramps from 0 to 120 V and vice versa are applied, first to the positive and the then to the negative actuation direction electrodes ($Vp+$ or $Vn+$) in each axis. The output voltages of

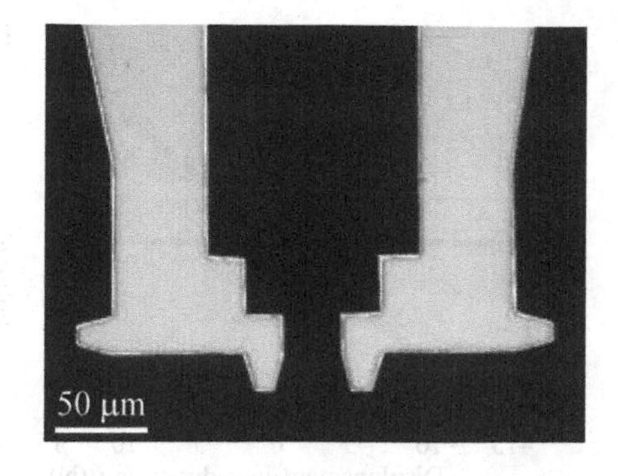

Fig. 4.16 Microscope picture of the tensile tester's end effectors with the geometrical model fitted to its outline, used for the visual rigid body tracking. Copyright 2010 *IEEE Journal of Microelectromechanical Systems* [24]

the two feedback sensors as well as the position given by the visual tracker are recorded and shown in Fig. 4.17a, c for the actuator calibration in the x-direction and in Fig. 4.17b, d for the y-direction. Clearly, there is no significant hysteresis observable in the actuator's characteristics, since the increasing and decreasing positions during the actuator calibration cycles overlap. Similar relationships are found for the force-sensing end effector arm calibration. In the case of the force sensor position feedback calibration, the movements are induced by the reference-force sensor during force calibration.

To calculate the calibration matrix and its uncertainties, the adaptive MCM described in the previous section is used. The results for the calibration of the actuator are shown in Fig. 4.17 as well as in Table 4.7.

The nondiagonal entries in the calibration matrix are almost zero, indicating almost completely independent movement in the x- and y-directions. However, as can be seen in Fig. 4.17b, c, a nonlinearity in the nondiagonal components of the calibration matrix can be detected. The error due to this nonlinearity is characterized by an additional uncertainty with a rectangular distribution, defined by a lower and a upper limit.

The results of the actuator position feedback and the force sensor position feedback calibration are summarized in Table 4.7. Together with the results presented in Table 4.6, these characteristics can be used to make force as well as position predictions along two axes from the four output voltages of the tensile tester. To calculate the uncertainties of the position and force predictions, the uncertainties in the calibration matrices and in the output voltages (as well as the nonlinearity) need to be taken into account.

4.3.6 Range Tunable Microforce Sensing

For most applications that require the measurements of microscale forces, it is challenging and often not intuitive to estimate the magnitude of the expected force range before actually measuring it. Therefore, designing a force sensor for a specific application is often an iterative and time-consuming process, since after the first measurement with such a sensor, it often needs to be redesigned for a different force

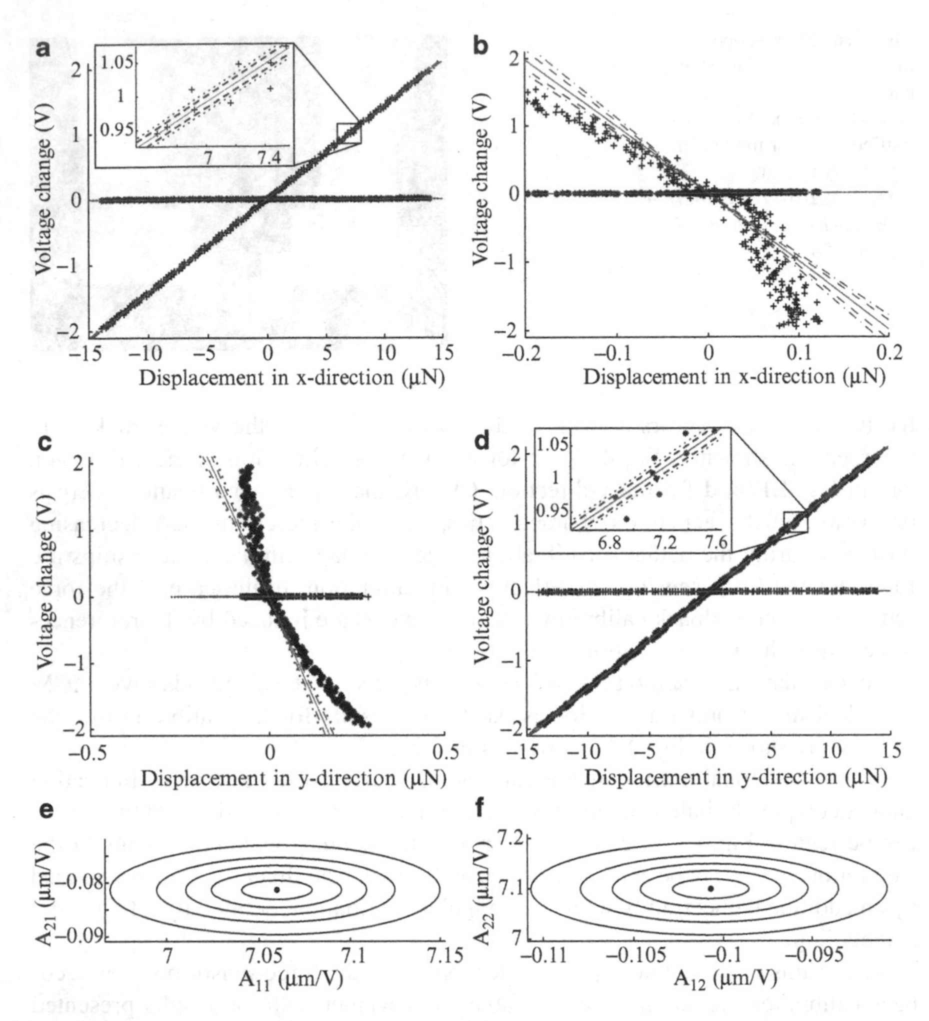

Fig. 4.17 Calibration results of the tensile tester's two-axis position feedback sensor in the actuated arm consisting of the calibration data (*plus, asterisk*) as well as the best fit (*solid line*), 68% (dashed line), and 95% (*dotted line*) coverage intervals of the calibration curve (**a, c**) in the *x*-direction and (**b, d**) in the *y*-direction, contour lines of the multivariate PDF of the calibration coefficients in (**e**) *Ax* and (**f**) *Ay* for coverage probabilities of 10%, 30%, 50%, 70% and 90%. Copyright 2010 *IEEE Journal of Microelectromechanical Systems* [24]

range or used within only a percentage of its range. For this reason, it would be desirable to have a force sensor that allows the tuning of its measurement range for a specific application while the measurements are taken, enabling a best possible signal-to-noise ratio.

As presented in Sect. 4.2.6, the capacity-to-voltage converter incorporates capabilities for the variation of some of its characteristic parameters, such as the integrator feedback capacitor C_{int} and the output amplifier gain "Gain" using a serial

Table 4.7 Calibration: two-axis position feedback sensors in the microtensile tester

Sensor type	Position feedback on force sensor	Position feedback on actuator
Input range (μm)		
D_x	2.2	16.1
D_y	1.1	16.2
Output range (V): V_x and V_y	0–4.5	0–4.5
Calibration matrix $\hat{\mathbf{A}}$ (μm/V)	$\begin{pmatrix} 0.93 & 0.12 \\ -0.03 & 0.36 \end{pmatrix}$	$\begin{pmatrix} 7.06 & -0.10 \\ -0.08 & 7.10 \end{pmatrix}$
Standard uncertainty $\mathbf{u}(\hat{\mathbf{A}})$ (μm/V)	$\begin{pmatrix} 0.02 & 0.01 \\ 0.02 & 0.01 \end{pmatrix}$	$\begin{pmatrix} 0.04 & 0.01 \\ 0.01 & 0.04 \end{pmatrix}$
Correlation coefficients		
r_{11-21}	−0.412	−0.001
r_{21-22}	−0.408	−0.001
u_{Noise} at 10 Hz (μm)		
D_x	0.003	0.02
D_y	0.001	0.02
u_{Drift} ($t = 30$s) (μm)		
D_x	0.003	0.02
D_y	0.001	0.02
$u_{\text{nonlinearity}}$ (μm/V) PDF: rectangular	–	$\begin{pmatrix} 0 & 0.06 \\ 0.06 & 0 \end{pmatrix}$

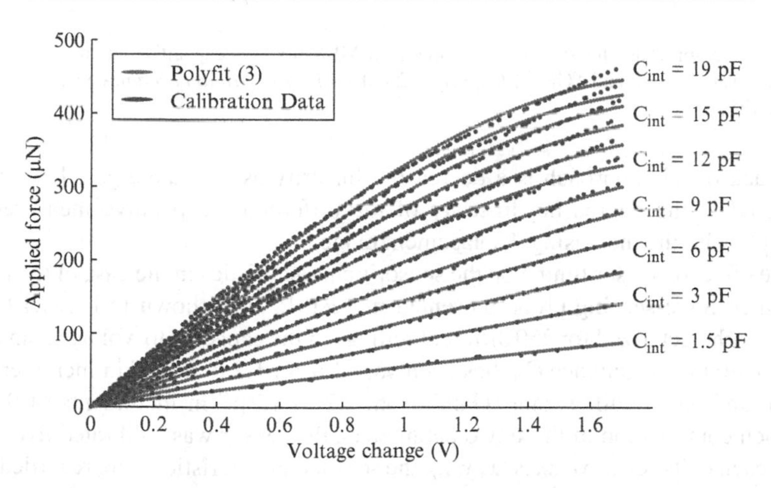

Fig. 4.18 Calibration curves for the single-axis MEMS-based capacitive microforce-sensing probe for varying C_{int}

interface. The effect of varying C_{int} can be seen in Fig. 4.18 for the case of the single-axis capacitive microforce sensor presented in Sect. 4.2.2. As C_{int} is increased, the range of the sensor increases, enabling the measurement of larger forces. The

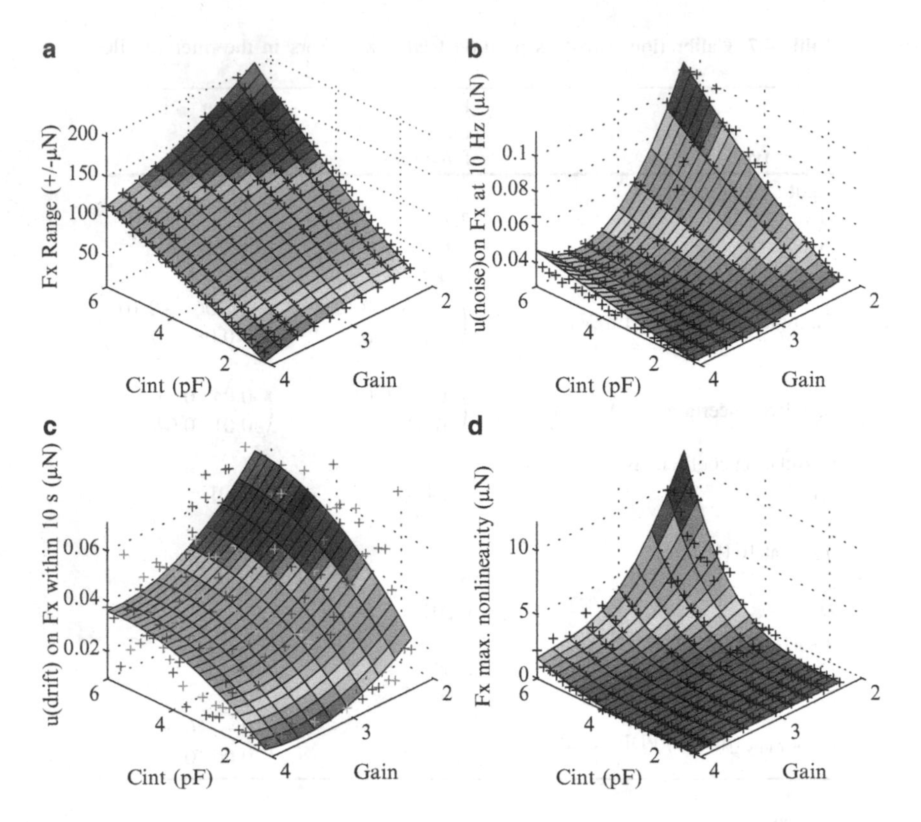

Fig. 4.19 Sensor characteristics for the three-axis MEMS-based capacitive microforce-sensing probe for varying C_{int} and "Gain". Copyright 2010 *Journal of Micromechanics and Microengineering* [25]

drawback of this approach is a decreasing linearity as the range gets larger; this is related to the decreasing linearity of the differential capacitive measurement principle with the increasing displacement range.

The effect of sensor tuning on the sensor's characteristics in the case of the three-axis microforce-sensing probe presented in Sect. 4.2.3 is shown in Fig. 4.19. The sensor is characterized for 250 different settings of the capacity-to-voltage converter. The integrator capacitance C_{int} has been varied from 1.2 to 6.0 pF in increments of 0.2 pF and the amplifier gain "Gain" from 2.2 to 4.0 pF in increments of 0.2 pF. For each combination of these two parameters, the sensor was calibrated five times along each of its sensitive axes (x, y, z), the sensor characteristics were recorded, and the corresponding measurement uncertainties calculated. The resulting data sets for the x-direction – from a total of 3,750 calibrations – are shown in Fig. 4.19. The raw characterization data are shown as $+$, and the surface plots show a fit using a second-order polynomial in two variables, fitted using a least squares algorithm. The range of the sensing probe can be changed from approximately ± 20 to $\pm 200\,\mu$N with a corresponding resolution from 30 to 110 nN.

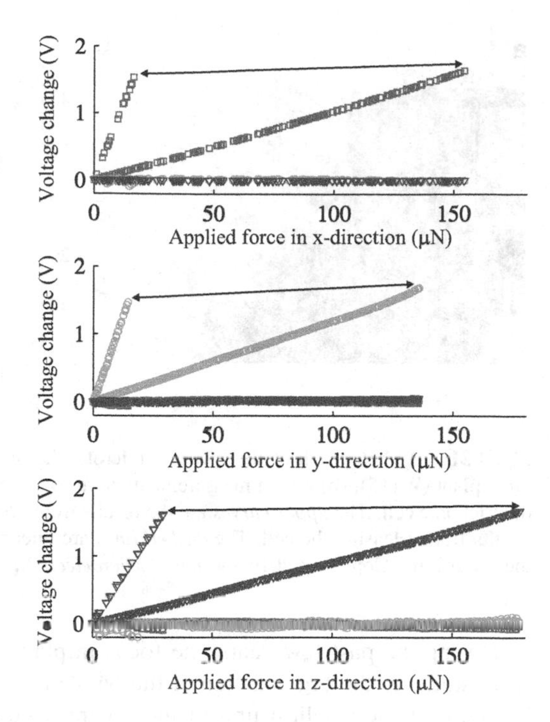

Fig. 4.20 The calibration curves of the three-axis MEMS-based microforce-sensing probe for two different CVC settings (*squares* indicate the raw calibration data V_x, *circles* V_y, and *triangles* V_z). Copyright 2010 *Journal of Micromechanics and Microengineering* [25]

The calibration curves for the minimum and maximum sensor range are shown in Fig. 4.20. The arrows between the main components indicate the range in which the calibration curves can be adjusted. These results further indicate that the goal of mechanically decomposing the forces has been successfully realized, and that the lower the range of the sensor is tuned, the smaller the forces that can be measured. Consequently, instead of redesigning the sensor, or using only a fraction of its range, it can be tuned to the requirements of an application, which will result in an optimal signal-to-noise ratio and minimum nonlinearity. Since the sensor has been precalibrated for all the different settings of the CVC, it can be tuned as the measurements are taken without requiring recalibration.

4.4 Micromechanical Investigation of Individual Plant Cells

As the world's primary producer of food and energy, photosynthetic organisms are of vital importance to human society. Making up over 99% of the earth's biomass, photosynthetic organisms also have a major impact on the global climate. Consequently, understanding how plants grow is of fundamental importance. Numerous studies and models of plant growth have been made based on limited and mainly quantitative knowledge of their mechanical properties.

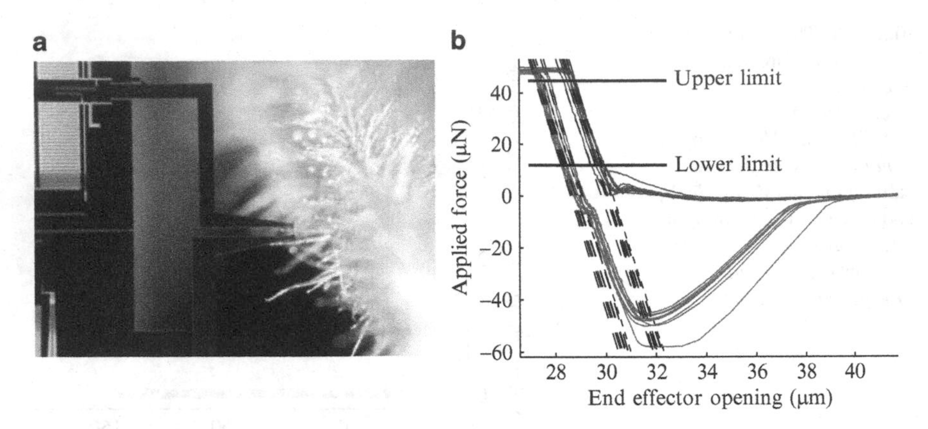

Fig. 4.21 (**a**) Photograph of the two-axis microtensile tester aligned with a trichome of the petunia plant (W115); (**b**) all ten measurement curves on one of the nine measurement locations on the trichome cell: The *upper curves* are the results from compressing and the *lower curves* are the results from releasing the cell. The *dashed lines* are linear fits to the data between the upper and the lower limit. Copyright 2010 *Journal of Microelectromechanical Systems* [24]

During the past few years, the focus in plant development biology has shifted from studying the organization of the whole body or individual organs toward the behavior of the smallest unit of the organism, the single cell [3]. Cell expansive growth is a mechanical process that balances internal and external stresses with the compliance to allow expansion. Various models to predict cell growth have been made, but to supply mathematical models with relevant and accurate input, quantitative values for a number of physical parameters need to be provided. Since educated guesses are often the only recourse [40], the aim of this application is to demonstrate the capabilities of the microtensile tester to quantitatively measure these properties in plant cells in their living states. This has the potential to provide the information needed to construct a new generation of mechanical models that are based on actual, measured properties.

The microtensile tester presented in Sect. 4.2.4 can be used for a wide range of applications in different fields, such as material science or mechanobiology. One of the key features of this device is its ability to be used as a compression tester. Much like the human hand, the gripper-like end effector design allows it to simply grasp an object, e.g., a cell in its living state, and "feel" its properties, such as its stiffness or size. This feature is demonstrated by measuring the mechanical properties of plant hairs (trichomes). These elongated epidermal plant cells are mechanically characterized while they are still attached to the living plant. The trichomes serve as an excellent model system to study various aspects of plant differentiation at the individual cell level, and are easily accessible because of their epidermal origin [3]. The glandular, non-branched multicellular trichomes from the wild-type petunia (W115) plant, as shown in Figs. 4.21a and 4.22a, are chosen as a sample. The goal is to measure the cell stiffness along one of the long inclining cells to find the

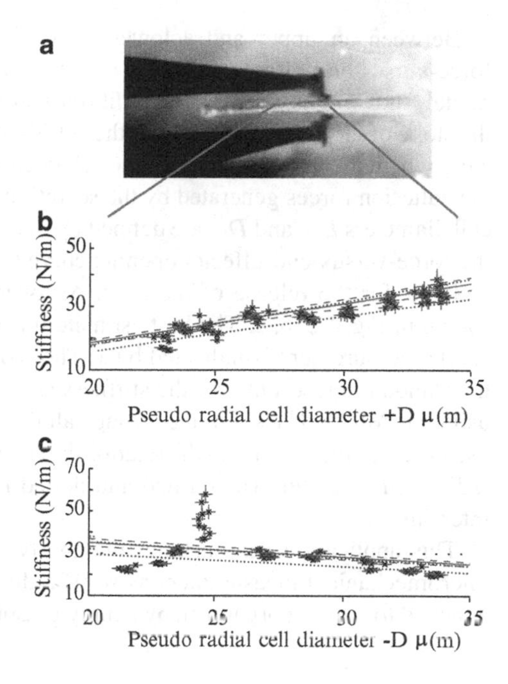

Fig. 4.22 Measurement results of the trichome cell characterization: (**a**) the microtensile tester's end effectors aligned with the trichome indicating the region of the stiffness measurements, (**b**) results of the stiffness-versus-cell-diameter measurement for the compression of the cell, (**c**) results of the stiffness-versus-cell-diameter measurement for the releasing of the cell. Copyright 2010 *Journal of Microelectromechanical Systems* [24]

relationship between the cell stiffness and its radial diameter. The tensile tester is mounted on a three-axis micropositioner with position encoders (SL-2040, SmarAct GmbH). The proper alignment of the tensile tester with the trichome cell is critical. Before the measurements, the end effectors are visually aligned at three locations along the cell. By interpolating between these locations, the trajectory along which the micropositioner will move the tensile tester during the automated measurement can be determined.

The stiffness of the trichome cell is measured by compressing the cell while measuring the force and the deformation, given by the sum of the displacements of the sensing and the actuating end effector. This is done in a fully automated fashion along the trichome cell over a length of 180 μm in 20 μm steps. At each measurement location, the cell is compressed ten times repeatedly. In Fig. 4.21b, the ten measurement curves from one of the nine measurement locations are shown. The measurements begin with the fully opened end effectors (right side in Fig. 4.21b). Then they are slowly closed at an average speed of 0.5 μm/s until the force sensor is saturated (upper curve in Fig. 4.21b). Subsequently, they are opened again to their initial position, shown in the lower curve in Fig. 4.21b. The large negative force on the lower curve is the adhesion force of the cell to the end effectors during the opening and indicates that the cell is under tension. The slow decrease in the adhesion force as the end effectors are separated is assumed to be related to the sticky secretion with which these glandular trichomes are covered. The extensive interpretation of the measurement data and the description of the material model are not within the scope of this work. However, to give an overview of the results, a simplified description of the cell using a linear model is shown in Fig. 4.22.

Between an upper and a lower limit of the force, shown in Fig. 4.21b, the force-versus-end-effector opening can be reasonably approximated using a liner model, shown with a least squares fit (dashed lines). The cell stiffness is defined as the slope of these curves. Defining the cell's radial diameter is difficult since it is not clear where the first contact with the cell occurred. This is suspected to be related to the adhesion forces generated by the secretion on the cell. Therefore, pseudo radial cell diameters $D+$ and $D-$ are defined by linearly extrapolating the linear regime of the force-versus-end-effector opening curve to zero force ($D+$ for the compression and $D-$ for the release of the cell). An overview of the measurement results is shown in Fig. 4.22b, c. The best estimate for the stiffness and cell diameter from a single measurement is indicated by x. The cross in each measurement point shows the standard uncertainty for the stiffness as well as for the cell diameter calculated using the MCM. The solid line through all the data indicates the best estimate of the least squares fit, describing the relationship between the cell stiffness and the pseudo radial cell diameter. The dashed and dotted lines are the 68% and 95% coverage intervals.

This application demonstrates the ability of the microtensile tester to perform micromechanical measurement as well as dimensional measurement on a sample, attached to a living organism, by simply grasping it.

4.5 Conclusion

In this chapter, the design of a single-axis, a two-axis, and a three-axis MEMS-based microforce-sensing probes, their fabrication, and their characterization is presented. In the case of the three-axis sensing probe, forces of up to ±200 μN along all its sensitive directions with a resolution as low as 30 nN can be measured. By combining capacitive sensing as well as electrostatic actuation elements on a single chip, the design of a monolithically integrated, two-axis MEMS-based microtensile tester has been shown, allowing the direct measurement of the mechanical as well as electrical properties of a sample along two directions. Due to the gripper-like design and the symmetrical force and displacement range, the tensile tester also can be used as a compression tester, allowing for the mechanical investigation of samples by simply grasping them.

To enable the use of these sensing tools for a wide range of applications without requiring a redesign when used in a different force range, a method for tuning their input range while taking measurements is incorporated into the sensors, enabling the best possible characteristics for a wide range of applications.

Driven by the unavailability of an SI-traceable reference standard in the nanonewton range, and the fact that none of the approaches by any of the NMIs to realize a primary small-force reference standard have matured enough to be implemented, a different approach for the calibration of microforce-sensing tools had to be found.

A combination of the highly accurate, compensated SI-traceable semi-microbalance, deadweights, and a custom macroscale reference sensor was used. With this blend of tools, it was possible to benefit from the advantages of the mature microbalance technology while avoiding their limitations, such as their slow output frequency. In combination with the implementation of the latest advancement in the field of multivariate uncertainty analysis using an MCM, SI-traceable microforce calibration in the nanonewton to micronewton range became achievable.

As an application example, the use of the microtensile tester chip is presented for the micromechanical investigation of individual plant cells – attached to the living plant – providing a method for extracting the properties needed to construct a new generation of mechanistic models of plant growth, based on actual measured properties.

This application demonstrates the tools and methodology necessary for the micromechanical investigation of small samples with a high accuracy.

References

1. U.S. Congress Office of Technology Assessment, "Miniaturization technologies," *U.S. Government Printing Office*, Nov 1991.
2. E.F. Schumacher, *Small is beautiful. A study of economics as if people mattered.* London: Vintage, 1993.
3. M. Hulskamp, "Plant trichomes: A model for cell differentiation," *Nature Reviews Molecular Cell Biology*, 5, 471–480, Jun 2004.
4. E. Peiner and L. Doering, "Force calibration of stylus instruments using silicon microcantilevers," *Sensors and Actuators A: Physical*, 123–24, 137–145, Sep 2005.
5. J. C. Doll, S. Park, A. J. Rastegar, N. Harjee, J. R. Mallon, G. C. Hill, A. A. Barlian, and B. L. Pruitt, "Force sensing optimization and applications," in *Advanced Materials and Technologies for Micro/Nano-Devices, Sensors and Actuators*, ed: Springer Netherlands, 2010, pp. 287–298.
6. M. S. Kim and J. R. Pratt, "SI traceability: Current status and future trends for forces below 10 micronewtons," *Measurement*, 43, 169–182, Feb 2010.
7. Y. Sun and B. J. Nelson, "MEMS capacitive force sensors for cellular and flight biomechanics," *Biomedical Materials*, 2, 16–22, Mar 2007.
8. C. F. Graetzel, S. N. Fry, F. Beyeler, Y. Sun, and B. J. Nelsons, "Real-time microforce sensors and high speed vision system for insect flight control analysis," *Experimental Robotics*, 39, 451–460, 2008.
9. J. C. Doll, S. Muntwyler, F. Beyeler, S. Geffeney, M. B. Goodman, B. J. Nelson, and B. L. Pruitt, "Measuring thresholds for touch sensation in *C. elegans*," in *International Conference on Microtechnologies in Medicine and Biology (MMB)*, Quebec City, Canada, Apr 2009.
10. X. J. Zhang, S. Zappe, R. W. Bernstein, O. Sahin, C. C. Chen, M. Fish, M. Scott, and O. Solgaard, "Integrated optical diffractive micrograting-based injection force sensor," in *Transducers: International Conference on Solid-State Sensors, Actuators and Microsystems* Boston, USA, Jun 2003.
11. T. Hugel and M. Seitz, "The study of molecular interactions by AFM force spectroscopy," *Macromolecular Rapid Communications*, 22, 989–1016, Sep 2001.
12. C. Gosse and V. Croquette, "Magnetic tweezers: Micromanipulation and force measurement at the molecular level," *Biophysical Journal*, 82, 3314–3329, Jun 2002.

13. Y. Sun, B. Nelson, D. P. Potasek, and E. Enikov, "A bulk microfabricated multi-axis capacitive cellular force sensor using transverse comb drives," *Journal of Micromechanics and Microengineering*, 12, 832–840, Nov 2002.
14. A. Sieber, P. Valdastri, K. Houston, C. Eder, O. Tonet, A. Menciassi, and P. Dario, "A novel haptic platform for real time bilateral biomanipulation with a MEMS sensor for triaxial force feedback," *Sensors and Actuators A: Physical*, 142, 19–27, Mar 2008.
15. A. Tibrewala, A. Phataralaoha, and S. Buttgenbach, "Simulation, fabrication and characterization of a 3D piezoresistive force sensor," *Sensors and Actuators A: Physical*, 147, 430–435, Oct 2008.
16. F. Beyeler, S. Muntwyler, Z. Nagy, C. Graetzel, M. Moser, and B. J. Nelson, "Design and calibration of a MEMS sensor for measuring the force and torque acting on a magnetic microrobot," *Journal of Micromechanics and Microengineering*, 18, 025004, Feb 2008.
17. F. Beyeler, S. Muntwyler, and B. J. Nelson, "A six-axis MEMS force-torque sensor with micro-newton and nano-newtonmeter resolution," *Journal of Microelectromechanical Systems*, 18, 433–441, Apr 2009.
18. J. R. Pratt, J. A. Kramar, D. B. Newell, and D. T. Smith, "Review of SI traceable force metrology for instrumented indentation and atomic force microscopy," *Measurement Science & Technology*, 16, 2129–2137, Nov 2005.
19. E. Peiner, L. Doering, M. Balke, and A. Christ, "Silicon cantilever sensor for micro-/nanoscale dimension and force metrology," *Microsystem Technologies: Micro-and Nanosystems Information Storage and Processing Systems*, 14, 441–451, Apr 2008.
20. A. Cherry, J. Abadie, and E. Piat, "Microforce sensor for microbiological applications based on a floating-magnetic principle," in *IEEE International Conference on Robotics and Automation (ICRA)*, Roma, Italy, Apr 2007.
21. L. D. Wang, J. K. Mills, and W. L. Cleghorn, "Development of an electron tunneling force sensor for the use in microassembly," in *Microsystems and Nanoelectronics Research Conference (MNRC)*, Ottawa, Canada, Oct 2008.
22. S. Fahlbusch and S. Fatikow, "Force sensing in microrobotic systems an overview," in *IEEE International Conference on Electronics, Circuits and Systems (ICECS)*, Lisbon, Portugal, Sep 1998.
23. P. Ruther, J. Bartholomeyczik, A. Trautmann, M. Wandt, O. Paul, W. Dominicus, R. Roth, K. Seitz, and W. Strauss, "Novel 3D piezoresistive silicon force sensor for dimensional metrology of micro components," in *IEEE Sensors*, Irvine, USA, Nov 2005.
24. S. Muntwyler, B. E. Kratochvil, F. Beyeler, and B. J. Nelson, "Monolithically Integrated Two-Axis Microtensile Tester for the Mechanical Characterization of Microscopic Samples," *Journal of Microelectromechanical Systems*, 19, 1223–1233, Oct 2010.
25. S. Muntwyler, F. Beyeler, and B. J. Nelson, "Three-axis micro-force sensor with sub-micro-Newton measurement uncertainty and tunable force range," *Journal of Micromechanics and Microengineering*, 20, 025011, Feb 2010.
26. Y. Zhu and H. D. Espinosa, "An electromechanical material testing system for in situ electron microscopy and applications," *Proceedings of the National Academy of Sciences of the United States of America*, 102, 14503–14508, Oct 11 2005.
27. J. J. Brown, J. W. Suk, G. Singh, A. I. Baca, D. A. Dikin, R. S. Ruoff, and V. M. Bright, "Microsystem for nanofiber electromechanical measurements," *Sensors and Actuators A: Physical*, 155, 1–7, Oct 2009.
28. H. Kahn, R. Ballarini, R. L. Mullen, and A. H. Heuer, "Electrostatically actuated failure of microfabricated polysilicon fracture mechanics specimens," *Proceedings of the Royal Society A: Mathematical Physical and Engineering Sciences*, 455, 3807–3823, Oct 1999.
29. M. T. A. Saif and N. C. MacDonald, "A millinewton microloading device," *Sensors and Actuators A: Physical*, 52, 65–75, Mar 1996.
30. H. Sato, T. Fukuda, F. Arai, K. Itoigawa, and Y. Tsukahara, "Parallel-beam sensor/actuator unit and its application to the gyroscope," *IEEE/ASME Transactions on Mechatronics*, 5, 266–272, Sep 2000.

31. F. Beyeler, A. Neild, S. Oberti, D. J. Bell, Y. Sun, J. Dual, and B. J. Nelson, "Monolithically fabricated microgripper with integrated force sensor for manipulating microobjects and biological cells aligned in an ultrasonic field," *Journal of Microelectromechanical Systems*, 16, 7–15, Feb 2007.
32. N. A. Burnham, X. Chen, C. S. Hodges, G. A. Matei, E. J. Thoreson, C. J. Roberts, M. C. Davies, and S. J. B. Tendler, "Comparison of calibration methods for atomic-force microscopy cantilevers," *Nanotechnology*, 14, 1–6, Jan 2003.
33. M. S. Kim, J. H. Choi, J. H. Kim, and Y. K. Park, "SI-traceable determination of spring constants of various atomic force microscope cantilevers with a small uncertainty of 1%," *Measurement Science & Technology*, 18, 3351–3358, Nov 2007.
34. Joint Committees for Guides in Metrology, "Guide to the expression of uncertainty in measurement," *International Organization for Standardization*, 2008.
35. Joint Committees for Guides in Metrology, "International vocabulary of metrology — Basic and general concepts and associated terms (VIM)," *International Organization for Standardization*, 2007.
36. Joint Committees for Guides in Metrology, "Supplement 1 to the guide to the expression of uncertainty in measurement - Propagation of distributions using a Monte Carlo method," *International Organization for Standardization*, 2008.
37. Joint Committees for Guides in Metrology, "Draft: Supplement 2 to the guide to the expression of uncertainty in measurement - Models with any number of output quantities," *International Organization for Standardization*, 2009.
38. I. Marson, H. Kahle, F. Chaperon, St Mueller, and F. Alasia, "Absolute gravity measurements in Switzerland: Definition of a base network for geodynamic investigations and for the Swiss fundamental gravity net," *Journal of Geodesy*, 55, 203 217, 1981.
39. B. E. Kratochvil, L. X. Dong, and B. J. Nelson, "Real-time rigid-body visual tracking in a scanning electron microscope," *International Journal of Robotics Research*, 28, 498–511, Apr 2009.
40. A. Geitmann and J. K. E. Ortega, "Mechanics and modeling of plant cell growth," *Trends in Plant Science*, 14, 467–478, Sep 2009.

Chapter 5
Cellular Force Measurement Using Computer Vision Microscopy and a Polymeric Microdevice

Xinyu Liu, Roxanne Fernandes, Andrea Jurisicova, Robert F. Casper, and Yu Sun

Abstract Assisted reproduction technologies (ART) require the reproductive quality of oocytes to be efficiently assessed. This chapter presents a cellular force measurement technique that allows for mechanical characterization of mouse oocytes during microinjection (i.e., in situ) without requiring a separate characterization process. The technique employs an elastic cell holding device and a sub-pixel computer vision tracking algorithm to resolve cellular forces in real time with a nanonewton force measurement resolution (2 nN at 30 Hz). The experimental results demonstrate that the in situ obtained force-deformation data are useful for distinguishing healthy mouse oocytes from those with aging-induced cellular defects. Biomembrane and cytoskeleton structures of the healthy and defective oocytes are also investigated to correlate the measured subtle mechanical difference to the cellular structure changes. The technique represents a promising means to provide a useful cue for oocyte quality assessment during microinjection.

Keywords Cellular force measurement • Cell manipulation • Sub-pixel visual tracking • Polymeric microdevice • Mechanical properties • Mouse oocytes • Oocyte quality assessment • Assisted reproduction technologies (ART) • In vitro fertilization (IVF)

Y. Sun (✉)
Department of Mechanical and Industrial Engineering, University of Toronto, 5 King's College Road, Toronto, ON, Canada M5S 3G8

Institute of Biomaterials and Biomedical Engineering, University of Toronto, 5 King's College Road, Toronto, ON, Canada, M5S 3G8
e-mail: sun@mie.utoronto.ca

C. Clévy et al. (eds.), *Signal Measurement and Estimation Techniques for Micro and Nanotechnology*, DOI 10.1007/978-1-4419-9946-7_5,

5.1 Introduction

An important procedure in assisted reproduction technologies (ART) is the assessment of the reproductive quality of oocytes for in vitro fertilization (IVF). The state-of-the-art morphology analysis method [1, 2] is often subjective and fails to provide definitive prediction for oocyte quality, causing low pregnancy rates and therefore, imposing extra difficulties on the follow-up reproductive studies.

Emerging techniques for oocyte quality assessment include genetic screening [3], spectroscopy-based metabolomic profiling [4], and polscope-based spindle imaging [5, 6]. Despite tremendous advances made by these techniques, they are limited by: (1) the invasive deoxyribonucleic acid (DNA) sampling procedure that may impair the oocytes and result in lower development competence [3]; or (2) the requirement of specific analysis equipment [3–6] and complex spectral data analysis [4]. ART demands efficient and low-cost methods for oocyte quality investigation.

It is well known that a variety of cellular functions, such as cell division, gene expression, signal transduction, and apoptosis, greatly depend on mediation and regulation of mechanical signals (i.e., forces and stresses), and on mechanical properties of cell membranes and intracellular proteins/fluid [7–9]. Similarly, a wide range of human diseases are also closely correlated with variations of the mechanical properties of cells [8, 10, 11], suggesting mechanical characterization of biological cells a possible candidate for disease state detection. In this context, quantification of mechanical properties (e.g., force-deformation data) of oocytes may reveal cellular defects, such as mitochondrial leakage and DNA damage. Hence, we were motivated to develop suitable mechanical characterization techniques for exploring the feasibility of using mechanical cues for oocyte quality evaluation.

For mechanical characterization of a living cell, the cell must be deformed in some way and the applied forces/stresses and cell deformations accurately measured. Experimental techniques for cellular force measurement include micropipette aspiration [12], optical tweezers [13], optical stretchers [14], atomic force microscopy (AFM) [15], magnetic bead measurement [16], and microelectromechanical systems (MEMS) transducer based measurement [17, 18]. Among these techniques, MEMS force transducers have certain advantages over other tools due to their cost-effectiveness and flexibility for system integration. However, the construction of the MEMS force sensors are typically based on silicon micromachining that requires much processing effort [17]. Furthermore, issues such as biocompatibility and operating in an aqueous environment for biological cells to survive often pose stringent challenges and intricacies in MEMS design, material selection, and microfabrication.

Instead of using silicon-based MEMS transducers, polymeric materials such as polydimethylsiloxane (PDMS) and polyacrylamide (PAM) have been widely employed as passive deformable force sensors for cellular force measurements, due to their high transparency, low stiffness, and good biocompatibility. Although PDMS or PAM flexible substrates have been used for characterizing cellular traction forces by visually tracking local deformations of the substrate [19, 20], the continuous deformation model of the substrate requires heavily complex computation to interpolate measured local discrete deformations into global continuous deformations.

Based on the same concept, researchers conceived innovative PDMS micro-post structures as force transducers, enabling measurements of local traction forces generated by adherent cells [21–23]. The devices can be easily constructed using the standard soft-lithography technique [24]. Image processing techniques were used for measuring the PDMS post deflection, and only a simple cantilever mechanical model is required for mapping post deflections into cellular forces.

Such micro-post devices were also modified by integrating magnetic nanowires into individual micro-posts so that external magnetic field induced forces can be applied to the cells [25]. The cellular retraction force response to the applied forces was examined, providing a new way to investigate cellular locomotion behavior under mechanical stimuli. Although the polymeric micro-post devices can both apply mechanical stimuli to adherent cells and simultaneously measure their traction forces, they are not suitable for characterizing mechanical properties of suspended cells such as oocytes/embryos.

Since in vitro oocyte fertilization is often achieved via intracytoplasmic sperm injection (ICSI), in which a single sperm is physically injected into an oocyte by a sharp micropipette, it would be convenient/efficient to mechanically characterize oocytes during the ICSI process (referred to as in situ in this work). This chapter presents a cellular force measurement technique and its application to in situ mechanical characterization of mouse oocytes during microinjection. A PDMS cell holding device (Fig. 5.1a) and a sub-pixel visual tracking algorithm are used together to visually resolve applied forces to a single oocyte with nanonewton force resolutions. Experimental results demonstrate that the in situ obtained force-deformation data are useful for distinguishing healthy oocytes from those with compromised cellular functions during microinjection. Follow-up structural imaging of the zona pellucida (ZP) and fluorescence analysis of filamentous actin (F-actin) contents verify that structural differences of the ZP and cytoskeleton exist between healthy and defective oocytes. These structural differences are speculated to result from oocyte defects.

5.2 Materials and Methods

5.2.1 *Mouse Oocyte Preparation*

In this study, oocytes from young (8–12 weeks old) and old (40–45 weeks old) imprinting-control-region (ICR) female mice (referred as young and old oocytes in the rest of this chapter) were used as a comparison model to investigate the feasibility of using mechanical cues to distinguish healthy oocytes from defective ones. 40–45 weeks old ICR female mice are near the end of their reproductive

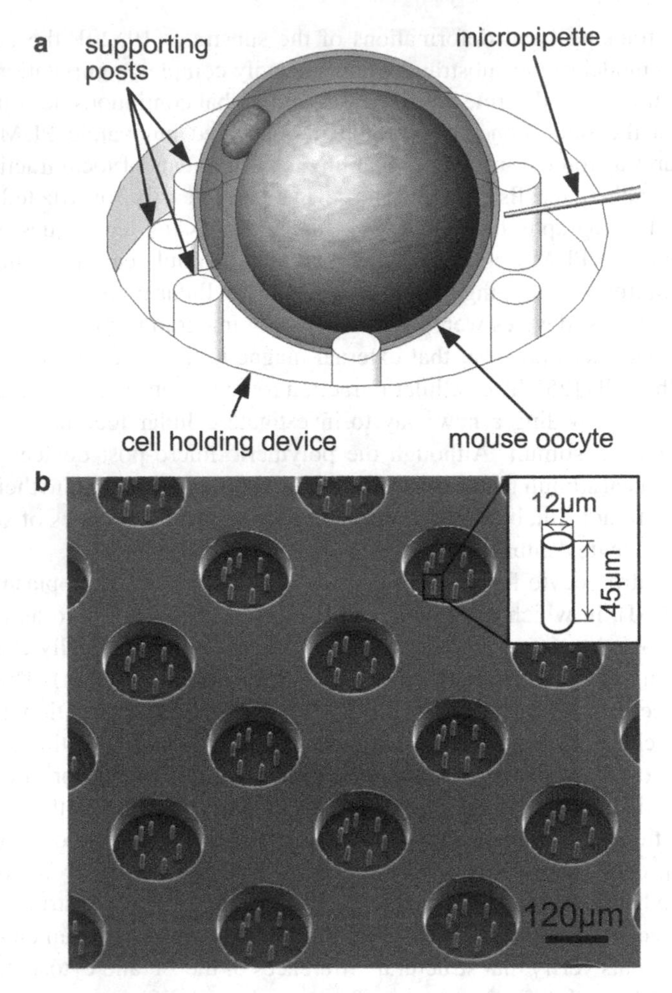

Fig. 5.1 (**a**) Schematic of cellular force measurement technique using low-stiffness elastic posts during microinjection. (**b**) Scanning electron microscopy (SEM) image of a PDMS cell holding device. Reproduced by permission. Copyright 2010 the Royal Society of Chemistry [43]

lifespan. Their oocytes and corresponding embryos reveal compromised developmental competence due to multiple cellular defects, such as meiotic irregularities and mitochondrial dysfunction [26]. The old mouse model has been widely used in reproductive biology as an analogue to human female infertility due to advanced maternal age (≥35 years) [27].

Defective oocytes often contain compromised mitochondria, insufficient maternal endowment of proteins, and/or transcripts leading to chromosomal aneuploidy, particularly evident with aging [26, 28]. Thus, it is anticipated that these molecular events may have impact on the cell membrane and cytoskeleton, and that these subtle changes induced by oocyte defects may lead to mechanical differences between defective and healthy oocytes. Therefore, cellular force-deformation measurements could provide useful information for detecting oocyte dysfunctions.

All experiments were conducted in compliance with federal laws and institutional guidelines and have been approved by the Mount Sinai Hospital Animal Care Committee in Toronto. Young and old ICR mice (Harlan Laboratories) were superovulated with 5IU of pregnant mare's serum gonadotropin (PMSG) (Sigma) and 48 h later with 5IU of human chorionic gonadotropin (hCG) (Sigma), by intraperitoneal injection. Mouse oocytes were collected from the superovulated female mice at 16 h post-hCG and cultured in potassium simplex optimization medium (KSOM, Specialty Media). The average diameter of the mouse oocytes was 96 μm.

5.2.2 Working Principle of the Cellular Force Measurement Technique

Vision-based force measurement techniques are capable of retrieving both vision and force information from a single vision sensor (CCD/CMOS camera) under microscopy observation [29,30]. For cellular force measurement during cell manipulation, this concept is realized by visually tracking structural deformations of an elastic cell holding structure, and subsequently, transforming material deformations into forces.

The cell holding device shown in Fig. 5.1b integrates an array of cavities (180 μm in diameter) for accommodating individual mouse oocytes. Inside each cavity, low-stiffness microposts (45 μm high and 12 μm in diameter) are arranged in a circular pattern to support the oocyte during microinjection. Figure 5.1a schematically illustrates the working principle of the cell holding device for vision-based cellular force measurement during oocyte injection. While the micropipette injects individual oocytes inside these cavities, applied forces are transmitted to the low-stiffness, supporting posts. In real time (30 Hz), a sub-pixel visual tracking algorithm measures post deflections that are fitted into an analytical mechanics model to calculate the force exerted on the oocyte.

We previously demonstrated this technique using a large-sized PDMS device on zebrafish embryos [31]. The study presented here focuses on miniaturizing the cell holding devices for studying mouse oocytes (100 μm in diameter vs. 1.2 mm zebrafish embryos), enhancing the force measurement resolution to the nanonewton level, and using the in situ obtained cell mechanical property information to distinguish healthy mouse oocytes from those with compromised developmental competence. Considering the high deformability of mouse oocytes, our present post arrangement employs the minimal number of supporting posts (e.g., three) for securely immobilizing an oocyte during microinjection (Fig. 5.1a) and for maximizing post deflections. In addition, the analytical mechanics model (Sect. 5.2.4) for calculating the cellular forces was modified based on a new boundary condition.

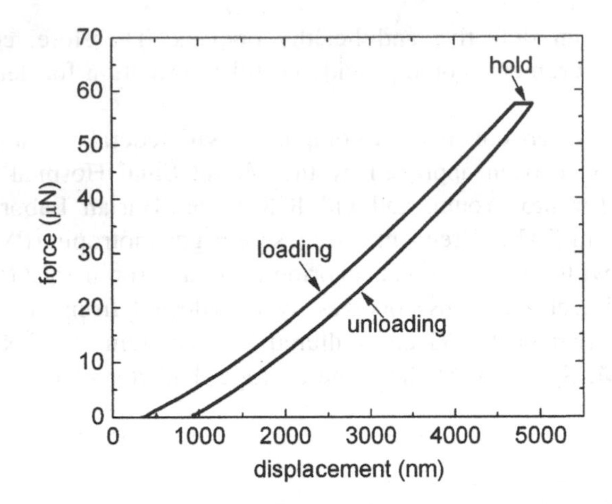

Fig. 5.2 A force–displacement curve of PDMS nanoindentation to calibrate the Young's modulus of cell holding devices. Reproduced by permission. Copyright 2010 the Royal Society of Chemistry [43]

5.2.3 Device Fabrication and Characterization

The cell holding device (Fig. 5.1b) was constructed with PDMS via standard soft lithography [31]. Briefly, PDMS prepolymer prepared by mixing Sylgard 184 (Dow Corning) and its curing agent with a weight ratio of 15:1, was poured over a SU-8 mold (SU-8 50, MicroChem) made on a silicon wafter using standard photolithography. After curing at 80°C for 8 h, the PDMS devices were peeled off the SU-8 mold. The depth of the cavity and protruding posts is 45 μm, and the diameter of the posts is 12 μm (Fig. 5.1b). In order to make the PDMS surface hydrophilic, the devices were oxygen plasma treated for 10 s before use.

The mechanics model for mapping post deflections into cellular forces, discussed in Sect. 5.2.4, requires the Young's modulus of the cell holding device to be accurately calibrated. Nanoindentation was used to determine the Young's modulus of the cell holding device and micro posts. Five devices were calibrated using a nanoindentation instrument (TI-750 Ubi nanomechanical test instrument, Hysitron). Figure 5.2 shows a calibration curve of applied forces vs. displacements. The determined Young's modulus value is 524.7 ± 22.1 kPa ($n = 5$), which is comparable to the previously reported data [32].

5.2.4 Force Analysis

Figure 5.3a shows a snapshot captured in the cell injection process. A micromanipulator controls an injection micropipette to exert an indentation force to a mouse oocyte, deflecting the three supporting posts on the opposite side. Post deflections,

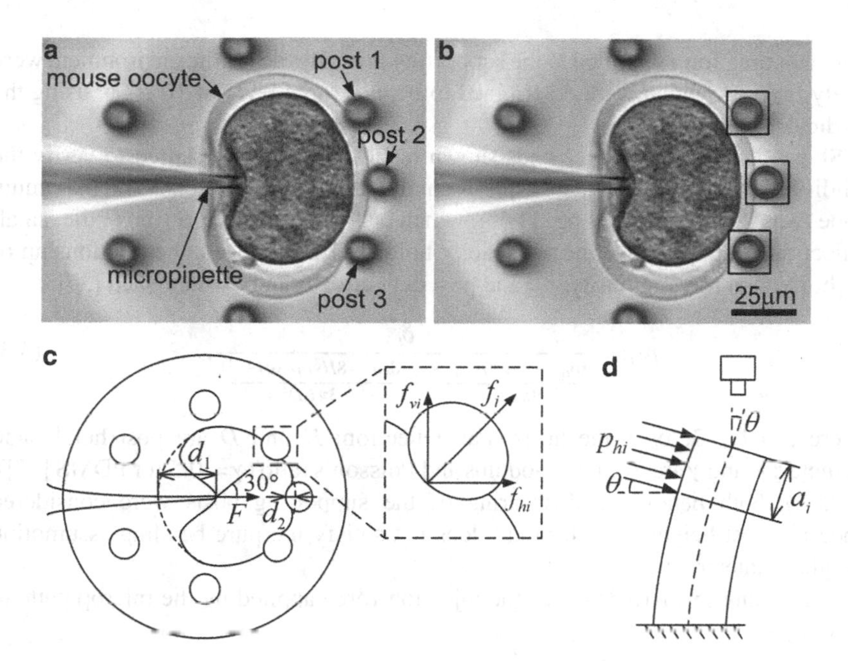

Fig. 5.3 (**a**) Indentation forces deform the mouse oocyte and deflect three supporting posts. (**b**) Image patches (*squares*) tracked by template matching and LSCD detected post *top circles*. (**c**) Force balance on the cell under indentation. (**d**) Post deflection model. Reproduced by permission. Copyright 2010 the Royal Society of Chemistry [43]

measured by a visual tracking algorithm (discussed in Sect. 5.2.5) are fitted to an analytical mechanics model to obtain contact forces between the oocyte and posts. Based on the contact forces, the indentation force applied by the micropipette on the oocyte is determined through the following force analysis.

The oocyte is treated as elastic due to the fact that quick indentation by the micropipette does not leave sufficient time for cellular creep or relaxation to occur. The injection force, F is balanced by the horizontal components, f_{hi} of contact forces between the oocyte and supporting posts (Fig. 5.3c),

$$F = \sum_{i=1}^{3} f_{hi}, \quad i = 1,2,3. \tag{5.1}$$

In the device configuration, the radius of the oocyte (~48 μm) is larger than the depth of the cavity and posts (45 μm), resulting in an initial point contact between the oocyte and supporting posts before post deflections occur. However, the high deformability of mouse oocytes makes cell membrane conform to the posts when an injection force is applied to the oocyte. It is assumed that the contact forces are evenly distributed over the contact areas. Thus, the horizontal components, f_{hi} are expressed by a constant force intensity, p_{hi} and a contact length, a_i (Fig. 5.3d)

$$f_{hi} = p_{hi} a_i, \quad i = 1,2,3. \tag{5.2}$$

Note that drag forces applied to the supporting posts by the fluidic environment were safely ignored, which were determined to be at a force level of 10^{-16} N using the fluidic drag model [33].

Slope θ of the posts' free ends shown in Fig. 5.3d was measured to verify the validity of linear elasticity that requires small structural deflections. The maximum slope was determined to be 11.1°, which satisfies $\sin\theta \approx \theta$ thus, the small deflection assumption of linear elasticity holds [34]. Therefore, the relationship of the horizontal force intensity, p_{hi} and post deflections can be established [34].

$$p_{hi} = \frac{\delta_i}{\frac{40a_i(1+\gamma)(2H-a_i)}{9\pi ED^2} + \frac{8(a_i^4+8H^3a_i-6H^2a_i^2)}{3\pi ED^4}}, \tag{5.3}$$

where $i = 1,2,3$; δ_i is the horizontal deflection; H and D are post height and diameter; E and γ are Young's modulus and Poisson's ratio ($\gamma = 0.5$ for PDMS [35]). In (5.3), both bending and shearing of the supporting posts were considered since the post height/diameter ratio does not satisfy the pure bending assumption (height/diameter ratio >5) [34].

Combining (5.1)–(5.3) yields the injection force applied by the micropipette to the oocyte.

$$F = \sum_{i=1}^{3} \frac{\delta_i a_i}{\frac{40a_i(1+\gamma)(2H-a_i)}{9\pi ED^2} + \frac{8(a_i^4+8H^3a_i-6H^2a_i^2)}{3\pi ED^4}}. \tag{5.4}$$

In (5.4), the unknown parameters are post horizontal deflections, δ_i and the contact length, a_i. Experimentally, imaging with a side-view microscope confirmed that the contact length, a_i increases at a constant speed, v_i for a given indentation speed. Hence, $a_i = v_i t$, where t denotes time.

Note that for a constant indentation speed of the micropipette, the variation speed of contact length a, v_i varies for different oocytes. At 60 µm/s used throughout the experiments, v_i of the tested mouse oocytes were measured to be 0.8–1.2 µm/s. Interestingly, the sensitivity of the mechanics model (5.4) to variations in v_i is low. The injection force varies only by 1% when v_i changes from 0.8 to 1.2 µm/s. Thus, the average value of the measured speeds, 1 µm/s was used to calculate injection forces for all the oocytes.

5.2.5 Visual Tracking of Post Deflections

In order to accurately track post deflections, a visual tracking algorithm with a resolution of 0.5 pixel was developed, which was described in detail previously [31]. A template matching algorithm tracks the motion of the supporting posts, providing processing areas for a least-squares circle detection (LSCD) algorithm to determine posts' center positions. The LSCD algorithm utilizes the Canny edge detector to obtain an edge image and then extracts a portion of the post top surface for circle fitting. The resolution of the visual tracking algorithm was determined by visually tracking the deflection of a stationary supporting post and calculating the standard deviation of the measured deflection data.

5.2.6 Electron Microscopy Imaging of Zona Pellucida

ZP is a unique extracellular membrane (6–8 μm thick in mouse oocytes/embryos) surrounding the oocyte/embryo, which is composed of three types of glycoproteins arranged in a delicate filamentous matrix [36]. The ZP structure significantly contributes to the mechanical stiffness of the oocyte [17, 37]. To understand the cause of possible subtle mechanical differences of healthy and defective oocytes, ZP surface morphology and glycoprotein structures were imaged using electron microscopy.

ZP surface morphologies of young and old oocytes were analyzed via scanning electron microscopy (SEM) imaging. Oocytes at 2 h post-collection were mounted on a Thermanox plastic coverslip (Fisher Scientific), fixed for 1 h with 2% glutaraldehyde in 1% sodium cacodylate buffer, postfixed for 1 h with 1% osmium tetroxide in 0.1 M sodium cacodylate buffer, and dehydrated in an acetone series of increasing concentration, according to a standard protocol [38]. After dehydration, the oocytes were CO_2 critical point dried in a polythene chamber, mounted on the specimen holder, coated with gold, and observed in an environmental SEM (XL-30, Philips).

Structural analysis of ZP glycoproteins was conducted via transmission electron microscopy (TEM) imaging of ZP cross-sections, following a standard method for sample preparation [39]. Young and old oocytes at 2 h post-collection were fixed for 1 h with 2% glutaraldehyde in 1% sodium cacodylate buffer, postfixed for 1 h with 1% osmium tetroxide in 0.1 M sodium cacodylate buffer, embedded in agarose, cut into 100 nm thick sections, and observed in a TEM (CM-100, Philips).

5.2.7 F-Actin Staining

Oocyte mechanical properties are also regulated by the cytoskeleton. In this study, F-actin contents of the young and old ooyctes were quantified by fluorescence microscopy. F-actin of the oocytes were stained by fluorescein isothiocyanate (FITC) conjugated phalloidin. Oocytes at 2 h post-collection were washed with phosphate buffered saline (PBS), fixed for 10 min with 4% formaldehyde in PBS, permeabilized with 0.1% Triton X-100 in PBS, and stained with a 5 μg/ml phalloidin-FITC solution in PBS for 60 min at room temperature. The concentration of 5 μg/ml and the staining time of 60 min were experimentally determined to guarantee the saturation of all F-actin binding sites for phalloidin-FITC [40].

Between consecutive steps, PBS washing was conducted three times. Nuclei of the oocytes were also stained with 4'-6-Diamidino-2-phenylindole (DAPI) for control purpose.

Finally, the stained oocytes were mounted with a 1:1 glycerol:PBS solution onto microscope slides, and analyzed on a deconvolution microscope (Olympus IX-70, Applied Precision Inc.) with a FITC filter. Ten 1-μm optical sections for each sample were obtained. Average fluorescent intensity of the ten optical sections was regarded as the F-actin content.

5.2.8 Statistical Analysis

Experimental data were analyzed using student's t-test (SigmaStat 3.5, Systat Software Inc.). Plots with error bars represent means ± one standard deviation (s.d.).

5.3 Results

All the experiments were conducted at 37°C inside a temperature-controlled chamber. With a 40× objective (NA 0.55), the pixel size of the imaging system was calibrated to be 0.24 × 0.24 μm. Micropipette tips used for indenting mouse oocytes were 5.3 μm in diameter.

The template matching algorithm and the LSCD algorithm together cost 22.3 ms for processing each frame of image. Figure 5.3b shows the tracked image patches and LSCD detected post top circles. The tracking resolution was determined to be 0.5 pixel (i.e., 0.12 μm).

5.3.1 Force-Deformation Measurement and Mechanical Characterization Results

Twenty young oocytes and twenty old oocytes were delivered onto the cell holding device using a transfer pipette and then indented. The micropipette was controlled to indent each oocyte by 25 μm at 60 μm/s. During the indentation process, force data were collected (30 data points per second). Figure 5.4a shows force-deformation curves of both young and old oocytes. The horizontal axis represents cell deformation, $d = d_1 + d_2$, where d_1 and d_2 were defined in Fig. 5.3c. The d values were calculated by subtracting the deflections of post 2 (Fig. 5.3a) from the displacements of the injection micropipette. The vertical axis is the cellular force. Force measurement resolution of the system is defined as the finest force

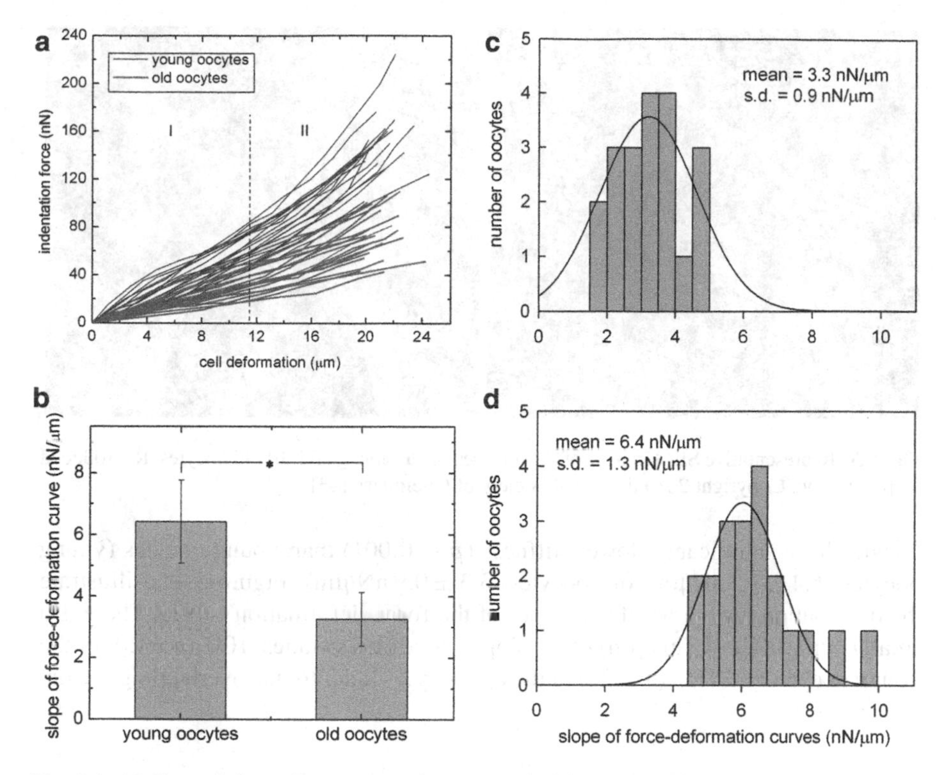

Fig. 5.4 (**a**) Force-deformation curves of young and old oocytes. (**b**) Means ± one standard deviations of force-deformation curve slopes from young and old oocytes ($^{*}p < 0.001$). (**c, d**) Distribution histogram of the slopes of the force-deformation curves from (**c**) young oocytes and (**d**) old oocytes. There is a small overlap (4.4–4.8 nN/μm) of slopes between young and old oocytes. Reproduced by permission. Copyright 2010 the Royal Society of Chemistry [43]

that a supporting post can measure, which is equal to the product of the stiffness of a suspended supporting post (16.8 nN/μm for current devices), and the tracking resolution of the post deflections (0.12 μm). The force measurement resolution was determined to be 2 nN at 30 Hz.

Most of the force-deformation curves of young and old oocytes separate themselves into two distinct areas with a slight overlap of a few curves. It was also observed that during microinjection, only the ZP was deformed with cell deformation less than a certain value (12.7 ± 3.4 μm, $n = 40$; no significant difference ($P = 0.487$) between young and old oocytes), corresponding to region I in Fig. 5.4a where ZP stiffness is dominant. After the cell deformation was beyond 12.7 ± 3.4 μm (region II in Fig. 5.4a), both ZP and cytoplasm were deformed. Thus, the force-deformation data in region II reflect the overall stiffness of the ZP and cytoplasm.

Slopes of the force-deformation curves were calculated using linear regression, which is considered as oocytes' overall stiffness. Figure 5.4b shows the means ± s.d. of the slopes of young and old oocytes. It was found that old

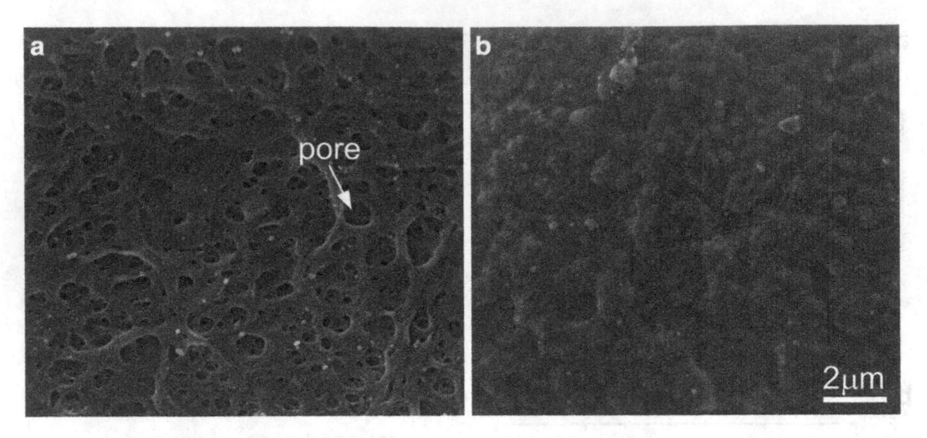

Fig. 5.5 Representative SEM images of ZP surfaces of (**a**) young and (**b**) old oocytes. Reproduced by permission. Copyright 2010 the Royal Society of Chemistry [43]

oocytes have significantly lower stiffness ($p < 0.001$) than young oocytes (young oocytes: 6.4 ± 1.3 nN/μm, old oocytes: 3.3 ± 0.9 nN/μm). Figure 5.4c,d illustrate the distribution histogram of the slopes of the force-deformation curves. There is a small overlap (4.4–4.8 nN/μm) of the slopes or stiffness values. 10% ($n = 20$) of the young oocytes and 15% ($n = 20$) of the old oocytes fall into this overlapping region.

5.3.2 *Zona Pellucida Structure Analysis*

In order to probe the cause of the detected mechanical changes in old oocytes, ZP thickness, surface morphology, and cross-sectional glycoprotein structures of young and old oocytes were analyzed by optical microscopy, SEM, and TEM imaging. Note that the in situ measured stiffness of young and old oocytes was not correlated with the ZP structures considering that indentation-induced global deformations of the ZP may change the ZP structures (e.g., ZP thickness and glycoprotein density). ZP thickness of 16 young oocytes and 10 old oocytes was measured under an optical microscope (Nikon TE-2000S) with 400× magnification. For each oocyte, ZP thickness of five different locations was measured, and the average was taken as the final thickness value. The young oocytes have a ZP thickness of 7.1 ± 0.3 μm (mean ± s.d.), which is not significantly different ($p = 0.098$) from that of young oocytes (6.8 ± 0.5 μm).

SEM imaging of ZP surfaces (Fig. 5.5) demonstrate that young and old oocytes reveal different surface morphologies. All observed young oocytes ($n = 8$) had a ‘spongy’ surface comprised of multiple layers of networked glycoproteins with numerous pores (Fig. 5.5a), while only 20% of the old oocytes had similar surfaces. The other 80% of the old oocytes showed a rough surface without pores(Fig. 5.5b).

The different ZP surface morphologies indicate different structures of ZP glycoproteins in young and old oocytes, which were then quantitatively analyzed via TEM imaging.

Figure 5.6a, b show TEM cross-sectional views of the ZP glycoprotein structures from young (Fig. 5.6a) and old (Fig. 5.6b) oocytes. The density of ZP glycoproteins was quantified using image processing. An adaptive thresholding algorithm [41] was used to recognize the glycoprotein structure areas (black areas in Fig. 5.6d). The area ratio of glycoprotein structures to the total image is defined as the relative density of glycoproteins. The final relative density value of each oocyte was obtained from five different ZP regions by averaging. As shown in Fig. 5.6e, the glycoproteins in old oocyte ZP are significantly sparser ($p < 0.001$) than those in young oocyte ZP. It is believed that the sparser ZP glycoproteins in old oocytes result in lower ZP stiffness than young oocytes, which mechanically differentiates young and old oocytes in region I of the force-deformation data (Fig. 5.4a).

5.3.3 *F-Actin Contents*

Figure 5.7a, b show the F-actin staining pictures of young and old oocytes, where the green and blue channels respectively represent F-actin and nucleus. Higher fluorescent intensity of the green channel indicates higher F-actin content. The fluorescence analysis results (Fig. 5.7c) show that old oocytes contain significantly less ($p < 0.001$) F-actin than young oocytes, which can be responsible for the stiffness difference in region II of the force-deformation data (Fig. 5.4a). Contour line plots (Fig. 5.7d, e) of F-actin fluorescence illustrate the F-actin distribution in the young and old oocytes. One can notice that the sub-cortical region of the old oocyte (labeled by white arrows in Fig. 5.7e) particularly lacks this cytoskeletal protein.

5.4 Discussion

Measurements of cell mechanical properties can be useful for predicting cellular response to mechanical stimuli and correlating mechanical properties to disease states [10, 42]. Characterizing mammalian oocytes during microinjection without a separate characterization process promises a useful and low-cost approach to detect potential oocyte defects and select high-quality oocytes for subsequent IVF and implantation. Targeting in situ distinguishing healthy oocytes from those with compromised cellular functions, a vision-based cellular force measurement technique was developed to resolve nanonewton-level cellular forces and characterize oocyte stiffness. Although mouse oocytes were characterized in this study, the technique can also be applied to characterize mechanical properties of mouse embryos by indenting those embryos with a micropipette without penetration.

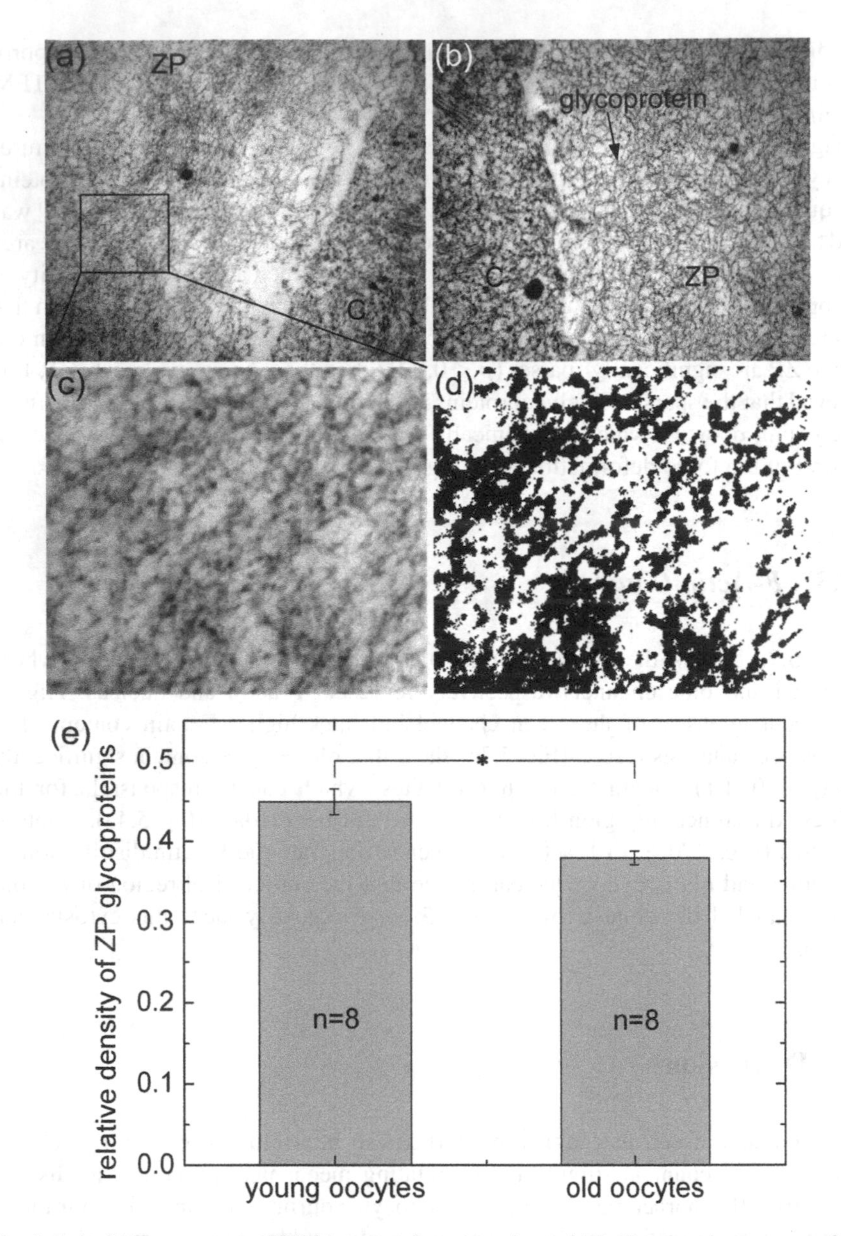

Fig. 5.6 (**a**, **b**) Representative TEM images of cross sections of ZP glycoprotein structures from (**a**) young and (**b**) old oocytes (C: cytoplasm). (**c**) A sub-region from (**a**) for image processing. (**d**) Binary image after adaptive thresholding of (**c**). (**e**) Relative density of ZP glycoprotein structures of young and old oocytes ($*p < 0.001$). Reproduced by permission. Copyright 2010 the Royal Society of Chemistry [43]

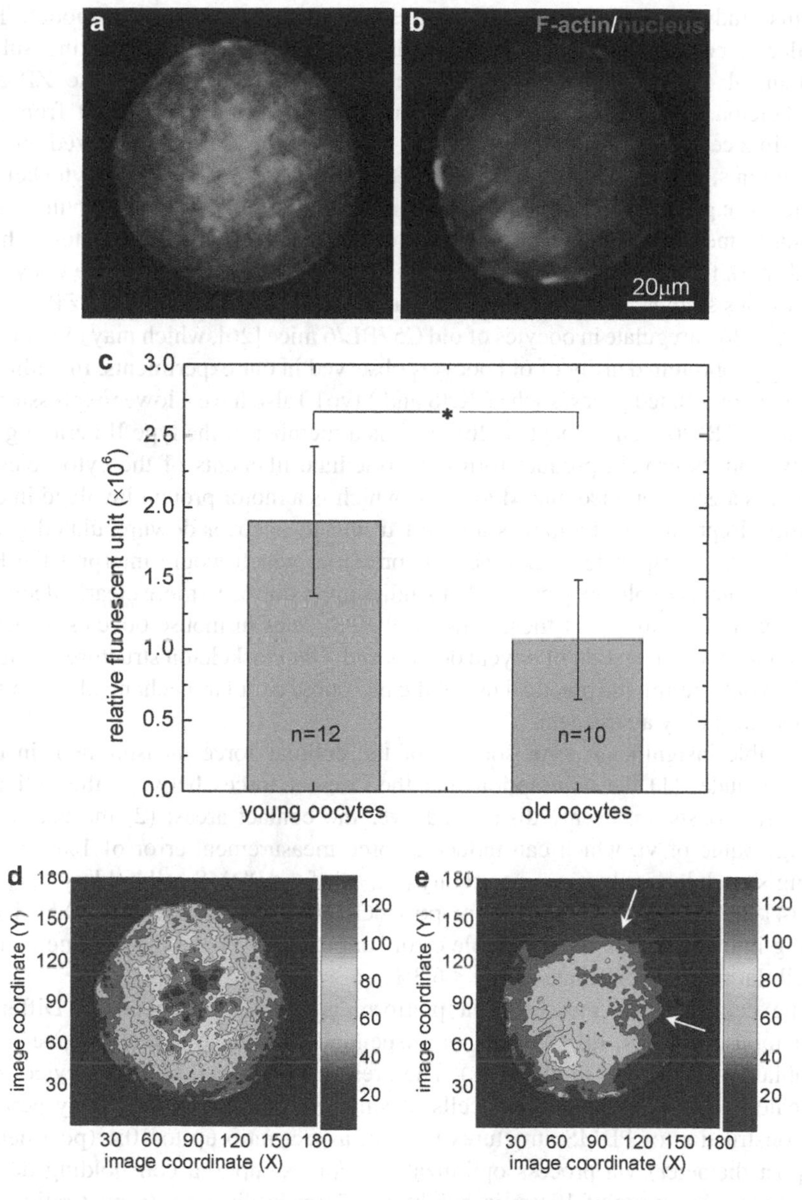

Fig. 5.7 F-actin content analysis. (**a**, **b**) F-actin staining images of (**a**) young and (**b**) old oocytes. (**c**) F-actin contents of young and old oocytes ($*p < 0.001$). (**d**) *Contour line* plot of F-actin fluorescence in (**a**). (**e**) *Contour line* plot of F-actin fluorescence in (**b**). Reproduced by permission. Copyright 2010 the Royal Society of Chemistry [43]

This study used young and old mouse ooyctes as a comparison model. The cellular force measurement technique was proven effective for resolving subtle mechanical changes of old oocytes, due to structural changes of the ZP and cytoskeleton. These cellular structure changes are speculated to result from the aging-induced defects of old oocytes. F-actin structures were analyzed in the experiments; however, it should be noted that structures of other cytoskeletal filaments (e.g., intermediate filament and microtubule) may also contribute to the measured mechanical differences between young and old mouse oocytes, which could spark further studies of cytoskeletal structure changes in old mouse oocytes.

Previous studies have demonstrated that all ZP glycoprotein genes (ZP1, ZP2, and ZP3) downregulate in oocytes of old C57BL/6 mice [26], which may explain the lower glycoprotein density of old oocytes observed in our experiments. In addition, cytoskeleton-related genes, such as Krt8 and Myo10 also have a lower expression in the old C57BL/6 mouse oocytes [26]. Krt8 is a member of the type II keratin gene family, and its protein product forms intermediate filaments of the cytoskeleton. Myo10 is a gene for encoding Myosin-X, which is a motor protein involved in cell motility. Reproductive biologists are still trying to uncover downregulated genes in old oocytes responsible for F-actin expression, which would interpret the low F-actin contents in old oocytes. Further studies are required to more clearly decipher the regulation pathways of these downregulated genes in mouse ooyctes to better understand the connection of ooycte defects and ZP/cytoskeleton structure changes, which would permit the practical use of the measured cellular mechanical properties for oocyte quality assessment.

Possible insignificant error sources of the cellular force measurement in this study include: (1) the assumption that the contact forces between the cell and supporting posts are evenly distributed over the contact areas; (2) the use of an average value of vi, which can induce a force measurement error of 1%; (3) the Young's modulus calibration uncertainty (4.2%) of the PDMS cell holding devices; and (4) the visual tracking error for post deflection measurements ($\leq$0.5 pixel). Taking into account all the countable error sources (2)–(4), the measurement error of cellular forces is calculated to be $\leq$6.3%.

The cellular force measurement platform is not scale dependent. Different from mouse oocytes, the majority of suspended cells have a smaller size (e.g., fibroblasts are ~15 μm in diameter). The present PDMS cell holding devices can be scaled down to accommodate cells of smaller sizes. Soft lithography permits the construction of PDMS structures with an aspect ratio up to 10:1 (post height vs. post diameter) via process optimization. For example, a cell holding device with supporting posts of 10 μm in height and 2 μm in diameter (aspect ratio: 5:1; mechanical stiffness of each post: 1.2 nN/μm), based on a 0.5 pixel visual tracking resolution obtained in this study, has the capability of visually resolving forces down to 145 pN with a 40 × objective. Thus, the device design and real-time visual tracking algorithm provide a cost-effective, useful experimental platform for single cell studies with a sub-nanoNewton force measurement resolution.

5.5 Conclusion

This chapter demonstrated a vision-based cellular force measurement technique and its application to in situ mechanical characterization of mouse oocytes. By visually tracking deflections of elastic, low-stiffness supporting posts on a PDMS cell holding device during microinjection, the technique measured cellular forces in real time (30 Hz) with a 2 nN resolution. An analytical mechanics model was developed to convert post deflections into cellular forces. Young's modulus of the cell holding devices were calibrated to be 524.7 ± 22.1 kPa ($n = 5$). Based on characterization experiments of 20 young mouse oocytes and 20 old mouse oocytes, it was found that the in situ obtained force-deformation data are useful for distinguishing healthy oocytes from defective ones. The follow-up analysis of oocyte structures also demonstrated that the subtle mechanical differences between young and old mouse oocytes may be due to structure changes of the zona pellucida and cytoskeleton.

Acknowledgements We thank Douglas Holmyard for the assistance with SEM/TEM sample preparation, Zahra Mirzaei for assistance with F-actin staining, Richard Nay from Hysitron for use of nanoindentation equipment, and the staff in microfabrication center at the Emerging Communications Technology Institute of University of Toronto. This work was supported by the Natural Sciences and Engineering Research Council of Canada, the Ontario Ministry of Research and Innovation, and the Ontario Centers of Excellence. We also acknowledge the financial support from the Canada Research Chairs program to YS, and the Ontario Graduate Scholarship program to XYL.

References

1. Xia, P. 1997. Intracytoplasmic sperm injection: correlation of oocyte grade based on polar body, perivitelline space and cytoplasmic inclusions with fertilization rate and embryo quality. Hum Reprod 12:1750-1755.
2. Meriano, J. S., J. Alexis, S. Visram-Zaver, M. Cruz, and R. F. Casper. 2001. Tracking of oocyte dysmorphisms for ICSI patients may prove relevant to the outcome in subsequent patient cycles. Hum Reprod 16:2118-2123.
3. Coutelle, C., C. Williams, A. Handyside, K. Hardy, R. Winston, and R. Williamson. 1989. Genetic analysis of DNA from single human oocytes: a model for preimplantation diagnosis of cystic fibrosis. Br. Med. J. 299:22-24.
4. Nagy, Z. P., S. Jones-Colon, P. Roos, L. Botros, E. Greco, J. Dasig, and B. Behr. 2009. Metabolomic assessment of oocyte viability. Reprod Biomed Online 18:219-225.
5. Moon, J. H., C. S. Hyun, S. W. Lee, W. Y. Son, S. H. Yoon, and J. H. Lim. 2003. Visualization of the metaphase II meiotic spindle in living human oocytes using the Polscope enables the prediction of embryonic developmental competence after ICSI. Hum Reprod 18:817-820.
6. Rienzi, L., F. Ubaldi, M. Iacobelli, M. G. Minasi, S. Romano, and E. Greco. 2005. Meiotic spindle visualization in living human oocytes. Reprod Biomed Online 10:192-198.
7. Bao, G., and S. Suresh. 2003. Cell and molecular mechanics of biological materials. Nat Mater 2:715-725.
8. Suresh, S. 2007. Biomechanics and biophysics of cancer cells. Acta Biomaterialia 3:413-438.
9. Janmey, P. A., and C. A. McCulloch. 2007. Cell mechanics: Integrating cell responses to mechanical stimuli. Annu Rev Biomed Eng 9:1-34.

10. Lee, G. Y., and C. T. Lim. 2007. Biomechanics approaches to studying human diseases. Trends Biotechnol 25:111-118.
11. Simmons, C. A. 2009. Aortic valve mechanics an emerging role for the endothelium. J Am Coll Cardiol 53:1456-1458.
12. Hochmuth, R. M. 2000. Micropipette aspiration of living cells. J Biomech 33:15-22.
13. Dai, J. W., and M. P. Sheetz. 1995. Mechanical-properties of neuronal growth cone membranes studied by tether formation with laser optical tweezers. Biophys J 68:988-996.
14. Guck, J., R. Ananthakrishnan, H. Mahmood, T. J. Moon, C. C. Cunningham, and J. Kas. 2001. The optical stretcher: a novel laser tool to micromanipulate cells. Biophys J 81:767-784.
15. Charras, G. T., P. P. Lehenkari, and M. A. Horton. 2001. Atomic force microscopy can be used to mechanically stimulate osteoblasts and evaluate cellular strain distributions. Ultramicroscopy 86:85-95.
16. Bausch, A. R., F. Ziemann, A. A. Boulbitch, K. Jacobson, and E. Sackmann. 1998. Local measurements of viscoelastic parameters of adherent cell surfaces by magnetic bead microrheometry. Biophys J 75:2038-2049.
17. Sun, Y., K. T. Wan, K. P. Roberts, J. C. Bischof, and B. J. Nelson. 2003. Mechanical property characterization of mouse zona pellucida. IEEE Trans Nanobiosci 2:279-286.
18. Yang, S., and T. Saif. 2005. Reversible and repeatable linear local cell force response under large stretches. Exp Cell Res 305:42-50.
19. Harris, A. K., P. Wild, and D. Stopak. 1980. Silicone-rubber substrata - new wrinkle in the Study of cell locomotion. Science 208:177-179.
20. Beningo, K. A., and Y. L. Wang. 2002. Flexible substrata for the detection of cellular traction forces. Trends Cell Biol 12:79-84.
21. Tan, J. L., J. Tien, D. M. Pirone, D. S. Gray, K. Bhadriraju, and C. S. Chen. 2003. Cells lying on a bed of microneedles: An approach to isolate mechanical force. P Natl Acad Sci USA 100:1484-1489.
22. du Roure, O., A. Saez, A. Buguin, R. H. Austin, P. Chavrier, P. Siberzan, and B. Ladoux. 2005. Force mapping in epithelial cell migration (vol 102, pg 2390, 2005). P Natl Acad Sci USA 102:14122-14122.
23. Zhao, Y., and X. Zhang. 2006. Cellular mechanics study in cardiac myocytes using PDMS pillars array. Sensor Actuat A-Phys 125:398-404.
24. Xia, Y. N., and G. M. Whitesides. 1998. Soft lithography. Angew Chem Int Ed 37:551-575.
25. Sniadecki, N. J., A. Anguelouch, M. T. Yang, C. M. Lamb, Z. Liu, S. B. Kirschner, Y. Liu, D. H. Reich, and C. S. Chen. 2007. Magnetic microposts as an approach to apply forces to living cells. P Natl Acad Sci USA 104:14553-14558.
26. Hamatani, T., G. Falco, H. Akutsu, C. A. Stagg, A. A. Sharov, D. B. Dudekula, V. VanBuren, and M. S. H. Ko. 2004. Age-associated alteration of gene expression patterns in mouse oocytes. Hum Mol Genet 13:2263-2278.
27. Perez, G. I., A. Jurisicova, T. Matikainen, T. Moriyama, M. R. Kim, Y. Takai, J. K. Pru, R. N. Kolesnick, and J. L. Tilly. 2005. A central role for ceramide in the age-related acceleration of apoptosis in the female germline. FASEB J 19:860-862.
28. Steuerwald, N. M., M. G. Bermudez, D. Wells, S. Munne, and J. Cohen. 2007. Maternal age-related differential global expression profiles observed in human oocytes. Reprod Biomed Online 14:700-708.
29. Luo, Y. H., and B. J. Nelson. 2001. Fusing force and vision feedback for manipulating deformable objects. J Robotic Syst 18:103-117.
30. Greminger, M. A., and B. J. Nelson. 2004. Vision-based force measurement. IEEE Trans Pattern Anal 26:290-298.
31. Liu, X. Y., Y. Sun, W. H. Wang, and B. M. Lansdorp. 2007. Vision-based cellular force measurement using an elastic microfabricated device. J Micromech Microeng 17:1281-1288.
32. Armani, D., C. Liu, and N. Aluru. 1999. Re-configurable fluid circuits by PDMS elastomer micromachining. In IEEE Int Conf. Micro Electro Mechanical Systems, Orlando, FL, USA. 222-227.

33. Hughes, W. F., and J. A. Brighton. 1999. Schaum's outline of theory and problems of fluid dynamics. McGraw Hill, New York.
34. Ugural, A. C., and S. K. Fenster. 2003. Advanced strength and applied elasticity. Prentice Hall PTR, Upper Saddle River, N.J.
35. Mark, J. E. 2009. Polymer data handbook. Oxford University Press, Oxford; New York.
36. Familiari, G., M. Relucenti, R. Heyn, G. Micara, and S. Correr. 2006. Three-dimensional structure of the zona pellucida at ovulation. Microsc Res Techniq 69:415-426.
37. Murayama, Y., J. Mizuno, H. Kamakura, Y. Fueta, H. Nakamura, K. Akaishi, K. Anzai, A. Watanabe, H. Inui, and S. Omata. 2006. Mouse zona pellucida dynamically changes its elasticity during oocyte maturation, fertilization and early embryo development. Hum Cell 19:119-125.
38. Nogues, C., M. Ponsa, F. Vidal, M. Boada, and J. Egozcue. 1988. Effects of aging on the zona pellucida surface of mouse oocytes. J In Vitro Fert Embryo Transf 5:225-229.
39. Martinova, Y., M. Petrov, M. Mollova, P. Rashev, and M. Ivanova. 2008. Ultrastructural study of cat zona pellucida during oocyte maturation and fertilization. Anim Reprod Sci 108:425-434.
40. Bereiter-Hahn, J., and J. Kajstura. 1988. Scanning microfluorometric measurement of TRITC-phalloidin labelled F-actin. Dependence of F-actin content on density of normal and transformed cells. Histochemistry 90:271-276.
41. Liu, X. Y., Y. F. Wang, and Y. Sun. 2009. Cell contour tracking and data synchronization for real-time, high-accuracy micropipette aspiration. IEEE Trans Autom Sci Eng 6:536-543.
42. Chen, C. S., M. Mrksich, S. Huang, G. M. Whitesides, and D. E. Ingber. 1997. Geometric control of cell life and death. Science 276:1425-1428.
43. Liu, X. Y., R. Fernandes, A. Jurisicova, R. F. Casper2, and Y. Sun. 2010. In-situ mechanical characterization of mouse oocytes using a cell holding device. Lab Chip 10:2154-2161.

Chapter 6
In Situ Characterizations of Thin-Film Nanostructures with Large-Range Direct Force Sensing

Gilgueng Hwang, Juan Camilo Acosta, Hideki Hashimoto, and Stephane Regnier

Abstract Thin-film nanostructures from various materials have a great potential to further miniaturize the devices like nanoelectronics and micromachines. Recently semiconductor nanofilms, mono or multiple atomic layer carbon nanofilms have been synthesized. However, the precise electrical and mechanical properties of these structures still need to be characterized in more detail. In this chapter, we introduce a large range force sensing tool that we recently developed. Three-dimensional piezorsistive helical nanobelts (HNB) will be described including their giant piezoresistivity. Their large force sensing range is characterized and calibrated by incorporating in situ scanning electron microscope (SEM) tuning fork sensors. This in situ characterization clearly revealed the non-constant stiffnesses of HNBs. Finally, as an application example, mechanical properties of nanowires are characterized by the HNBs. The proposed large range force characterization system is useful and promising toward creating thin-film micro and nanodevices.

Keywords In situ characterization • Scanning electron microscope • Nanomanipulation • Thin-film nanostructures • Nanoelectromechanical systems (NEMS) • Helical nanobelts • Force sensing • Tuning fork

6.1 Introduction

Micro/nanomanipulation and assembly technologies are important toward the development of micro-nanoelectromechanical systems (MEMS/NEMS) [1–3]. However these manipulations are currently being performed manually by human operators which makes the task very stressful and time-consuming. It is mainly because

G. Hwang (✉)
Laboratory for Photonics and Nanostructures, CNRS, route de Nozay, 91460 Marcoussis, France
e-mail: gilgueng.hwang@lpn.cnrs.fr

C. Clévy et al. (eds.), *Signal Measurement and Estimation Techniques for Micro and Nanotechnology*, DOI 10.1007/978-1-4419-9946-7_6,

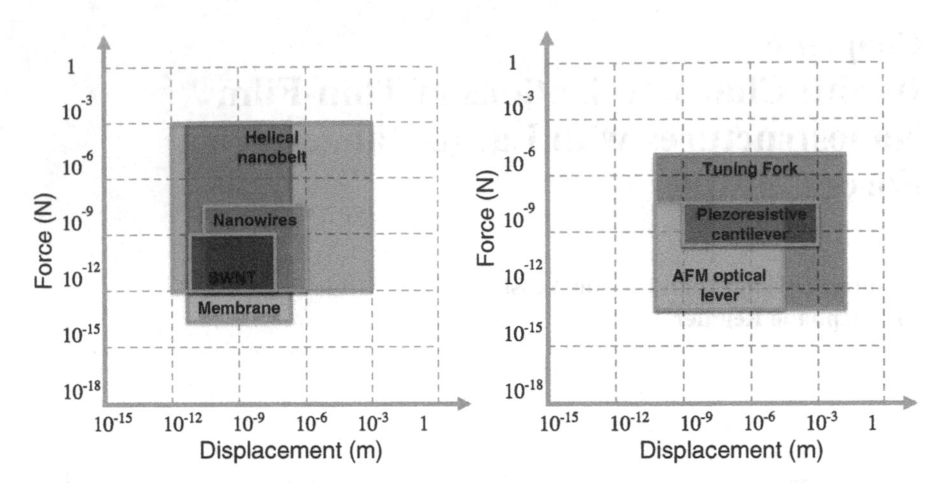

Fig. 6.1 Comparison in force and displacement sensing range. (**a**) Ultra flexible nanostructures, (**b**) Force sensors

of the lack of proper sensing tools to measure different physics in micro/nano scale. Currently available sensors cannot measure such different physical transitions because of their limited sensing resolution and range (Fig. 6.1).

For an automated micro/nanomanipulation, it is necessary to have both visual and force feedback. In-situ scanning electron microscope (SEM) nanomanipulation was proposed to combine good enough visual feedback but the force sensing is still in their infancy. Therefore, we summarize our recent works on the high resolution, wide range force and stiffness sensing with helical nanobelt (HNB) and tuning forks that can be utilized in the in-situ electromechanical property characterizations. The developed force sensing mechanisms can also be utilized to characterize the mechanical properties of other ultra-flexible nanostructures. Mechanical property characterizations of ultra-flexible three-dimensional nanostructures are important during the development of NEMS. It also requires ultra-high precision and wide range force calibration systems. For even much higher resolution of sensing, optical tweezers [4], magnetic bead [5], and atomic force microscopy (AFM) [6] have been mostly studied, and AFM among them is the mostly used. However these methods are limited in large displacement measurement and principal difficulty to measure 3-D forces and suffer from several limitations such as their limited force and displacement ranges. Considering the force resolution, optical method can easily reach piconewtons to the contrary of the other methods which are in the upper range of nanonewtons [7]. Furthermore, cantilevers need strong calibration and are difficult to be used as a sensor or actuator for several dimensions (the cantilevers torsion is difficult to characterize) [8]. Optical tweezers are useful for a large variety of object to characterize [9]; however, they are strictly limited in the piconewton range of the forces to measure and its in liquid environment [10]. For studies that require larger forces, atomic force microscope cantilevers are more suitable choices because they are stronger (less compliant) than optical tweezers. Finally, their potential integration with experimental setup is challenging and expensive because of necessary external laser optics, especially in vacuum. And they are also limited in the range depending on the stiffness of cantilever. The studied whole systems are limited in their force range or in their displacement range (Fig. 6.1). To build a sensor which is able to measure piconewton as well as millinewton and capable of picometer displacements as well as millimeter displacement still remains a challenging research area. The newly proposed sensors would then be useful in a wide variety of objects to characterize such as nanowires, tissues, viruses, bacteria, living cells, colloidal gold, and even DNA. The further advantage consists in its simple and straightforward integration that will lead to the conception of complex MEMS (nano-translators with high displacement range, nano-sensors with high force range) for future complex bio- or non-organic applications.

The aim of this chapter is to introduce our recent works in development of large range force sensing tools (micronewton range with the resolution of hundreds of piconewton and the stiffness resolution in N/m) for in situ property characterizations of ultra flexible nanostructures. For this purpose, the sensors should be integrated to electron microscopes. Thus, the sensors should also be vacuum and electron beam compatible. For this aim, we have developed two different types of force and stiffness sensors. In this chapter, the first part will describe the 3-D piezoresistive HNB force sensors and their applications. Then the tuning fork and HNB based force sensors will be shown. Using these tools, we therefore can study electromechanical properties of ultra-flexible nanostructures by in-situ SEM nanomanipulation with direct force characterization by incorporating nanomanipulators.

6.2 Large Range Direct Force Sensing Devices

6.2.1 Piezoresistive HNBs as Force Sensors

HNBs with metal pads attached on both sides were fabricated using microfabrication techniques (Fig. 6.2) [11]. With the metal connectors, good electrical contact can be achieved. Besides the electromechanical characterization, such connectors also allow for the integration of these structures into more complex assemblies. Nanomanipulation inside an SEM was used for their electromechanical property characterization. The experimental results showed that the structures exhibit a unusually high piezoresistive response. Moreover, electrostatic actuation was used to excite the structures at their resonance frequency and investigate their resistance to fatigue. With their low stiffness, high strain capability, and good fatigue resistance, the HNBs can be used as high-resolution and large-range force sensors. By variation of design parameters, such as the number of turns, thickness, diameter, or pitch, a HNB with the required stiffness can be designed through simulation. The fabrication process is suitable for further miniaturization. Nanometer scale diameter and wire width can be achieved through changes in the later design and by using electron–beam lithography, respectively.

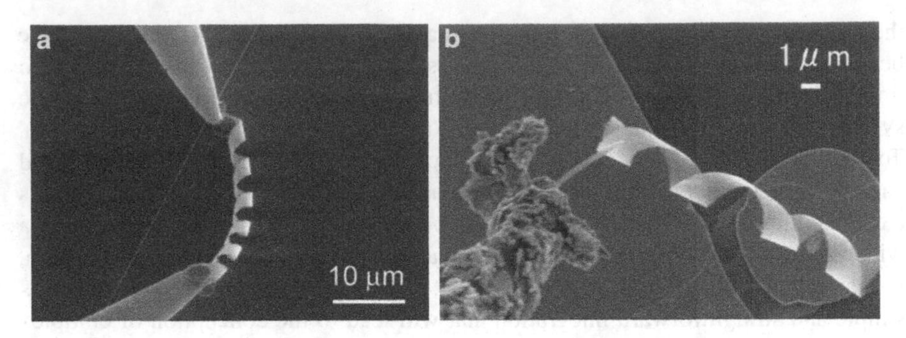

Fig. 6.2 Metal connector attached 3-D HNBs and their electromechanical property characterization with two metal probes. (**a**) The piezoresistivity was characterized in the longitudinal tensile elongations. (**b**) Mechanical property of silicon nanowire is characterized by HNBs. Copyright 2010 ACS Nano Letters [11]. The [11] refers to the complete citation in the reference section

6.2.2 Large Range Mechanical Property Characterization and Calibration of HNBs by Tuning Fork

As another force sensing tool, tuning fork was incorporated to HNB to be able to function in dynamic force gradient measurement applications. The tuning fork as force interaction tools have been proposed mainly to replace the AFM optical cantilevers for imaging [12]. However, the large range force sensing was not attempted due to their principle force gradient sensing. In our work, the wide range force sensor based on frequency modulation AFM tuning fork and attached HNBs was developed. HNBs attached to the tuning fork tip enable to measure the force by measuring. The force is obtained from the stiffness of HNBs measured by tuning fork gradient force sensor and by the displacement measurement of HNBs with SEM imaging. Attaching a high aspect ratio probe tip onto tuning fork enabled amplification to the amount of stress applied to the tuning fork body. For example, the stiffness measurement resolution of tuning fork is 0.009 N/m. Considering the SEM imaging resolution to measure the displacement of HNBs as 200 nm, the resolution of force sensing is 1.8 nN. If we use theoretical stiffness measurement resolution of tuning fork and high resolution SEM (1 nm resolution), the force resolution can be improved to 600 fN. And the upper measurement range was measured up to the range of μN. Furthermore, the measured frequency shift of tuning fork does not have to be calibrated. was calibrated with as-calibrated cantilever in the nano-newton range force and HNBs with the stiffness estimation from the model for the sub-nanonewton range (Fig. 6.3). The demonstrated force sensors are easily integrated to SEM thus the demonstrated technology have strong potential in nanomanipulations of various nanostructures.

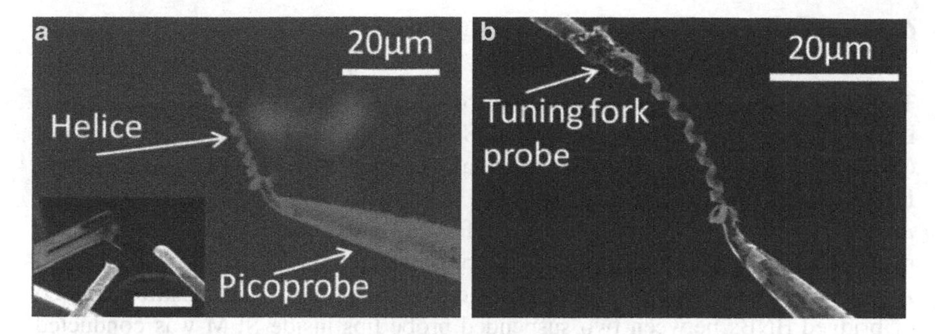

Fig. 6.3 Tuning fork force calibration using HNB. (**a**) HNB picked with probe of manipulator (*inset* figure depicts the two probes of manipulators and tuning fork for manipulation and calibration of HNB). (**b**) the picked HNB was attached to the probe of tuning fork

6.2.3 Potential Applications

The static nanomechanical characterization of ultra-thin film nanostructures is promising application of the proposed HNB force sensors. Since the device is based on smart sensing with simple current monitoring to measure the force, any complicated external read-out devices are not required. We have applied these sensors to characterize the mechanical properties of nanowires with even smaller dimensions than HNBs (Fig. 6.2). Visual tracking of the deformation is also promising to measure the force using HNBs in case of wet applications such as biological manipulation. Furthermore, tuning fork and HNB force sensors can be applied to reveal mechanical properties of the dynamic systems such as resonators based on thin membrane nanostructures. By incorporating with cryogenic or biological manipulation setup, it can also be utilized to measure the dynamically varying biological nanostructures.

6.3 Piezoresistive HNB Force Sensors

The objective of this section is to characterize the piezoresistivity of InGaAs/GaAs HNBs as the sensing mechanism of the proposed force sensor. Since the piezoresistivity of HNBs was not so far known, our concern of this section is to make a clear piezoresistivity characterization of as-fabricated HNBs as the force sensor elements. This material is promising for the use in electromechanical sensors. It is necessary to characterize individual nanostructure's properties in order to use it as assembled devices. In this section, clear characterizations of piezoresistivity with the sensitivity parameters such as gauge factors and piezoresistance coefficients of InGaAs/GaAs HNBs was experimentally achieved to show its high potential as the nano building blocks to nano electromechanical devices such as force sensors [11].

6.3.1 HNBs as Force Sensors

This section presents the characterization of piezoresistive HNBs for an improved interconnection assembly to create NEMS. The HNB fabrication process is based on microfabrication techniques to create a planar pattern in a 27 nm thick, n-type InGaAs/GaAs bilayer that self-forms into 3-D structures during a wet etch release. As the HNBs have lower dopped thin and flexible layers small metal connectors are necessarily attached to both sides for assuring stable ohmic contact with electrodes. A 3-D nanorobotic assembly including electron-beam-induced-deposition (EBID) of bridged HNBs between two suspended probe tips inside SEM was conducted to probe the structures for electromechanical characterization. With their strong piezoresistive response, low stiffness, large-displacement capability, and good fatigue resistance they are well suited to function as sensor elements in high-resolution, large-range electromechanical sensors.

Stable electromechanical characterizations of piezoresistivity requires the HNBs to be assembled with proper soldering method between independent electrodes with high precision actuation. Attaching metal electrodes to nanostructures is useful for their assembly and better making an electrical ohmic contact. Magnetic field assisted assembly of carbon nanotubes with ferromagnetic metal (Ni) ends proved to be useful for both electrical property characterization and device assembly [13, 14]. Dexterous robotic manipulation with multiple end-effectors can be more useful for the stable dynamic property characterization [15]. HNBs can serve as a mechanism to transduce force to displacement. The deformation is detected through piezoresistance. This sensing mechanism was demonstrated previously with other structures on the micro- and nano-scale. For example, piezoresistive microcantilevers have been used to characterize low force electrical contacts [16]. Moreover, an increased piezoresistivity could be demonstrated for Si nanostructures [17, 18]. In this section, InGaAs/GaAs HNBs were fabricated with one end fixed to non-scrolling supports for control over their position. Moreover, we present an extension to the fabrication process described in [20]. We demonstrated the fabrication of HNBs with small metal connectors attached on both ends. These connectors help to achieve good electrical contacts which are required for the electromechanical characterization, but can also be useful for the assembly of more complex devices with the self-sensing HNBs as components. The electromechanical response of the piezoresistive HNBs is characterized using nanomanipulation in SEM. Their stiffness and stress distribution of each bilayer are simulated using the model that was validated for similar structures in [20]. Moreover, the fatigue behavior of the structures is investigated during oscillation at resonant frequency.

6.3.2 The Principle of Piezoresistivity

Piezoresistivity is the fractional change in bulk resistivity induced by small mechanical stresses applied to a material. Most materials exhibit some piezoresistive effect [21]. The gauge factor of semiconductors is much higher than the one of metals. The deformation of the energy bands by an applied stress causes piezoresistivity. The deform of the energy bands affects the effective mass and mobility of the electrons and holes. The fractional change on resistivity is first order linearly dependent on the stress components parallel and perpendicular to the direction of the current flow.

$$\frac{\Delta\rho}{\rho} = \pi_{\parallel}\sigma_{\parallel} + \pi_{\perp}\sigma_{\perp}, \tag{6.1}$$

where ρ is the resistivity of the material, $\pi_{\parallel}$ and $\pi_{\perp}$ are the parallel and perpendicular piezoresistivity coefficients, $\sigma_{\parallel}$ and $\sigma_{\perp}$ are the parallel and perpendicular stress components. The resistivity coefficients can be obtained when the resistivity tensor of the material is transformed in these directions:

$$\pi_{\parallel} = \pi_{11} - 2(\pi_{11} - \pi_{12} - \pi_{44})(l_1^2 m_1^2 + l_1^2 n_1^2 + m_1^2 n_1^2), \tag{6.2}$$

where $(\pi_{11}, \pi_{12}, \pi_{44})$ are the three piezoresistive coefficients of material and (l_1, m_1, n_1) are the set of direction cosines between the parallel resistor direction and the crystal axes of the material, (l_2, m_2, n_2) is the set of direction cosines between the perpendicular resistor direction and the crystal axes of the material.

6.3.3 Two Point Contact Resistance Measurements

Since there are contacts between HNB and manipulator probe, the contact resistances need to be considered for the total measured resistance through the HNB. Contact resistance measurement techniques fall into four main categories: two-contact two-terminal, multiple-contact two-terminal, four-terminal, and six terminal methods. Four-point probe methods have been widely used to characterize the contact resistance [17, 22]. However, our HNB measurement should basically use only the both ends of HNBs which prevents to use the 4-point probe methods. The main reason why two point contact was used for our experiments is that the HNBs require larger displacement range from their large elastic range. For the experiments, the HNBs were picked up from the substrate to prevent the interference from the substrate, since the 3D morphology is easier to get interrupted when in-plane measurement configuration.

Still, none of these methods is capable of determining the specific interfacial resistivity ρ_i. Instead they determine the specific contact resistivity ρ_c which is not the resistance of the metal-semiconductor interface alone, but it is a practical quantity describing the real contact. It is, therefore, always difficult issue to compare theory with experimental results because theory cannot predict ρ_c accurately and experiment cannot determine ρ_i unambiguously.

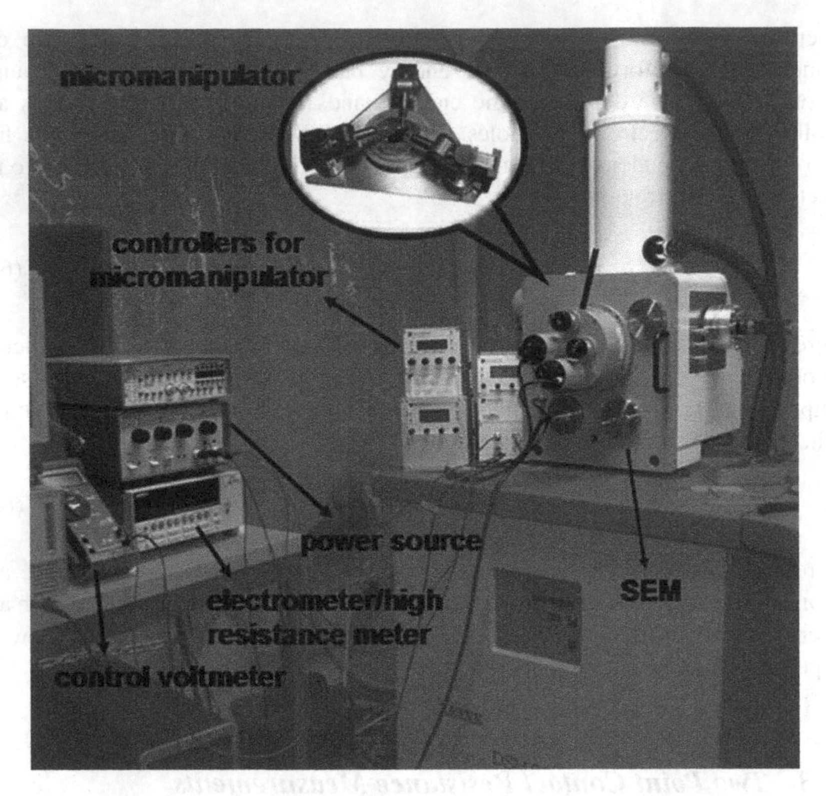

Fig. 6.4 Picture of the measurement setup and the SEM used to characterize the HNB piezoresistivity

In this section, we limit our discussion on the contact resistivity to the two-contact two-terminal method. And the contact resistance will be considered by the transmission line model (TLM) described in the [23]. The detailed method of TLM is not covered in this chapter.

6.3.4 Nanorobotic Manipulation System for Piezoresistivity Measurement

As was discussed to measure properly the piezoresistivity of HNBs, at least two probes with actuation capability with high precision positioning accuracy are required. Also the whole characterization processes should be recorded with SEM for the later image analyses on the deflected HNBs. Therefore, multiple nanorobotic manipulators are installed inside the SEM chamber. Total measurement setup including a SEM (Carl Zeiss DSM 962) is shown in the Fig. 6.4.

Fig. 6.5 Nanorobotic manipulation system inside a SEM (DSM 962). Three MM3A micromanipulators with Picoprobes and homemade microgripper are installed

The nanorobotic manipulation system shown in Fig. 6.5 has been used for the manipulation of the as-fabricated HNBs inside a SEM. Three nanorobotic manipulators (Kleindiek, MM3A) are installed inside the SEM; each has three degrees of freedom, and 5, 3.5, and 0.25 nm resolution in X, Y, and Z directions at the tip. The basic nanorobotic manipulation tasks using the system was tested as shown in the Fig. 6.6.

The system is eligible for the HNB manipulations inside SEM by the probe tip functionalization such as using the vacuum compatible glue, etc. For the electrical measurement, the low current electrometer (Keithley 6517A) was used to generate the input voltage and record the measured current data. The probes of each nanomanipulator are connected to the electrometer. The detailed measurement schematics and piezoresistivity experiments using the built nanorobotic characterization system will be covered in the later sections.

6.3.5 Axial Piezoresistivity Characterizations

In this section, the axial piezoresistivity of HNBs was measured using the nanorobotic manipulation systems. Figure 6.7 shows the schematics of the axial piezoresistivity measurements. Both ends of HNB metal connectors are attached with the probe tips of nanomanipulators. The contact was made by the EBID by

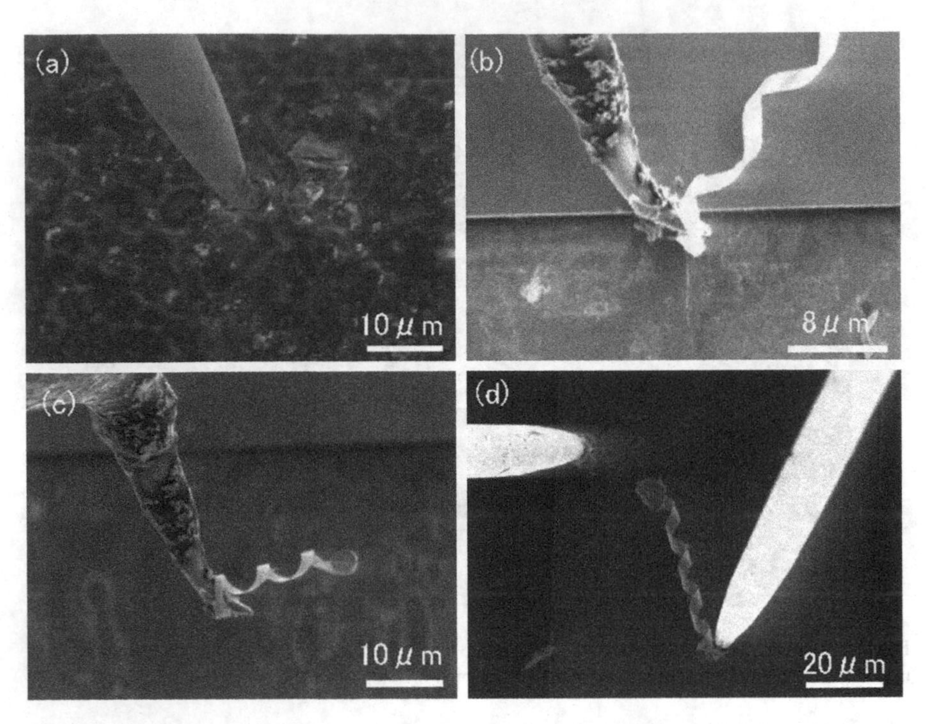

Fig. 6.6 Nanorobotic manipulation of a HNB onto in-plane electrodes. (**a**) Dipping with sticky tape. (**b**) Breaking a HNB with sticky probe. (**c**) Picking up the HNB. (**d**) Releasing using a non-sticky probe (Photos were not taken from a single sequential manipulation)

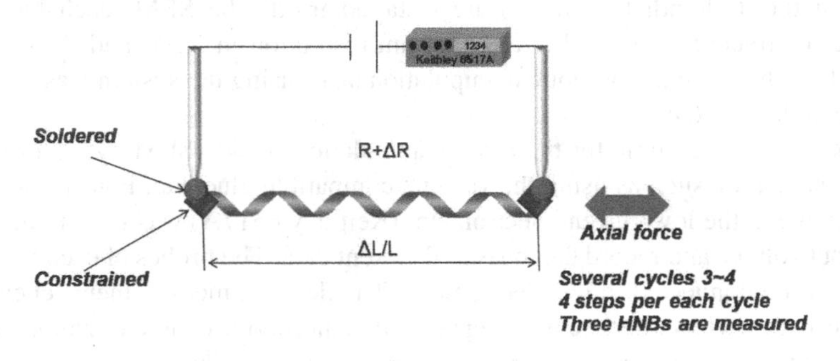

Fig. 6.7 The schematic configuration of axial piezoresistivity measurement of HNBs

applying the W(CO)6 precursor in gas phase. One side of the probe was fixed and the other side of the probe is actuated in axial direction. The reason of this measurement configuration can be explained by following reasons. It's a 3-D structure and it is easily interrupted from the surface tension since it is especially the flexible structures. So, proper force should be measured by the picked up

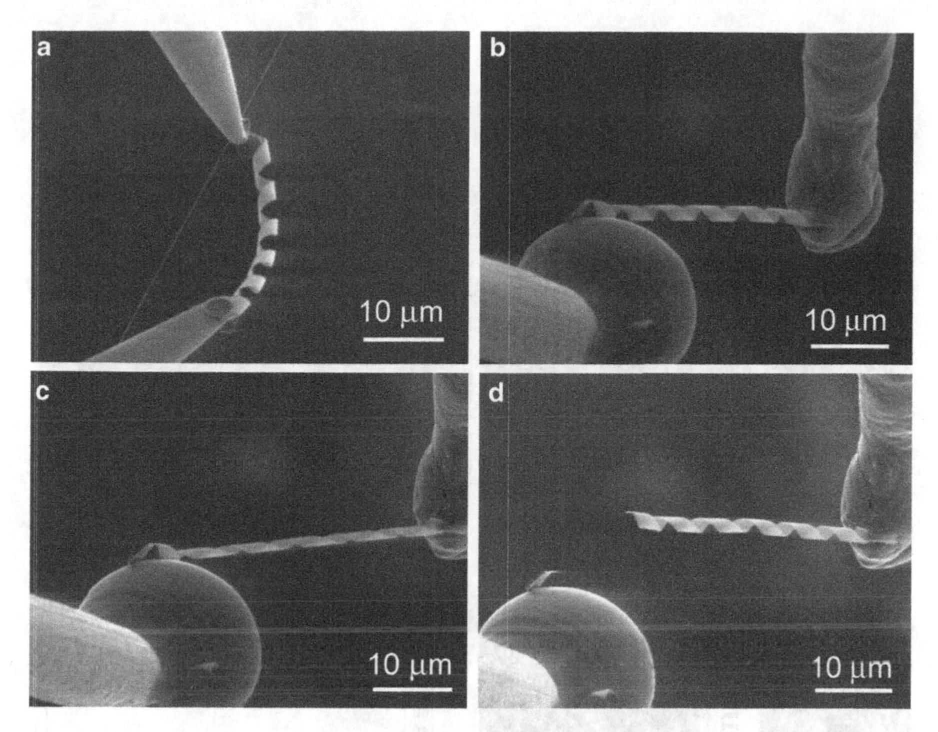

Fig. 6.8 HNBs with two metal pads during manipulation. (**a**) Picking up. (**b**) During measurement with zero-deformation. (**c**) During measurement with maximum deformation. (**d**) After the structure breaks with too much force. Copyright 2010 ACS Nano Letters [11]. The [11] refers to the complete citation in the reference section

structure ideally. Another reason to pick up is to utilize the large displacement range of force sensing on the benefit of the flexible structures. (comparing with the force sensing range and resolution of CNT force sensors [24, 25]).

Using this measurement configuration, three HNBs were characterized in 3–4 cycles and 4 steps in each cycle.

Results for the electromechanical characterization experiments, two micromanipulators (Kleindiek, MM3A), each with two metal probes (Picoprobe, T-4-10-1mm) with a tip radius of 100 nm attached, were installed inside an SEM (Zeiss, DSM 962).

The experimental procedure is illustrated in Fig. 6.7. The SEM photos during this experiment are shown in the Fig. 6.8. One manipulator was used to break and pick up a HNB on one side. For this purpose the HNBs were fabricated with a small length between the support and the first metal pad. The other manipulator was used to make contact to the other side. In order to achieve good electrical contacts on both sides of the HNBs EBID with W(CO)6 precursor was used [26, 27]. This way a voltage could be applied on both sides of the HNBs and the current could be measured with a low-current electrometer (Keithley 6517A).

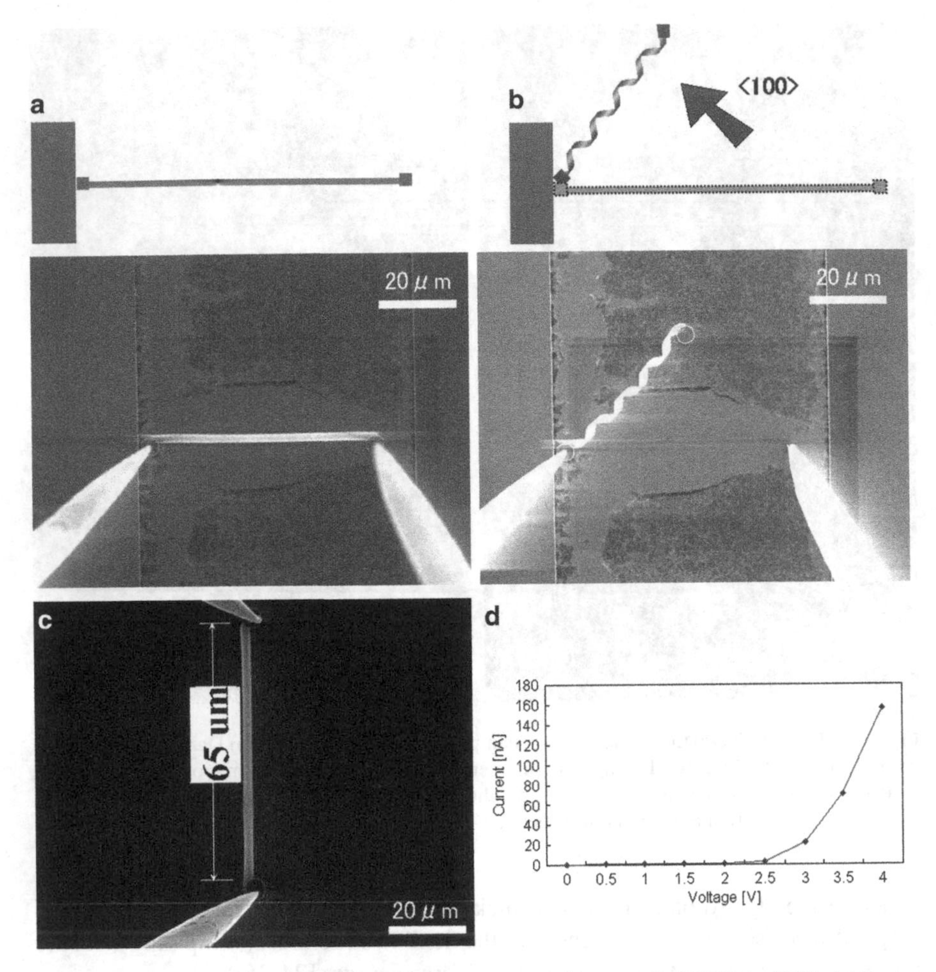

Fig. 6.9 (**a, c**) Fully extended HNB before scrolling. (**b, d**) the scrolled HNB in ⟨100⟩. (**c**) Resistance measurement of extended HNB before scrolling. (**d**) Its I–V curve

The InGaAs/GaAs bilayer resistance of extended HNB before scrolling was measured using the less under-etched HNB as shown in Fig. 6.9c. Its measured I–V curve (Fig. 6.9d) reveals that the resistance of extended HNB before scrolling is much higher than the scrolled HNBs. It shows the intrinsic beam type InGaAs/GaAs bilayer resistance with zero torsion before scroll is much higher than the total resistance measurement of the scrolled HNB. It assumes the negative piezoresistivity by the torsion force increase. In this case, the probes are firmly applying enough pressure to make ohmic contacts. Therefore, the contact resistance of HNB (Fig. 6.8) does not take a big role in the total measured resistance of HNB piezoresistivity at least in zero deformation.

Table 6.1 Specifications of HNBs used in experiments and simulations

HNB	1	2	3
Thickness $In_{0.15}Ga_{0.85}As$		11.6 nm	
Thickness GaAs		15.6 nm	
Thickness Cr		~10 nm	
Thickness Au		~100 nm	
Diameter		2.1 μm	
InGaAs/GaAs Doping		6×10^{18}–1×10^{19} cm^{-3}	
Pitch	5.9 μm	6.6 μm	7.1 μm
Stripe width	2.9 μm	3.0 μm	2.9 μm
Number of turns	6	5.5	4.5
Stiffness (Simulation)	0.010 N/m	0.024 N/m	0.030 N/m

For the HNB piezoresistivity measurement, we instead used fresh metal probe and EBID as shown in Fig. 6.8. After a HNB was attached as described above, a tensile force was applied to it by moving one probe away from the other in the axial direction (Fig. 6.8b–d). Continuous frames of images were taken to detect the deformation and I–V curves were recorded for the different positions. The characterization was carried out for three different HNBs. Their dimensions are summarized in Table 6.1. It was checked from the images that the boundary conditions did not change significantly during the experiments. In Fig. 6.8d, it can be seen that even after the HNB broke the attachment to the probe did not change. The SEM images were analyzed to extract the HNB deformation for a certain I–V measurement. From these data, different plots can be generated. In Fig. 6.10, different I–V curves are shown for different elongations. The current increases for an increasing axial elongation.

After the self-scrolling of the structures the outer InGaAs layer is under compressive stress while the inner GaAs layer is under tensile stress. Moreover, across the thickness of each layer the stress is inhomogeneous. When a tensile force is applied to the structure, the compressive stress in the InGaAs layer decreases and the tensile stress in the GaAs layer increases. Moreover, there is shear stress all of which make it difficult to model the exact stress state of these structures under a tensile load.

Moreover, the piezoresistive response due to changes in this stress state is even more complex. For GaAs there are four physical mechanisms that cause changes in electrical resistance due to stress [28]. Firstly, the electron effective mass changes under stress leading to a change in mobility. Secondly, changes in stress cause the transfer of electrons between the high-mobility band gap minimum and low-mobility minima due to change of their relative energy. Thirdly, stress can cause electrons to freeze to deep-level impurities. Fourthly, piezoelectric charges induced by stress gradients also change the resistivity. Besides these mechanisms for the piezoresistive behavior of the InGaAs/GaAs HNBs, size effects influence the magnitude of the piezoresistive response, as has been reported previously for other types of nanostructures [17, 18].

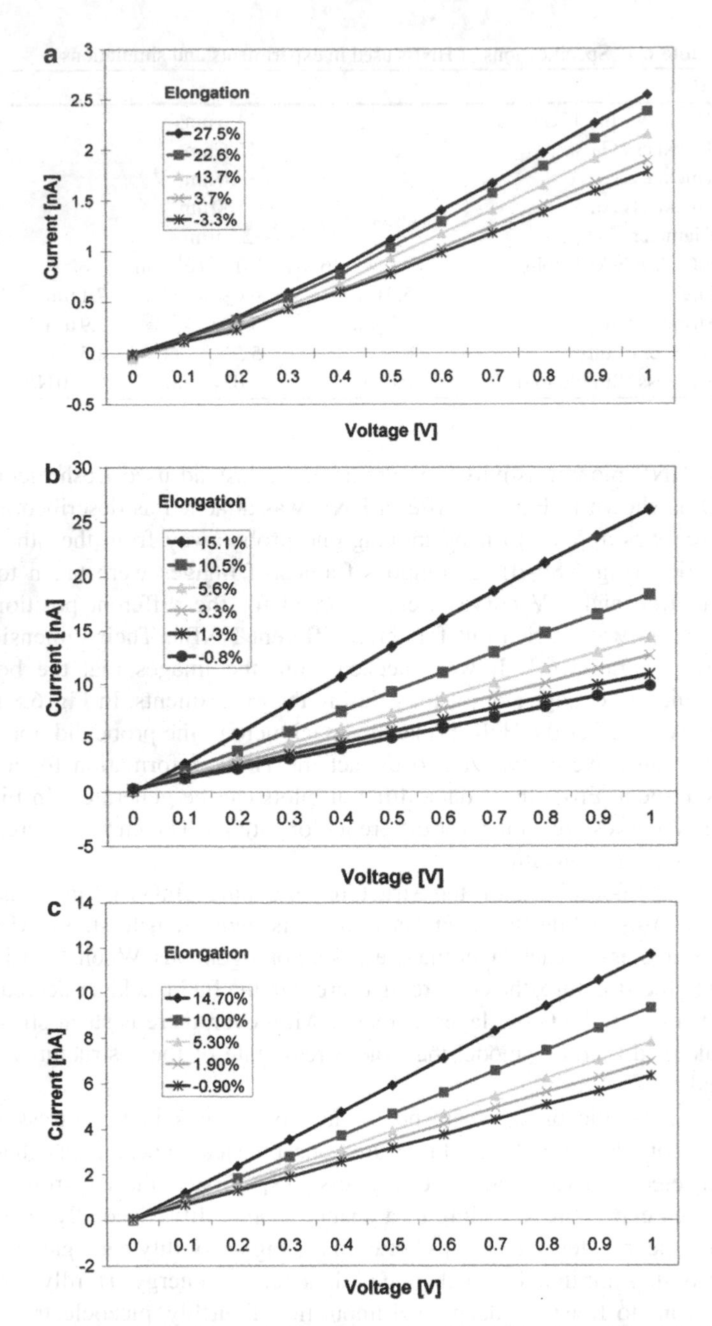

Fig. 6.10 I–V curves for different deformations of the HNBs listed in Table 6.1. (**a**) HNB 1. (**b**) HNB 2. (**c**) HNB 3. Copyright 2010 ACS Nano Letters [11]. The [11] refers to the complete citation in the reference section

Table 6.2 Experimentally obtained quantitative parameters in full range: resistance (MΩ),gauge factor

HNB	1	2	3
Resistance (MΩ)	394–568	42–102	84–158
Gauge Factor $\beta_{GF}(\varepsilon)$	1.1	4.3	3.3

For all of these reasons the complete modeling of the changing stress state and piezoresistive response of the HNBs in this work is very complex and has not been attempted yet. From Fig. 6.10 it can be seen that both the resistance and the piezoresistive response are different for the three HNBs in Table 6.2.

The resistance range and the gauge factor of HNBs are summarized in Table 6.2. The piezoresistive gauge factor $\beta_{GF}(\varepsilon)$ of the HNBs is given by

$$\beta_{GF}(\varepsilon) = ((\Delta R(\varepsilon)/R_0))\varepsilon^{-1}, \tag{6.3}$$

where $\Delta R(\varepsilon)$ is the resistance change, R_0 is the initial resistance at zero strain, and ε^{-1} is the strain. To compare the response of the measured HNB piezoresistivity, the resistance range and the gauge factors were shown in Table 6.2. HNB 1 has a much higher resistance at zero elongation as well as a lower piezoresistive response. The chip that was used to fabricate this HNB was taken from a different wafer. Besides its higher resistance the HNB with lower doping concentration also exhibits a lower piezoresistive response. HNBs 2 and 3 were made from the same wafer. The difference in resistance between these two structures can be due to the difference in contact resistance between the probes and the HNBs. The piezoresistance of the structures can be further increased if Al is incorporated in the bilayer [28]. Increasing the Al content, however, reduces the etch selectivity with the $Al_{0.8}Ga_{0.2}As$ sacrificial layer [41]. In Fig. 6.11, the change in resistance is plotted versus the change in length for all HNBs. All three results have one trend in common: During the repeated cycles of pulling and pushing the resistance at zero deformation increases. There are two ways to explain this. The most probable reason is that under repeated loading and unloading the contact resistance increased because the attachment between the probes and the HNB changed. From the TLM, the resistance at zero deformation depicts the contact resistance if we assume the initial contact resistance as static zero [23]. The linearity of resistance versus deflection at each cycle proves that contact resistance at each deflection was maintained as static. Another reason can be that charging and joule heating in the SEM caused an increase in the resistance of the HNBs. As the initial contact was made by EBID, HNBs have initial high temperature but cools down during the measurement.

To prove the e-beam effect inside SEM, another experiment was performed. We made the same configuration of HNB between two probes and measured the resistance change. The resistance was measured at the same interval of piezoresistivity measurement. HNBs show the temperature dependence. However, the increase of the average resistance is not caused by some intrinsic property of the HNBs,

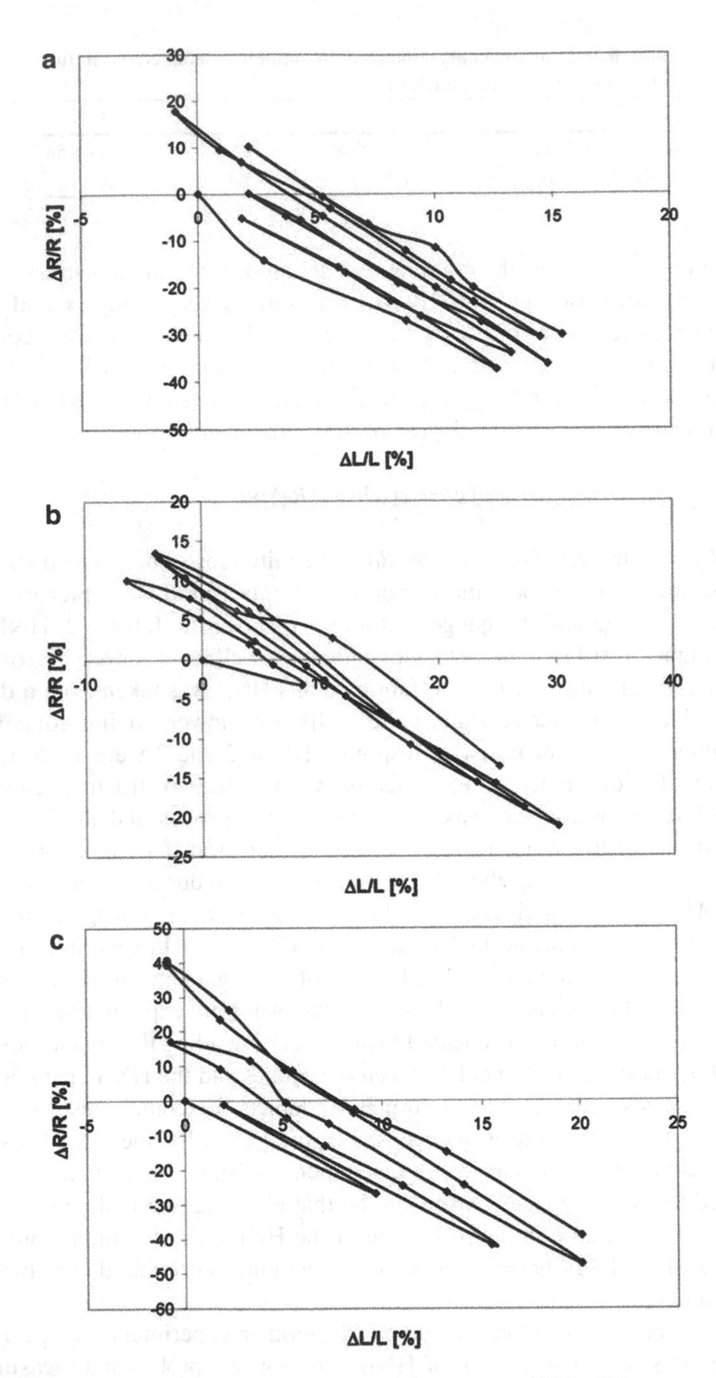

Fig. 6.11 Plots showing the length of the HNBs and the current at 1 V for different measurements. **(a)** HNB 1. **(b)** HNB 2. **(c)** HNB 3. Copyright 2010 ACS Nano Letters [11]. The [11] refers to the complete citation in the reference section

Table 6.3 Experimentally obtained axial piezoresistance coefficients (π_l^ρ) in full range

HNB	1	2	3
Piezoresistance coefficient (π_l^ρ)($\times 10^{-8}$Pa^{-1})	−35.6	−24.5	−9.96
Pitch (μm)	5.9	6.6	7.1
Stiffness (K)(N/m)	0.010	0.024	0.03
Torsional coefficient:T_c (pitch × stiffness)	0.059	0.15	0.213

but by the entire measurement system. For HNBs that are integrated with stable electrical connections into some system outside of the SEM, a repeatable output can be expected. From the results in Fig. 6.11 it can also be seen that the response is almost linear at most points. For the stable electromechanical sensor applications, the dynamic contact resistance analysis is important as we can easily calibrate the initial contact resistance as long as it is static.

Since each cycles of piezoresistivity curve in Fig. 6.11 were almost linear to the maximum elongations, we could evaluate the sensitivity of HNB's piezoresistivity by the piezoresistance coefficients for the comparison with other nanostructures. The longitudinal piezoresistance coefficients (π_l^σ) of three HNBs were calculated from the measured I-V characteristics. (π_l^σ) is defined as the relative change in conductivity per unit stress:

$$\pi_l^\sigma = \frac{1}{X}\frac{\Delta\sigma}{\sigma_0}, \tag{6.4}$$

where σ_0 is the conductivity under zero stress, $\Delta\sigma$ is the conductivity change and X is the stress. The piezoresistance coefficient was defined with resistivity as $\pi_l^\rho = (\Delta_\rho/\rho_0)$. The conversion is $\pi_l^\sigma = -\pi_l^\rho$ for small σ. Uniaxial stresses were applied on HNB along their lengths by the method shown in the Fig. 6.7. Since the linearity of each cycle from the measured data is almost constant, the widest measured region of the cycles was chosen to calculate the piezoresistance coefficient. Geometry information from the Table 6.1 was used as the parameters.

Table 6.3 summarizes the obtained results of three different HNBs and mechanical parameters such as pitch and stiffness to see the relation between them. We did not consider the doping concentrations on each HNBs because of no precise doping concentrasions are specified for each HNB. Therefore we only consider the effect from the geometry here for the simplification of the discussion. Three coefficients are from -35.6×10^{-8}Pa^{-1}, -24.5×10^{-8}Pa^{-1} and -9.96×10^{-8}Pa^{-1}.

We can assume that the piezoresistivity of HNBs are from the torsional effect when they are elongated through the length axis. Mechanical stiffness of axial direction of helical nano structures are known to be dominated by the torsional force [29] compared to the shear, bending and tension. Since the torsional effect is inversely proportional to the pitch and the stiffness, pitch×stiffness was defined as the torsional coefficient (T_c) in this case for the comparison (Table 6.3). T_c of HNB2 is 2.685 times higher than the one of HNB1. And the T_c of HNB3 was 3.61 times higher than the one of HNB1. The piezoresistance coefficient of HNB2 was 1.453 times higher than the one of HNB1. And the HNB1 is 3.57 times higher than

Table 6.4 Comparison of axial piezoresistance coefficients (π_l^p)

	Si Bulk	Bn-Si	SiNW	CNT	HNB
$\pi_l^p (\times 10^{-10} \text{Pa}^{-1})$	−1.7 to −9.4	−4	3.5 to 355	−400	−996 to −3,560

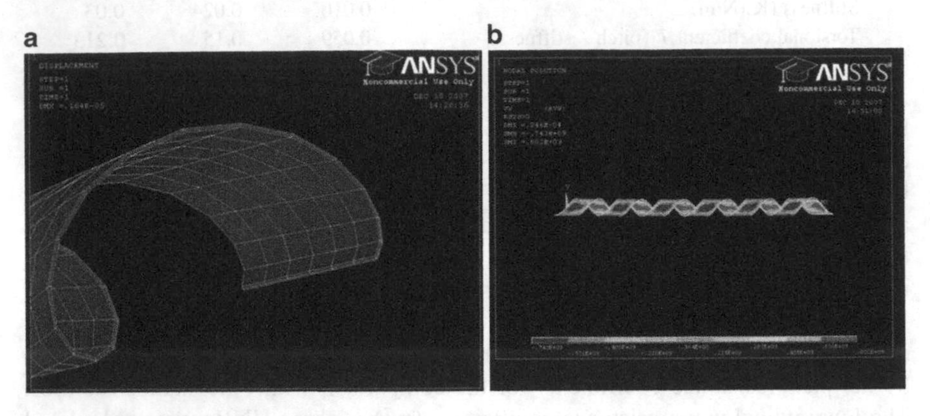

Fig. 6.12 Structural simulation for HNB 1. (**a**) Elements close-ups showing both layers. (**b**) Bilayer stress distribution in GPa for axial tensile force of 500 nN on one end and fixed on other end. Copyright 2010 ACS Nano Letters [11]. The [11] refers to the complete citation in the reference section

the one of HNB3. Therefore the piezoresistance coefficient is increased when the T_c is decreased but not so inversely proportional between the HNB1 and HNB3. It can be explained by the fact that HNBs used for experiments have different doping concentrations. Therefore the comparison of the precise doping concentrations of each sample should further be performed to prove the mechanism further precisely.

These obtained piezoresistance coefficients of HNBs are compared with other elements as the piezoresistors. Each reported piezoresistance coefficients of Bulk Si [30], boron doped Si [31], and Si-NW [17], and CNT [32] were summarized in Table 6.4. Especially Si-NW was recently reported as the giant piezoresistive effect in the [17]. The piezoresistance coefficients of the HNBs are even much higher than the one of Si-NW. These results are straight forward from the benefit of HNB's flexibility considering the (6.4) since the tiny input stress (X) can cause the resistance change $\left(\frac{\Delta\sigma}{\sigma_0}\right)$. At least the torsional effect seems to affect much to the high piezoresistivity response of the HNBs. As was discussed on the difficulty in exact modeling of the piezoresistivity of HNBs, mechanism behind of such a high piezoresistive response was not understood yet. However, there is the fact that the HNBs are the promising nanostructures to be used as the elements of ultra high resolution electromechanical devices such as force sensors.

Finite element simulation was used to estimate the stiffness of the HNBs and the bilayer stress distribution (Table 6.1). In Fig. 6.12, a plot of the displacement along the axial direction is shown for HNB 1 from Table 6.1 with an applied axial force

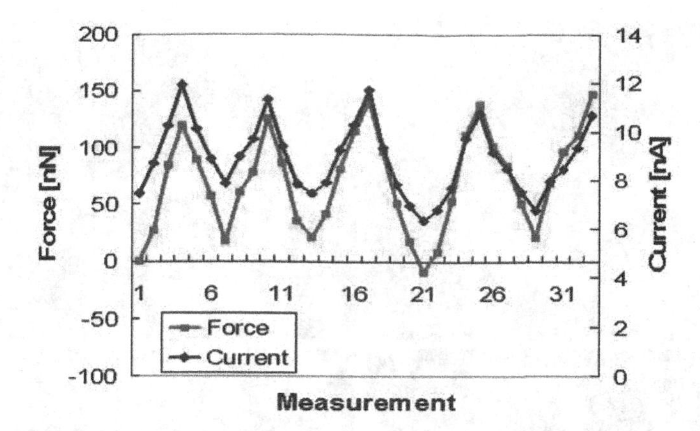

Fig. 6.13 Electromechanical measurements. Current at 1 V of a HNB 3 corresponds with the applied force at each point. Measurement in X axis depicts the measurement sequence. Copyright 2010 ACS Nano Letters [11]. The [11] refers to the complete citation in the reference section

of 0.1 μN. The displacement of the end where the axial force is applied is 5 μm. Therefore the axial stiffness of the structure is 0.02 N/m.

In Fig. 6.12a, the close-ups show both layers in the model which includes an inner layer (GaAs) and an outer layer (InGaAs). To show how the bilayer stress distribution when a tensile force is applied to the structure. A stress state simulation was conducted. A stress state distribution in each layer when a tensile force of 500 nN is applied to the structure is shown in Fig. 6.12b. An initial stress of the structure in zero strain was assumed to be zero. The tensile stress of the inner GaAs layer was increased to 0.802 GPa while the compressive stress of the outer InGaAs layer decreases to −0.743 GPa. The stresss mismatch between the inner and the outer layers definitely increase the torsional effect and is believed to dominantly contribute to the higher piezoresistivity of HNBs. As previously mentioned, the further analysis based on the HNB piezoresistivity model considering many different aspects is out of scope in the literature.

Force calibration of the single HNB in axial direction using as-calibrated atomic force microscope cantilever can be referred from the previous work [20]. Based on the stiffness estimation, force versus current curve of the HNB 3 is shown in Fig. 6.13. The current at 1 V corresponds with the applied force at each point for 4.5 times of repeated compressive and tensile cycle loadings.

In order to show that the HNBs can operate as sensor elements over a long time another experiment was carried out. A HNB without metal connectors was picked up with one probe while another probe was moved close to the free-standing end. When a sinusoidal voltage with an amplitude of 5 V was applied between the two probes the HNB started to oscillate due to the electrostatic force between the HNB and the other probe. Depending on the position of the second probe relative to the HNB axis different modes of oscillation could be induced. A HNB was excited at its resonance frequency both in bending mode (4.9 kHz) and in the axial displacement mode (10.5 kHz).

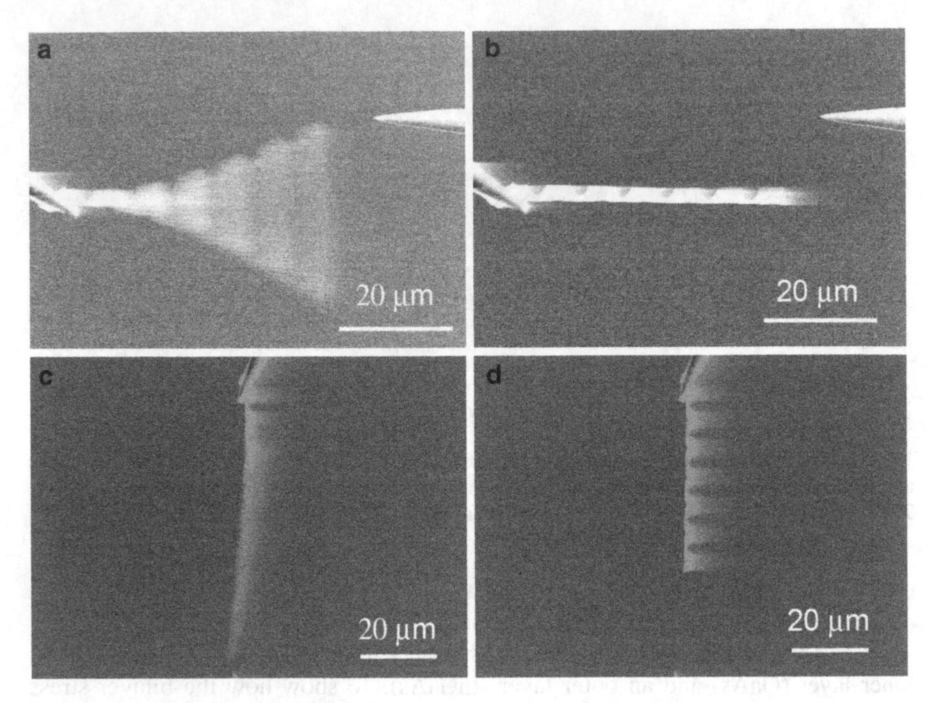

Fig. 6.14 SEM image of InGaAs/GaAs HNB. (**a**)–(**b**) in bending resonance (4.9 kHz). (**c**)–(**d**) in axial displacement resonance (10.5 kHz). Copyright 2010 ACS Nano Letters [11]. The [11] refers to the complete citation in the reference section

Since the resonance frequency did not change during more than 10^6 cycles in both modes it can be assumed that the mechanical properties of the bilayer material do not change with multiple loading. Images from these experiments are shown in Fig. 6.14.

6.3.6 Summary

Longitudinal piezoresistivity of HNBs was clearly characterized. For the characterization of individual HNB, EBID and nanorobotic assembly was used. Longitudinal piezoressitivity showed the giant piezoresistivity compared to the conventional materials including bulk Si, SiNW and CNT. Piezoresistance coefficient in axial direction was -9.95×10^{-8} to -35.6×10^{-8} and it is at least 249–890 times higher than the boron doped Si piezoresistors [33]. It is even higher than the recent report on the giant piezoresistivity of SiNW [17]. During the piezoresistivity characterization, the most important issue was how to properly assemble HNBs between probes assuring the stable electrical conductivity.

6.4 Large Range Mechanical Property Characterization and Calibration of HNBs by Tuning Fork

The objective of this section is to characterize and calibrate the mechanical property of InGaAs/GaAs HNBs in full range by incorporating tuning fork stiffness sensors and SEM imaging. Although mechanical property of HNBs have been attempted to measure using AFM cantilever inside SEM, the measurement range was still limited especially revealing the severe fluctuations within the displacement range of 10 μm due to the limited resolution and range of in situ characterization tools [20]. However it is inevitable to calibrate the full range mechanical property of HNBs for their force sensing applications. Therefore, our concern of this section is to make a clear mechanical property characterizations of HNBs using in situ SEM tuning force stiffness sensors [19].

6.4.1 Principles of Force Gradient Measurement with Tuning Fork for HNB Characterization

This section is divided into three parts. First, the principles of force gradient measurement with tuning fork are presented followed by the HNB model and mechanical properties measured from previous works [20]. Finally a unified model for tuning fork probe attached to HNB is presented.

Force measurement with a tuning fork is possible with either amplitude and phase modulation (AM/PM) or frequency modulation (FM) AFM techniques. For the first, a lock-in amplifier can be utilized in order to separate amplitude and phase from the original signal. From these two signals, the force can be obtained through analytical conversion formule [34, 35]. For the second, an Automated Gain Controller (AGC) and a Phase-Locked Loop (PLL) controller are used to control respectively amplitude and phase, thus frequency shift is obtained. With the frequency shift the gradient of the force can be obtained [36, 37]. Choosing AM or FM depends mainly on the required reaction time τ of the tuning fork. For AM-AFM regulation the reaction time $\tau = Q/(\pi f_0)$ (where f_0 is the resonant frequency and Q the quality factor) is highly dependent on the quality factor. Furthermore, the quality factor is higher in vacuum conditions of the SEM. In consequence, the bandwidth analysis will be limited. FM-AFM removes the time constant dependency [38] of the analysis allowing wide bandwidth with high quality factor which makes it the primary selection for this work [19].

The tuning fork frequency shift can be expressed as [39]:

$$\frac{\Delta f}{f_0} = \frac{1}{A \cdot K_{\text{TF}}} \int_0^{1/f_0} F_{\text{int}}(\omega \cdot t) \cdot \cos(\omega \cdot t) \mathrm{d}t, \tag{6.5}$$

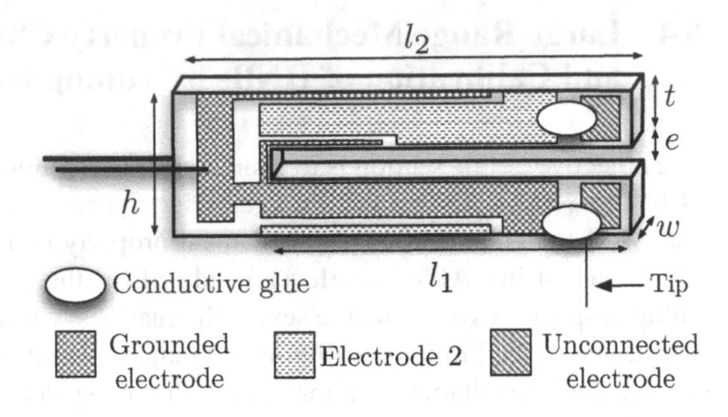

Fig. 6.15 Schematic of Tuning Fork electrodes with glued probe tip

Where K_{TF} is the stiffness of the tuning fork, A the tuning fork mechanical oscillation amplitude, F_{int} the interaction force between the tip of the tuning fork and the sample, f_0 the resonant frequency of the tuning fork and Δf the frequency shift. ω is the angular frequency of the tuning fork.

The stiffness of the tuning fork can be obtained using a geometrical model. The tuning fork dimensions can be measured with a microscope or SEM, thus using a geometrical method is feasible.

$$K_{TF} = \frac{E \cdot w \cdot t^3}{4 \cdot l_1^3}. \tag{6.6}$$

In (6.6), w, t and l_1 are the width, height and the length of tuning fork prong (geometrical parameters of tuning fork can be seen on Fig. 6.15) and $E = 78.7$ GPa, is the Young modulus of the quartz crystal of tuning forks. The stiffness of the tuning fork can be assumed to be constant [40]. In the following, the stiffness model of HNB is presented.

6.4.2 *Stiffness Modeling of HNBs*

InGaAs/GaAs bilayer HNBs were utilized for the experiments. HNBs were fabricated by the process described in [11]. The interaction force F_{int} at the end of the HNB can be expressed with Hooke's law for tiny elongation as follows:

$$F_{int} = K_{HNB} \cdot x, \tag{6.7}$$

Where K_{HNB} is the stiffness of the HNB and x is the elongation. Finite element method (FEM) simulation was used to estimate the deflection by the applied force onto HNBs, thus, obtain the rest (at no elongation) stiffness. The dimensions of the HNBs used are summarized in Table 6.5. The simulation result has previously been validated with experimental results for similar structures [20]. Simulation was carried out in the linear elastic range (small displacements). Values of the materials

Table 6.5 HNB specifications

	HNB 1	HNB 2
Thickness of InGaAs/GaAs (nm)	11.6/15.6	11.6/15.6
Length (μm)	25.4	53.4
Pitch (μm)	3.9	8.9
Number of turns	6.5	6
Stripe width (μm)	1.5	2.5
Diameter (μm)	2	2.5
Longitudinal spring (FEM) K_{long}(N/m)	0.009	0.011

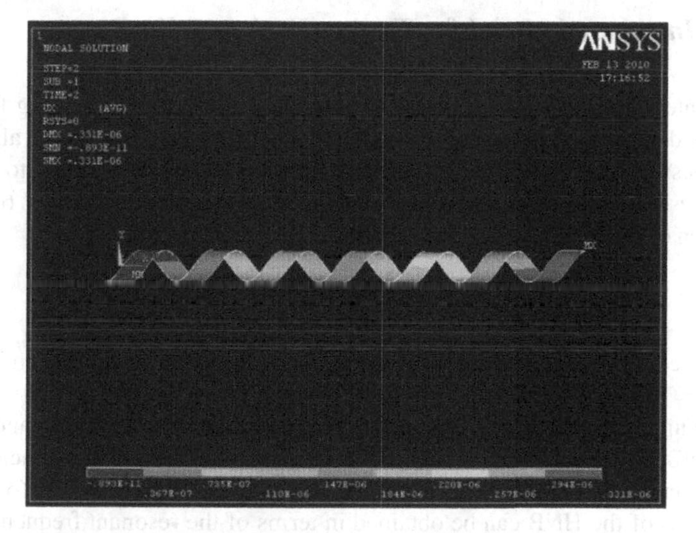

Fig. 6.16 Finite element simulation of HNB for displacement along the longitudinal direction

properties in the model were taken from [20] with the rule of mixture applied for the InGaAs layer. Both ends of the helix were constrained from rotation around all three axes. Moreover, on one end it was constrained from all translational movements, and on the other end it was constrained from translational movement perpendicular to the axis. On this end, a force in the longitudinal (X-axis) direction was applied to compute the displacement. In Fig. 6.16, a plot of the displacement along the longitudinal direction of the HNB is shown. From the simulation, longitudinal stiffness is estimated to be 0.009 N/m for HNB 1 and 0.011 N/m for HNB 2 as summarized in Table 6.5.

This simulation demonstrates rest position stiffness of the HNB. Nevertheless, non-constant behavior of the stiffness for upper elongation range was demonstrated by previous experimental works [20] with AFM cantilever under SEM. However, in previous works, full range measurement was not attempted due to the lack of wide range force sensing. As K_{HNB} is not constant, the stiffness obtained with FEM can only be used for very small displacements. Thus, it is important to characterize HNB in full range. In the following, an unified model including tuning fork and HNB is presented.

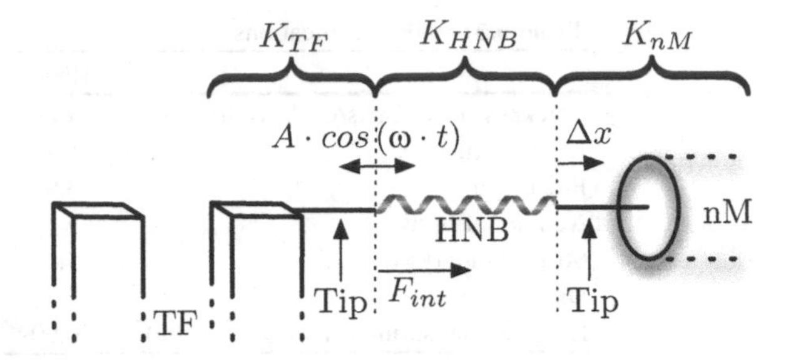

Fig. 6.17 Schematic of the interacting elements during the experiments. TF and nM stand for Tuning Fork and nanoManipulator

6.4.3 *Integrated Model for Tuning Fork and HNB*

For the integrated model for tuning fork and HNB, the tip of the tuning fork is in contact and aligned with one end of the HNB (Fig. 6.17). Because of the alignment, the stiffness of the tuning fork tip and manipulator tip are not taken into account. Further description on the mechanical configuration can be seen on Sect. 6.4.4. The elongation of the HNB for the integrated system can be expressed as:

$$x = A \cdot \cos(\omega \cdot t) + \Delta x, \tag{6.8}$$

Where $A \cdot \cos(\omega \cdot t)$ is contribution from the oscillation amplitude of the tuning fork probe, $f = \omega/(2 \cdot \pi)$, and Δx is, the linear elongation of the HNB.

In the unified model, due to the contact between the tuning fork tip and the end of the HNB, the interaction force F_{int} of HNB model (6.7) and displacement x (6.8), can be replaced in the frequency shift of the tuning fork (6.5) resulting in (6.9) where the stiffness of the HNB can be obtained in terms of the resonant frequency of the tuning fork, the frequency shift and the stiffness of the tuning fork.

$$K_{\text{HNB}} = \frac{2 \cdot \Delta f \cdot K_{\text{TF}}}{f_0}, \tag{6.9}$$

As A, K_{TF} and f_0 are constant, it is noticeable from previous equation that the measured frequency shift will reveal the stiffness behavior of the HNB.

6.4.4 *Tuning Fork Mechanical Characterization System*

In situ SEM tuning fork mechanical property characterization system is presented. For the oscillation control of the tuning fork and data acquisition, an OC4-Station from SPECS-Nanonis was used. This station has the advantage of having a lock-in amplifier, PLL, AGC, data acquisition hardware and software and real time operating system. The electronic preamplifier for the tuning fork was specially

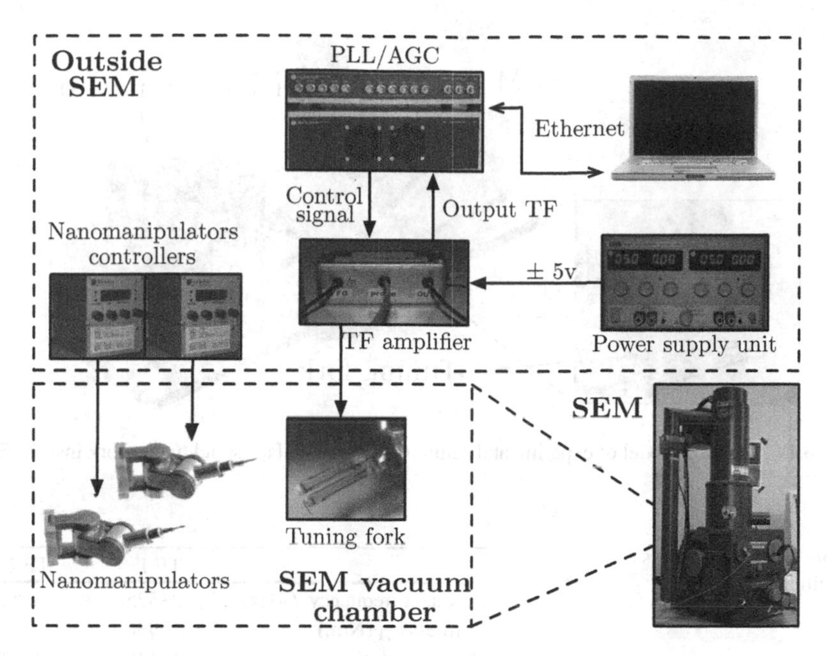

Fig. 6.18 System configuration of used hardware inside and outside the SEM

designed for use in SEM imaging conditions. A forthcoming article will describe in details this electronics. A TT*i* EX752M multi-mode Power supply unit was used with fixed +/−5v for the tuning fork electronic preamplifier. The detailed experimental setup is shown in Fig. 6.18. The main advantage of this system is that all the electronics for tuning fork and for the manipulators are outside the SEM chamber, avoiding influence from electron beam and space occupation.

For vacuum environment and visual feedback an SEM (Leica stereoscan 260 cambridge instruments) is used. Two nanomanipulators (MM3A-EM,Kleindiek) were used for manipulation of HNBs. Further details on the role of the manipulators and the manipulation procedure is described in Sect. 6.4.6. Each has 3 degrees of freedom and respectively 5 nm, 3.5 nm, and 0.25 nm resolution at the tip in X, Y and Z axis. Each axis is actuated with piezo stick-slip principle and is controlled via open loop piezo controller. Configuration of the manipulators and tuning fork inside the SEM chamber can be seen in Fig. 6.19.

The tuning forks were manufactured by Citizen America – CFS206 32.768 KDZB-UB. A tip is attached to the tuning fork in order to fix the HNB. Tips were glued to the tuning fork with conductive EPOTEK glue for grounding the tips (Fig. 6.15). Picoprobes, tungsten tips (T-4-10-1 mm, tip radius: 100 nm, GGB industries) and tips made with platinum iridium Pt90/ir10 wires were used for the nanomanipulator and tuning fork, respectively. In the following, the preparation of tuning forks as well as the procedure for attaching probes to them is presented.

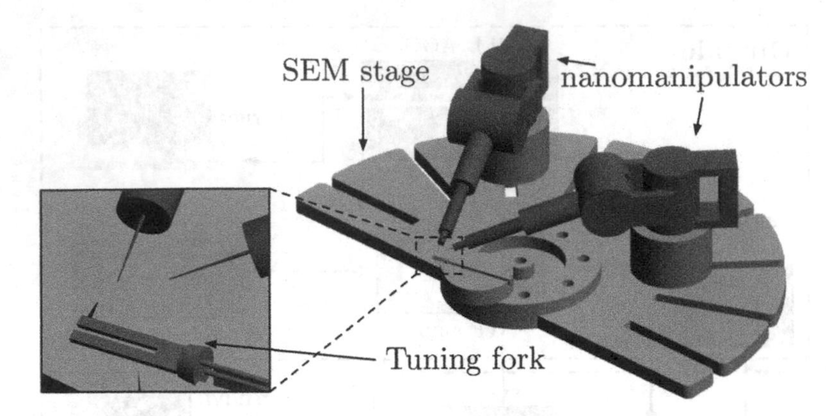

Fig. 6.19 3D CAD model of experimental setup of nanomanipulators and tuning fork inside SEM chamber

Table 6.6 Tuning forks specifications

	TF 1	TF 2
Resonant frequency f_0(Hz)	28325.5	30895.2
Stiffness k_{TF}(N/m)	7936	7936
Quality factor	11145	19800
Prong length l_1 (μm)	3204	3204
Prong height t(μm)	382	382
Prong width w(μm)	238	238

6.4.5 Tuning Fork Probe Preparation

Several things have to be considered before adding the tip. The quality factor of the tuning fork should remain as high as possible to obtain the higher sensitivity. It is based on the weight balancing between the two prongs. Any weight added on one of the prongs should be compensated by the other one to avoid decreasing of the quality factor [40]. As shown in Fig. 6.15, for grounding with prong of the tuning fork, conductive glue is used to attach the tip, thus avoid electrostatic charging by electron beam inside the SEM. Glue needs to be added for weight compensation on the other prong of the tuning fork, it can be done with either conductive or non conductive glue. Using conductive glue avoids charging raise by electron beam, however it will increase the risk of making short circuit between the interdigitated electrodes during deposition. Nevertheless, as the electron beam is mainly focused and zoomed to the tip of the probe, the other prong of the tuning fork has few risk of charging. The geometry information and the estimated stiffness of the two tuning forks are summarized in Table 6.6.

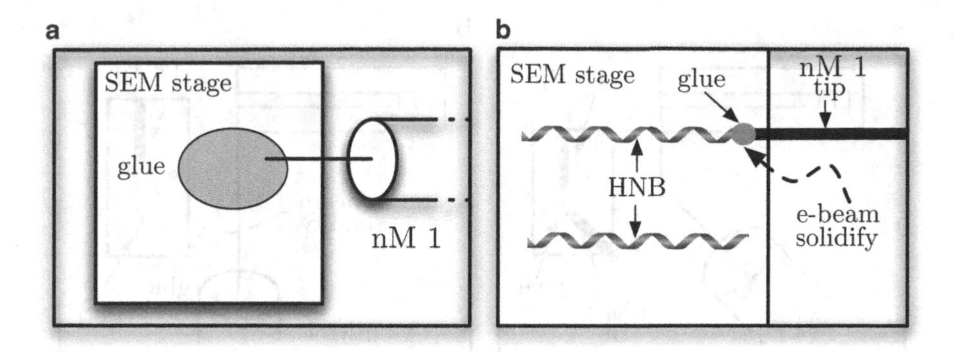

Fig. 6.20 Protocol for picking the HNB with manipulator tip. (**a**) Adding glue to the probe of the nanomanipulator. (**b**) Picking HNB from subtract with the nanomanipulator probe and soldering with e-beam. nM1 and nM2 are nanomanipulator 1 and 2 respectively

6.4.6 Assembly of HNB and Tuning Fork

Once the tip is attached to the tuning fork, the next step consists in assembling the HNB between one manipulator (for motion) and the tuning fork (sensing tool). As tuning fork is very sensitive to motion, it was mechanically fixed on top of the SEM stage. In order to pick up the HNB, the free end of the HNB is attached to the tip of the manipulator. For this purpose, the probe of the manipulator is dipped into EPO-TEK H21D (Epoxy Technology) glue (Fig. 6.20a), then it is approached to contact the end of the desired HNB (Fig. 6.20b). The SEM electron beam is focused onto the glue to solidify it.

Due to the big difference between the HNB lateral and longitudinal stiffness, compression is very challenging. In consequence, the HNB has to be elongated. For this purpose, two different experiments were carried out. For the first, the HNB is fixed to the tip of the TF. For this, glue is added to the tip of the tuning fork with the second nanomanipulator (Fig. 6.21a). Then, the HNB is approached to contact the tip of the tuning fork. The SEM electron beam is used to solidify the glue (Fig. 6.21b). With this technique, full range characterization of the HNB can be done, however, the HNB has to be destroyed eventually to disconnect system. For the second experiment, electrostatic and van der Waals forces are used to maintain the HNB attached to the probe of the tuning fork. This is basically the same configuration without the application of chemical glue between the tuning fork probe and HNB. In addition, it is a non destructive technique for the HNB and can be repeated several times. The experiments and results from these two methods are shown.

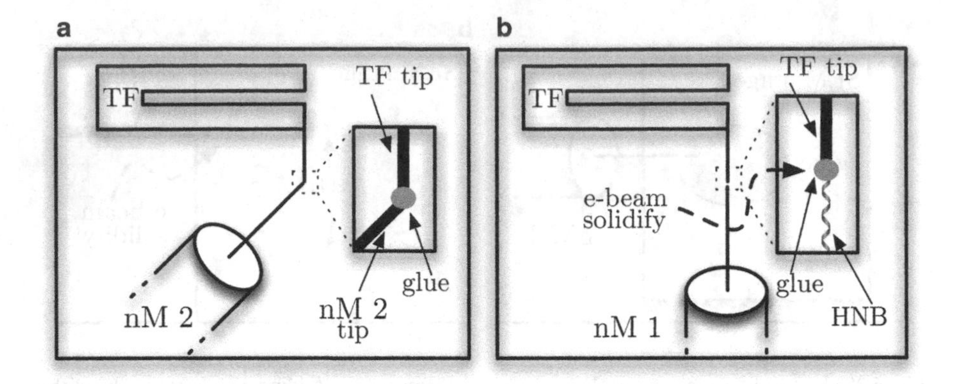

Fig. 6.21 Attachment of the HNB to the tuning fork's tip. (**a**) Adding glue to tip end of the tuning fork probe (**b**) Attaching of HNB to tuning fork tip and soldering with e-beam. nM1 and nM2 are nanomanipulator 1 and 2 respectively

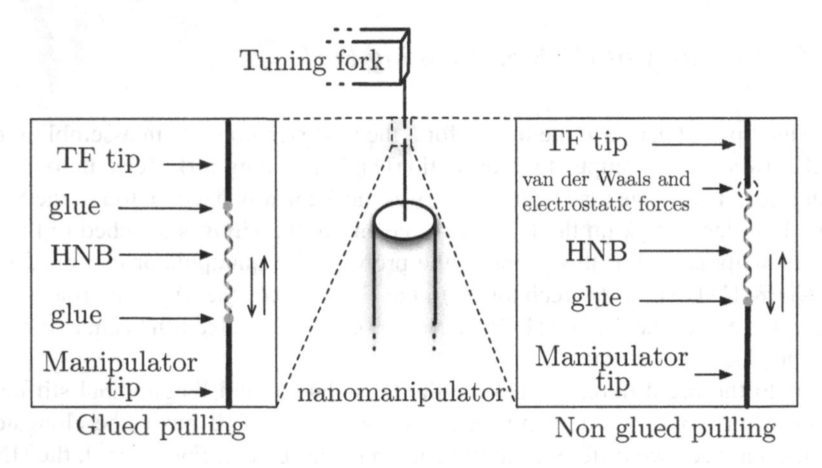

Fig. 6.22 Experimental configurations for longitudinal pulling. (**a**) HNB glued to the tip of tuning fork and (**b**) HNB not glued to the tip of tuning fork

6.4.7 *Full Range Characterization: HNB Attached to the Tuning Fork Probe*

For the first experiment, the nanomanipulator is moved by steps in the longitudinal direction of the HNB (which is aligned with the tuning fork and nanomanipulator tips), the procedure is described in Fig. 6.22a. The movement of the tip of the nanomanipulator generates a force in the axial direction of the HNB which elongated it until it breaks. This way full range for stiffness characterization can be obtained. The data flow can be seen in detail in Fig. 6.23. The tuning fork frequency shift is recorded during the entire experiment. Raw data and selected points for experiments can be seen in Fig. 6.24a. This figure stands the fact that the

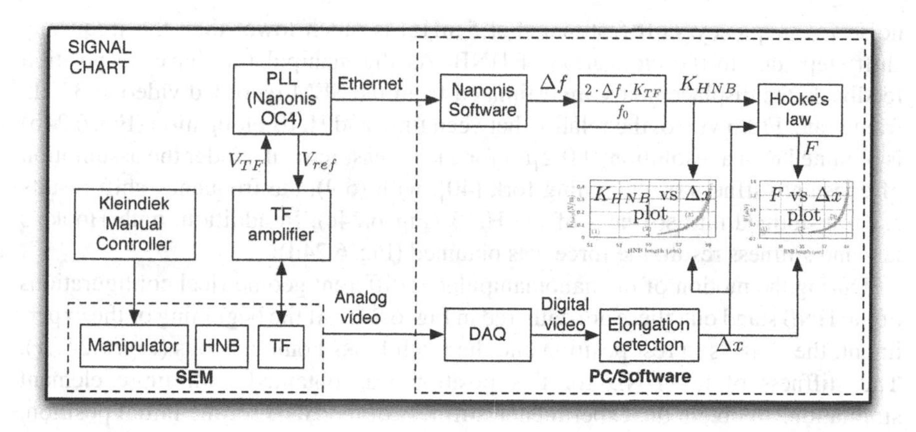

Fig. 6.23 Experiments data flow for plot generation. The tuning fork fork frequency shift obtained thanks to the PLL with amplitude controller on, is then transformed to stiffness with (6.6), which is transformed to force with Hooke's law. SEM video feedback is imported to a computer with a data acquisition card. The elongation of the HNB is estimated from video acquisition

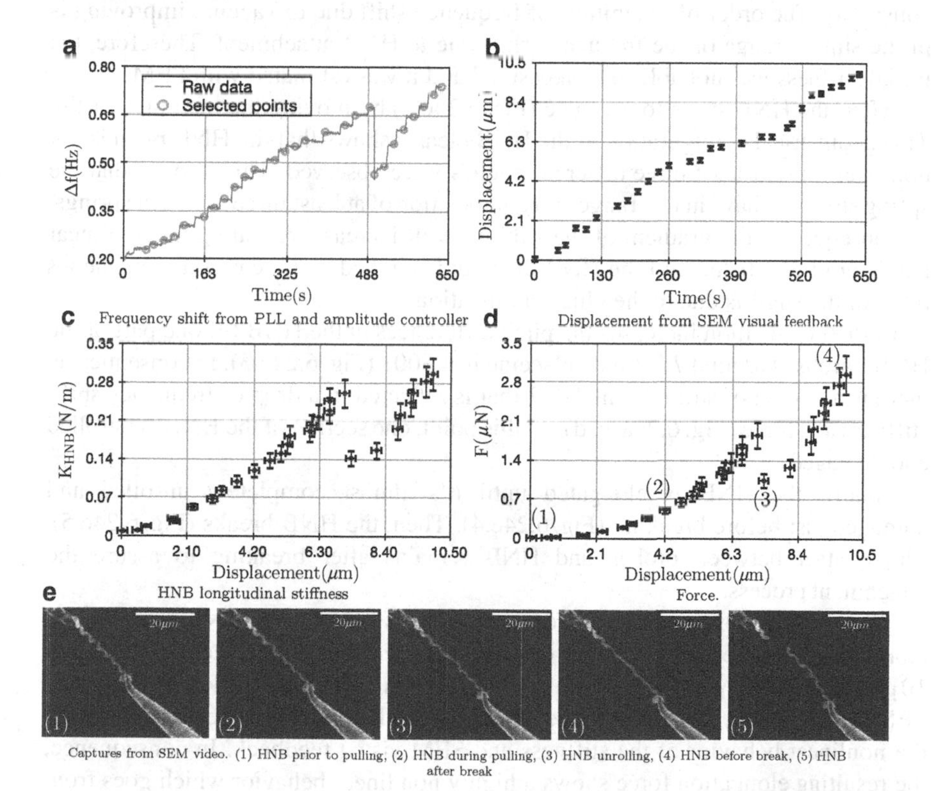

Captures from SEM video. (1) HNB prior to pulling, (2) HNB during pulling, (3) HNB unrolling, (4) HNB before break, (5) HNB after break

Fig. 6.24 Full range longitudinal pulling experiments HNBs: (**a**) frequency shift from phase-locked loop and amplitude gain controller, (**b**) measured displacement of HNB, (**c**) longitudinal stiffness of HNB, (**d**) measured force, and (**e**) screen shots of longitudinal pulling of HNBs

noise for frequency shift (estimated at 5 mHz) is much lower than the frequency shift steps due to the elongation of HNB. As the manipulators have no position feedback, the displacements are estimated from the SEM recorded video at 33 Hz frame rate. From video, the relation between time and HNB elongation (Fig. 6.24b) is obtained with a resolution of 0.2 μm for each measurement. Under the assumption of constant stiffness of the tuning fork [40], with (6.9), the frequency shift results are transformed into stiffness of the HNB (Fig. 6.24c). In addition, with Hooke's law and stiffness result, the force was obtained (Fig. 6.24d).

During the motion of the nanomanipulator, different geometrical configurations of the HNB stand out, these are gathered in Fig. 6.24e. At the beginning of the experiment, the HNB is in rest position and the pitch looks homogeneous (Fig. 6.24e.1). The stiffness of the HNB for this position was obtained with finite element simulation. To obtain the experimental stiffness of the HNB for the initial position, the difference between tuning fork resonant frequency before and after HNB attachment needs to be obtained with glue already applied to the tip of the tuning fork. However one of the main problems for this measurement was that the vacuum condition of the SEM improved during time making the resonant frequency increase constantly. The order of magnitude of frequency shift due to vacuum improving is in the similar range of the frequency shift due to HNB attachment. Therefore, the initial stiffness was not able to be measured and it was estimated with FEM.

After, the HNB starts to elongate (Fig. 6.24e.2) by moving the manipulator tip. This geometrical configuration of the HNB clearly shows that the HNB pitch is not homogeneous, at least three different pitches were observed. This implies that the spring start to behave in this range as a composition of at least three different springs. In consequence, the gradient of the stiffness will increase, revealing the nonlinear behavior of the stiffness of the HNB. It should be noted that the non-homogeneous pitch of the HNB is due to the glue solidification.

Further elongation increases the pitch differences in the HNB till one part of the HNB unrolls at around 7.3 μm displacement at 500 s (Fig. 6.24e.3). In consequence, there is a release of strain in the HNB that is reflected in a drop of frequency shift, stiffness and force (Fig. 6.24a, c, d). At this point, one section of the HNB is unrolled and damaged.

Finally, the HNB is elongated until it's almost completely unrolled and damaged just before breaking (Fig. 6.24e.4). Then, the HNB breaks (Fig. 6.24e.5). The contact between probes and HNB remains after breaking to assure the attachment process.

These results confirm the non-constant stiffness behavior of HNB in full range elongation. This behavior was not clearly measured for displacement of less than 10 μm in previous works [20] with atomic force microscope cantilevers inside the SEM. Furthermore, the non-homogeneous pitch of this HNB has been revealed with the nonlinear behavior of the stiffness and SEM visual feedback. In consequence, the resulting elongation force shows a highly non linear behavior which goes from 14.5 nN for the smallest step done, to 2.95 μN before breaking, showing the wide range sensing of the system.

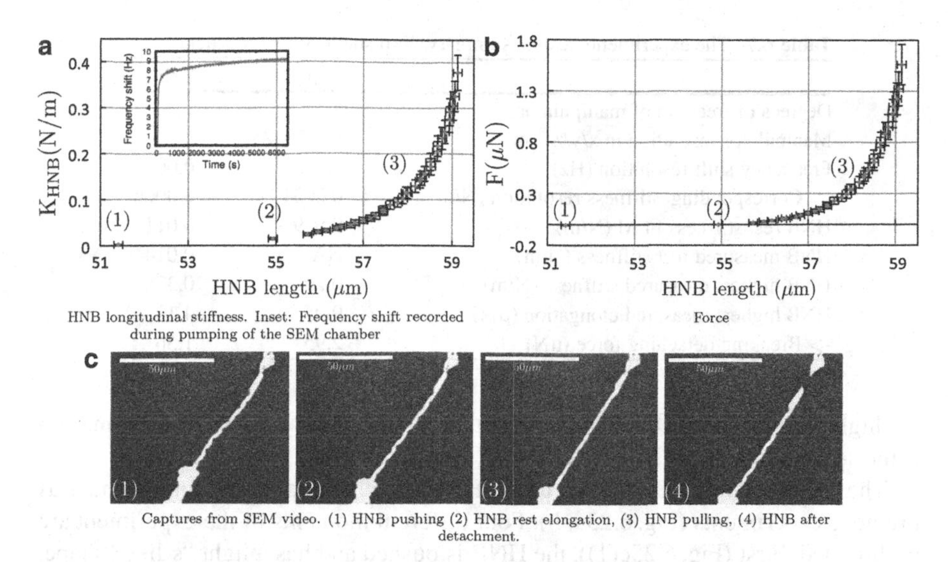

Fig. 6.25 Non destructive longitudinal pulling experiments of HNBs: (**a**) longitudinal stiffness of HNB, (**b**) measured force, and (**c**) screen shots of longitudinal pulling of HNB [19]

6.4.8 Non-Destructive Characterization

The previous experiment had mainly three limitation, we aim to solve in this second experiment. First the vacuum conditions were constantly affecting the resonant frequency of the tuning fork with an increasing offset through time. Second, it is a destructive method where the HNB is destroyed after the experiment, furthermore, the tips of the tuning fork and manipulator can be contaminated with glue and remained parts of HNB. In addition to this, the attaching of the HNB in both ends prevents its rotation during tensile elongation, eventually damaging it. Finally, the SEM video is analysed manually. This means that for every frame of interest, the elongation of the HNB is estimated with manually placed points in video.

In order to solve the first problem, vacuum characterization is done prior to the experiment to identify the saturation time after pumping where the variation of vacuum conditions will not affect the experiment. For this purpose, frequency shift is recorded during the pumping process (Inset Fig. 6.25a). After ninety minutes, the frequency shift drift due to vacuum conditions is small enough for a ten minute experiment.

To overcome the second problem, as despited in Sect. 6.4.4, electrostatic and van der Waals forces are used to maintain the HNB attached to the tip end of the tuning fork. For this, the HNB surface charge was increased with the electron beam. Furthermore, only one manipulator is needed for this experiment. Frequency shift is recorded in the same way as previous experiment and the higher quality factor of the second tuning fork makes the frequency shift noise decrease from 5 to 1 mHz. For the elongation of the HNB, offline visual tracking software. This, in addition to

Table 6.7 The experiments results summary. Exp stands for experiment

	Exp 1	Exp 2
Degrees of freedom of manipulator	3	
Manipulator resolution in x/y/z (nm)	5/3.5/0.25	
Frequency shift resolution (Hz)	0.005	0.001
-> Corresponding Stiffness resolution (N/m)	0.0031	0.0006
HNB rest stiffness FEM (N/m)	0.009	0.011
HNB measured rest stiffness (N/m)	NA	0.014
HNB highest measured stiffness (N/m)	0.297	0.378
HNB highest measured elongation (μm)	9.95	4.13
-> Breaking/detaching force (μN)	2.95	1.56

the high contrast used in the SEM, makes the error decrease from 0.2 to 0.1 μm even if the scale passed from 20 to 50 μm.

The stiffness and force applied to the HNB are obtained in the same manner as previous experiments (Fig. 6.25a, b). Four different moments of the experiment are highlighted. First (Fig. 6.25c(1)), the HNB is pushed and has a light "s-like" shape. This is mainly due to the much lower lateral stiffness than longitudinal stiffness of the HNB. In consequence the stiffness measured is composed of both lateral and longitudinal and its absolute value is lower than longitudinal rest position stiffness. The elongation being negative results in a negative force vector. After, the HNB is elongated to the rest position (Fig. 6.25c(2)). and then, elongated (Fig. 6.25c(3)) till it detach from the probe of the tuning fork (Fig. 6.25c(4)).

It is noticeable in comparison to the previous experiment the higher non-linearity. This is due to the non-constrained rotation of the end of the HNB in contact with the tuning fork. In consequence the HNB freely adjust its number of turns through elongation, avoiding damages and looses of helicoidal shape. Furthermore, the estimated force before releasing, 1.56 μN, correspond to the addition of van der Waals and electrostatic forces. The experiments results are gathered in Table 6.7.

6.4.9 Summary

The full range mechanical characterization and calibration of HNBs are inevitable to their force sensing applications. In situ SEM force and stiffness characterizations have been demonstrated by incorporating tuning forks and HNBs. The measurement system consists of SEM, two micromanipulators, a fixed tuning fork and attached HNBs. Tuning fork is used to mechanically characterize the stiffness shifts of HNBs. Then the deflections of HNBs are measured from SEM imaging to obtain the direct force. Two experiments have been demonstrated. For the first experiment, it was fixed between the tips of the tuning fork and manipulator for full range characterization. For the second experiment, electrostatic and van der Waals forces are used to maintain attached the HNB to the tip of the tuning fork, this way non destructive characterization can be done. The non-constant stiffness behavior of

HNBs during their controlled tensile elongation was clearly revealed in full range for the first time to the best of our knowledge. Furthermore, the revealed non-linear behavior of the stiffness with SEM visual feedback shows the capability of the proposed system to understand the mechanical properties of the nanostructure due to geometry deformation.

The obtained stiffness of HNB ranges from 0.009 to 0.297 N/m during full elongation and 0.011–0.378 N/m for the non-destructive method. It was transformed with Hooke's law into forces as high as 2.95 μN for the first experiment and 1.56 μN for the second. The minimum steps of frequency shift measured are more than five times higher than noise levels, in consequence, the stiffness measurement resolution of the system with this specific tuning fork is around 0.0031 N/m for the first, and 0.0006 N/m for the second. Changing the tuning fork with a less stiff one can dramatically improve the resolution. Even though, the main limit of the system is the resolution of SEM visual analysis. Taking pictures and not videos from SEM to estimate with more accuracy the elongation of the HNB and obtain stiffness for smaller displacements can be done. However, for long scanning times, the HNBs are too much exposed to e-beam thus will deposit to increase mass of HNB due to contaminants (mainly carbon) deposition. Furthermore, the nanomanipulator can be installed on top of a close loop controlled xyz piezo nanostage to obtain the displacement with more accuracy and increase the dexterity and resolution of the system.

Dynamic mechanical characterization of other ultra flexible nanostructures like nanowires, nanotubes and graphene membranes for example are possible in the future with the proposed system. Moreover, the dynamic measurement in addition to the dexterity of the system make it ideal for example for measuring the dynamic oscillation mode of membranes for optical micro-mirror applications. Furthermore, by incorporating environmental electron microscopes (ESEM) or fluorescence optics, flexible and elastic biological nanostructures such as DNA, proteins, cells, tissues are also in the scope of this new system.

6.5 Application Example: Nanowire Characterizations

As an application of HNBs as force sensor, we demonstrate mechanical property characterization of newly synthesized tungsten nanowires. Tungsten nanowires are defined as the 1 dimensional nanostructures (100 nm to 1 μm long and 20–100 nm in diameter) synthesized from sputtered or chopped tungsten material with approximately 800°C and 1×10^{-2}Pa ambient pressure [42,43]. Tungsten nanowires are expected to be used as emitters of field emission display from their easiness of synthesizing by just heating the tungsten substrate and high aspect ratio. Tungsten nanowire is a very promising potential building blocks for the future NEMS such as parallel aligned nanowire field-emitters [45], vertically aligned nanowire tweezer, and other electromechanical sensors. However, these tungsten nanowires vary in length, direction, and density during their growth. Therefore, it is necessary to control these parameters to apply tungsten nanowires to devices.

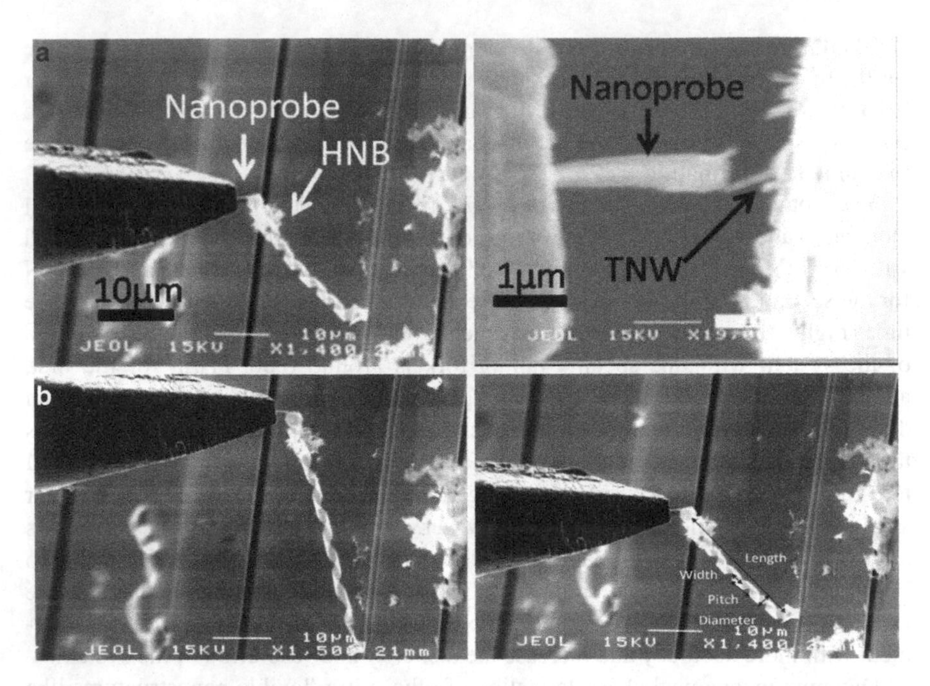

Fig. 6.26 Mechanical property characterization of tungsten nanowire using HNB and nanosoldering: (**a**) experimental view of nanorod calibration by HNB and nanowire calibration by nanorod, (**b**) full tensile elongation of HNBs and neutral position of HNBs with parameters

The first target is aimed to control the growth direction. Then, it is inevitable to characterize the mechanical property of single tungsten nanowire to control the growth direction of their array. However, the extremely small 1-dimensional nanowires grown from the flat chip surface is difficult to calibrate the single nanowire. There is no nanostructure or electromechanical sensors with well-known mechanical properties in this scale. Therefore, our approach is to use HNBs as a calibration tool for nanowires. Since both nanostructures are different in scale, we proposed an indirect way of calibration as shown in Fig. 6.26. We used the focused ion beam structured nanorod interfacing between the nanowire and HNBs. For the fabrication of nanorod, a tungsten rod (Nilaco co.) with 0.3 mm diameter was electro-polished to make a sharpen probe tip. Then, the probe was applied with focused ion beam to make the tip in geometry with $200 \times 200 \times 2{,}100$ nm cube. Figure 6.26 shows the nanorod is calibrated first by the elongation of HNBs. The probe was mounted onto the built-in pizeo actuated manipulator (SMM-7801, Sanyu co.) inside the SEM (JSM-6301F, JEOL co.). The manipulator has 3 degrees of freedom actuation with 1 nm position resolution. To measure the stiffness of the probe, it is necessary to use as-calibrated or known structure with the capability to measure the deflections of both structures inside SEM. Therefore, both experimentally and modeled InGaAs/GaAs HNBs were used for these

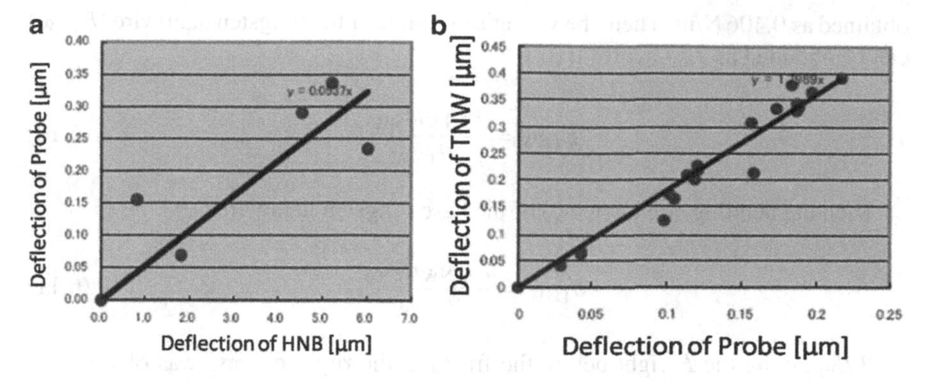

Fig. 6.27 Measurement of nanowire mechanical property, (**a**) nanorod calibration result using HNB, (**b**) nanowire calibration result using as-calibrated nanorod

experiments [20]. To simplify the experiments, the assembled force sensor from double HNBs was not used. However, for the real time measurement of the direct force while the calibration, the assembled HNB force sensor can be used for the experiment.

The chip with HNBs was mounted to the SEM sample holder. When the nanorod was attached to the end of HNBs, they were soldered together. The maximum deflection was tested to assure that the stable mechanical interconnection was established by manipulating the attached nanorod and the HNB together (Fig. 6.26). Then each deflection of the HNB and the nanorod was recorded to estimate the stiffness of the probe. When the stiffness of the nanorod was estimated, the same nanorod was used to calibrate the tungsten nanowire. Another type of tungsten probe (Picoprobe, T-4-10-1 mm, tip radius: 100 nm) was used to synthesize the tungsten nanowires on its surface. The picoprobe was sonicated in an ultra sonic bath for 3 min with aceton and isoprophanol, respectively. When the vacuum chamber was pumped down to 1.0×10^{-2} then the oxygen gas was injected. The temperature was increased to 800°C for 10 min then naturally cooled down. The synthesized tungsten nanowires was observed in SEM then another image was taken to obtain the 3 dimensional coordinates by rotating the sample stage. Then as-calibrated nanorod was attached to one of the tungsten nanowires on the tungsten probe for the calibration (Fig. 6.26). Figure 6.27 shows the linear curve fitted result plot of the nanorod versus HNBs deflection and the tungsten nanowire versus the nanorod deflection respectively. In this experiment, the tungsten nanowire was taken off the nanorod after its deflection of 0.217 μm. From this result, the mechanical property of the tungsten nanowire was obtained. External force, the stiffness coefficient, deflection, length, geometric moment of inertia and young's modulus were F, k, x, l, I and E respectively. Prior to consider the nonlinear mechanics of HNBs, the stiffness of HNB (k_{HNB}) was estimated as 0.021 N/m at rest position from the simulation. From the experiment and Hook's law, the stiffness of nanorod (k_{NR}) was obtained as 0.406 N/m. Then, the young's modulus of the tungsten nanowire (E_{TNW}) can be obtained as 7.5 Gpa from (6.10).

$$E_{\mathrm{TNW}} = \frac{k_{\mathrm{TNW}} l_{\mathrm{TNW}}^3}{2 I_{\mathrm{TNW}}}. \quad (6.10)$$

Then the bending stress (σ_{TNW}) of the base tungsten nanowire is,

$$\sigma_{\mathrm{TNW}} = \frac{F l_{\mathrm{TNW}} d_{\mathrm{TNW}}}{2 I_{\mathrm{TNW}}}. \quad (6.11)$$

By applying the F right before the fracture, the rupture stress was obtained as 18 Gpa. Then young's modulus of the bulk tungsten is 410 Gpa and the one of the tungsten nanowire was reported as 100–300 Gpa [44]. It is quite different from the one obtained by the experiment. It can be explained by two facts. First, the nonlinear mechanics of HNBs in tensile elongation could increase their stiffness drastically from the neutral one at no elongation. It should be noted that the stiffness of HNBs (k_{HNB}) was estimated at rest condition. Our experiments on other HNBs in Sect. 6.4 from same material revealed their stiffness increases from 0.011 to 0.378 N/m at maximum elongation of 4.13 μm. Considering these non-constant stiffnesses of HNBs, the stiffness of nanorod (k_{NR}) is obtained as 13.951 N/m thus the young's modulus of the tungsten nanowire (E_{TNW}) can be obtained as 257.7 Gpa at such elongation. The large range (100–300 Gpa) of tungsten nanowire mechanical property is in consistency to the reported results [44]. Furthermore, the mechanics of the tungsten nanowire can also be affected by the deposition of any contaminants inside SEM during the elapsed time taken to finally measure the probe. This was also considered to increase the stiffness k_{NR} higher than the initial condition. To prove it, SEM observation of the nanorod was done and its width was increased around 30% between two experiments. To minimize the error, exposure time under electron beam should be reduced. Figure 6.27 also shows that the tungsten nanowire was in elastic deflection from the facts that it was recovered back to the original position after experiment and the values are all in the linear curve. Therefore, it can be said that the tungsten nanowire was in elastic deflection till the base of tungsten nanowire was taken off. From these facts, tungsten nanowires experience a brittle fracture and not going to plastic deformation before the fracture. Therefore, doing direction control only by applying external force is difficult to achieve. In summary, Young's modulus and the rupture stress of tungsten nanowires are measured and they are the structure showing a brittle fracture.

6.6 Conclusions

In situ characterizations of thin-film nanostructures have been investigated for their NEMS applications. Piezoresistivity of HNBs have been characterized and demonstrated their effectiveness for the tools of handling, structuring, and characterizing other nanostructures such as nanowires. Furthermore, non-constant stiffness of 3-D HNBs have been clearly revealed within the displacement range of 10 μm using the tuning fork force gradient sensing. The demonstrated large range force sensors provides the possibility for in situ active property characterization, structuring, assembly of nanomaterials and nanostructures. These sensors based on smart sensing mechanisms in combination with nanomanipulation enable the construction of NEMS based on ultra-flexible nanostructures and their calibrations and property characterizations.

References

1. Chaillet, N., Regnier, S.: Microrobotics for Micromanipulation. John Wiley and Sons (2010)
2. Dong, L., Nelson, B.: Tutorial - Robotics in the small part II: Nanorobotics. Rob. Auto. Mag. IEEE. **14(3)**, 111–121 (2007)
3. Nelson, B., Dong, L., Subramanian, A., Bell, D.: Hybrid nanorobotic approaches to NEMS. Rob. Res. 163 174 (2007)
4. Nagato, K., Kojima, Y., Kasuya, K., Moritani, H., Hamaguchi, T., Nakao, M.: Local synthesis of tungsten oxide nanowires by current heating of designed micropatterned wires. Appl. Phys. Exp., **1**, 014005 (2008)
5. Novoselov, K. S., Geim, A. K., Morozov, S. V., Jiang, D., Zhang, Y., Dubonos, S. V., Grigorieva, I. V., Firsov, A. A.: Electric field effect in atomically thin carbon lms. Science, **306(5696)**, 666–669 (2004)
6. Sun, Y., Nelson, B. J., Greminger, M. A.: Investigating protein structure change in the zona pellucida with a microrobotic system. Int. J. of Rob. Res., **24(2-3)**, 211–218 (2005)
7. Sun, Y., Nelson, B. J.: MEMS for cellular force measurements and molecular detection. J. Info. Acq., **1(1)**, 23–32 (2004)
8. Xie, H., Vitard, J., Haliyo, S., Regnier.: Calibration of lateral force measurements in atomic force microscopy with a piezoresistive force sensor. Rev. Sci. Inst., **79**, 033708 (2008)
9. Ashkin, A., Dziedzic, J. M., Bjorkholm, J. E., Chu, S.: Observation of a single-beam gradient force optical trap for dielectric particles. Opt. Lett., **11**, 288-290 (1986)
10. Pacoret, C., Bowman, R., Gibson, G., Haliyo, S., Carberry, D., Bergander, A., Regnier, S., Padgett, M. Touching the microworld with force-feedback optical tweezers. Opt. Exp., **17(12)**, 10260 (2009)
11. Hwang, G., Hashimoto, H., Bell, D. J., Dong, L. X., Nelson, B. J., Schn, S.: Piezoresistive InGaAs/GaAs Nanosprings with Metal Connectors. Nano Lett., **9(2)**, 554–561 (2009)
12. Karrai, K., Grober, R. D.: Piezoelectric tip-sample distance control for near field optical microscopes. Appl. Phys. Lett. **66**, 1842 (1995)
13. Niyogi, S., Thamankar, R. M., Chaiang, Y., Kawakami, R., Myung, N. V., Haddon, R. C.: Magnetically Assembled Multiwalled Carbon Nanotubes of Ferromagnetic Contacts. J. of Phys. Chem. B, **108**, 19818–19824 (2004)
14. Bentley, A., Trethewey, J., Ellis, A., Crone, W.: Magnetic manipulation of copper-tin nanowires capped with nickel ends. Nano Lett., **4**, 487–490 (2004)
15. Hwang, G., Hashimoto, H.: Development of a Human-Robot-Shared Controlled Teletweezing System. IEEE Trans. Cont. Sys. Tech., **15**, 960–966 (2007)
16. Park, S. J., Goodman, M. B., Pruitt, B. L.: Analysis of nematode mechanics by piezoresistive displacement clamp. Proc. of the Natl. Aca. of Sci., **104**, 17376–17381 (2007)
17. He, R., Yang, P. D.: Giant piezoresistance effect in silicon nanowires. Nat. Nanotech., **1**, 42–46 (2006)

18. Toriyama, T., Funai, D., Sugiyama, S.: Piezoresistance measurement on single crystal silicon nanowires. J. of Appl. Phys., **93**, 561–565 (2003)
19. Acosta, J. C., Hwang, G., Polesel-Maris, J., Regnier, S.: A tuning fork based wide range mechanical characterization tool with nanorobotic manipulators inside a scanning electron microscope. Rev. Sci. Instrum., **82**, 035116 (2011)
20. Bell D. J., Dong, L., Zhang, L., Golling, M., Nelson, B. J., Gruetzmacher, D.: Fabrication and characterization of three-dimensional InGaAs/GaAs nanosprings. Nano Lett., **6**, 725–729 (2006)
21. Madou, M. J.: Fundamentals of Microfabriation. CRC press, (2006)
22. Timo, S., Choi, T., Schirmer, N., Bieri, N., Burg, B., Tharian, J., Sennhauser, U., Poulikakos, D.: A dielectrophoretic method for high yield deposition of suspended, individual carbon nanotubes with 4-point electrode contact. Nano Lett., **7**, 3633–3638 (2007)
23. Schroder, D. K.: Semiconductor Material and Device Characterization. New Jersey: Willy-Interscience (2006)
24. Stampfer, C., Helbling, T., Obergfell, D., Scholberle, B., Tripp, M. K., Jungen, A., Roth, S., Bright, V. M., Hierold, C.: Fabrication of Single-Walled Carbon-Nanotube-Based Pressure Sensors. Nano Lett. **6**, 233–237 (2006)
25. Stampfer, C., Jungen, A., Linderman, R., Obergfell, D., Roth, S., Hierold, C.: Nano-electromechanical displacement sensing based on single-walled carbon nanotubes. Nano Lett., **6**, 1449–1453 (2006)
26. Molhave, K., Madsen, D., Dohn, S., Boggid, P.: Constructing, connecting and soldering nanostructures by environmental electron beam deposition. Nanotech., **15**, 1047–1053 (2004)
27. Dong, L. X., Arai, F., Fukuda, T.: Electron-beam-induced deposition with carbon nanotube emitters. Appl. Phys. Lett., **81**, 1919–1921 (2002)
28. Hjort, K., Soderkvist, J., Schweitz, J. A.: Gallium-arsenide as a mechanical material. Journal of Micromechanics and Microeng., **4**, 1–13 (1994)
29. Chen, X., Zhang, S., Dikin, D. A., Ding, W., Ruoff, R. S., Pan, L., Nakayama, Y.: Mechanics of a Carbon Nanocoil. Nano Lett., **3**, 1299–1304 (2003)
30. Kanda, Y.: A graphical representation of the piezoresistance coefficients in silicon. IEEE Trans. on Elect. Dev., **29**, 64–70 (1982)
31. Harley, J. A., Kenny, T. W.: High-sensitivity piezoresistive cantilevers under 1000 A thick. Appl. Phys. Lett., **75**, 289–291 (1999)
32. Tabib-Azar, M., Wang, R., Xie, Y., You, L.: Self-welded metal catalyzed carbon nanotube piezoresistors with very large longitudinal piezoresistivity of $\sim 4 \times 10^{-8} Pa^{-1}$. Proc. of the 1st IEEE Intl. Conf. on Nano/Micro Eng. and Mol. Sys., 1297–1302 (2006)
33. Saya, D., Belaubre, P., Mathieu, F., Lagrange, D., Pourciel, J., Bergaud, C.: Si-piezoresistive microcantilevers for highly integrated parallel force detection applications. Sens. and Act. A, **123-124**, 23–29 (2005)
34. Katan, A. J., Van Es, M. H., Oosterkamp, T. H.: Quantitative force versus distance measurements in amplitude modulation AFM: a novel force inversion technique. Nanotech., **20(16)**, 165703 (2009)
35. Hu, S., Raman, A.: Inverting amplitude and phase to reconstruct tipsample interaction forces in tapping mode atomic force microscopy. Nanotech., **19**, 375704 (2008)
36. Giessibl, F. J.: Forces and frequency shifts in atomic-resolution dynamic-force microscopy. Phys. Rev. B, **56**, 16010 (1997)
37. Sader, J. E., Jarvis, S. P.: Accurate formulas for interaction force and energy in frequency modulation force spectroscopy. Appl. Phys. Lett., **84**, 1801 (2004)
38. Albrecht, T. R., Grutter, P., Horne, D., Rugar, D.: Frequency modulation detection using high-Q cantilevers for enhanced force microscope sensitivity. J. of Appl. Phys., **69**, 668 (1991)
39. Sader, J. E., Uchihashi, T., Higgins, M. J., Farrell, A., Nakayama, Y., Jarvis, S. P.: Quantitative force measurements using frequency modulation atomic force microscopytheoretical foundations. Nanotech., **16**, S94 (2005)
40. Castellanos-Gomez, A., Agrat, N., Rubio-Bollinger, G.: Dynamics of quartz tuning fork force sensors used in scanning probe microscopy. Nanotech., **20** (2009)

41. Voncken, M., Schermer, J., Bauhuis, G., Mulder, P., Larsen, P.: Multiple release layer study of the intrinsic lateral etch rate of the epitaxial lift-off process. Appl. Phys. A: mat. sci. and proc., **79**, 1801–1807 (2004)
42. Gu, Y., Kwak, E. S., Lensch, J. L., Allen, J. E., Odom, T. W., Lauhon, L. J.: Near-field scanning photocurrent microscopy of a nanowire photodetector. Appl. Phys. Lett. **87**, 43111 (2005)
43. Kojima, Y., Kasuya, K., Ooi, T., Nagato, K., Takayama, K., Nakao, M.: Effects of Oxidation during Synthesis on Structure and Field-Emission Property of Tungsten Oxide Nanowires. Jap. J. Appl. Phys. **46**, 6250 (2007)
44. Liu, K. H., Wang, W. L., Xu, Z., Liao, L., Bai, X. D., Wang, E. G.: In situ probing mechanical properties of individual tungsten oxide nanowires directly grown on tungsten tips inside transmission electron microscope. Appl. Phys. Lett. **89**, 221908 (2006)
45. Chen, J., Dai, Y. Y., Luo, J., Li, Z. L., Deng, S. Z., She, J. C., Xu, N. S.: Field emission display device structure based on double-gate driving principle for achieving high brightness using a variety of field emission nanoemitters. Appl. Phys. Lett. **90**, 253105 (2007)

Chapter 7
A Mechanism Approach for Enhancing the Dynamic Range and Linearity of MEMS Optical Force Sensing

Gloria J. Wiens and Gustavo A. Roman

Abstract Optical-based force sensors can provide the desired resolution and maintain relatively large sensing ranges for cell manipulation and microneedle injections via a force-sensing method that uncouples the conflicting design parameters such as sensitivity and linearity. Presented here is a mechanism approach for enhancing the performance of a surface micromachined optical force sensor. A new design is presented, which introduces a special mechanism, known as the Robert's mechanism, as an alternate means in which the device is structurally supported. The new design's implementation is achievable using an equivalent compliant mechanism. Both analytical pseudo-rigid-body modeling and FEA methods are used for determining the geometric parameters of the compliant Robert's mechanism, optimized to obtain a sensor with improved linearity and sensitivity. Overall, the force sensor provides higher sensitivity, larger dynamic range, and higher linearity compared to a similar optical force sensor that uses a simple structural supporting scheme. In summary, the effectiveness of using a mechanism approach for enhancing the performance of MEMS sensors is demonstrated.

Keywords MEMS optical force sensor • MEMS transducer • Compliant mechanism • Microsensors • Micro-manipulation • Robert's mechanism • XBob

7.1 Introduction

A wealth of research has been conducted in microelectromechanical systems (MEMS) to develop physical sensors and actuators. As a consequence, MEMS technology has provided great advances in the area of force sensing. The miniaturization

---

G.J. Wiens (✉)
Department of Mechanical and Aerospace Engineering, University of Florida,
Gainesville, FL 32611-6250, USA
e-mail: gwiens@ufl.edu

C. Clévy et al. (eds.), *Signal Measurement and Estimation Techniques for Micro and Nanotechnology*, DOI 10.1007/978-1-4419-9946-7_7,

has allowed sensors to be packaged in sub-cubic millimeter volumes. This opened the door for the placement of sensors in previously impossible situations and locations. For example, the tedious task of microassembly has greatly benefited from the integration of force feedback into microgrippers and tweezers for pick and place operations. The outcome has led to improvements in yield and reliability [1–6]. Similarly, the incorporation of force feedback information from part interactions during the assembly process leads to more reliable and robust assembled devices [4–7]. For diverse areas in biomedical research and clinical medicine, the versatility of microsensors and actuators is enabling ever-greater functionality and cost reduction in smaller devices for improved medical diagnostics and therapies.

Another advantage of MEMS sensors and actuators is the fact they can incorporate multiple physical domains. Numerous combinations from the mechanical, electrical, thermal, magnetic, and optical realms are available to MEMS designers for both sensing and actuation. A conventional optical-mechanical-based design is an optical force sensor developed by Zhang et al. [8]. It is a 1 DOF sensor using a linear optical encoder. The authors of [8] report a resolution of less than 1 μN and a range of about 10 μN. The general operating principle of this device is common to many force sensors. That is, they measure the displacement of a compliant structure with a known stiffness and calculate the force required to induce that displacement. Where these force sensors vary is the detection scheme used to determine the displacement. Another common method is capacitive sensing. Sun et al. [9] have developed a capacitive force sensor to measure forces up to 490 μN with a resolution of 0.01 μN in x, and up to 900 μN with a resolution of 0.24 μN in y. Other researchers have explored piezoresistive sensors for sub-Newton applications [3, 4, 6, 10]. Dao et al. [10] developed a six-degree-of-freedom piezoresistive sensor with mN range. This sensor used a compliant platform approach with 16 embedded piezoresistors arranged in a half-active Wheatstone bridge fashion. A design similar to this was further explored by Roman et al. [11].

Many of the above devices use simple compliant beams to support the sensing element of the sensor. There have been attempts to use more complicated compliant mechanisms in place of simple beams to improve sensor characteristics, such as linearity, resolution, and dynamic range [12–15]. The thrust of this chapter is to explore a mechanism approach for enhancing the performance of MEMS sensors. Presented herein, a compliant mechanism comprised of Robert's mechanisms is introduced as a design basis for replacing the simple beams in the above device of Zhang. For this design, an optimization was conducted to improve the performance characteristics of the sensor.

7.2 Enhancement of Optical Force Sensor: Modeling

As a state-of-the-art base design, the MEMS optical force sensor first presented by Zhang et al. [8] was selected and is presented in detail. The basic operating principle of the sensor is first discussed to provide the fundamental understanding of the physics. All aspects of the sensor design will remain the same as the base design, except for the manner in which the index grating is structurally supported. Particular attention is then placed on the sensor's structural mechanics and the identification of critical design parameters, such as material choice, stiffness, and geometry. Parameters that cannot be modified or assumed predetermined are also detailed. In addition, the performance characteristics of the force sensor are reported and the results of FEA analysis are shown.

As stated above, the main goal is to improve the linear relationship between force and displacement over the dynamic range of the sensor via design changes in the support structural mechanics of the index grating. A candidate design toward achieving this goal is a compliant mechanism, termed "XBob," developed in [16] using a series and parallel combination of Robert's mechanisms to acquire straight-line motion. This serial-parallel mechanism pair is introduced into the current force sensor design and optimized to achieve the desired design objectives. Using the pseudo-rigid-body model (PRBM) method [17], standard rigid-body kinematics can be used to characterize equivalent compliant mechanisms. Thus, using the PRBM, analytical equations are developed to initially model the design. A more detailed and accurate model is then constructed using FEA methods. Using the FEA model, optimization of critical parameters is done to further improve the design with respect to desirable characteristics. A sensitivity study is performed to test the design's robustness subject to fabrication errors. Finally, comparison of the new design with the old highlights the improvements and benefits gained from integrating into the base design mechanisms with special force and/or displacement characteristics.

7.2.1 *Prior Art: Optical Force Sensor*

The MEMS force sensor presented in [8] was fabricated using a surface micromachining process at the Stanford Nano-fabrication Laboratory. The purpose of this sensor is to allow for the characterization of microinjection forces. The authors specifically design their sensing device for the injection of genetic material into biological model systems, such as Drosophila embryos. The device is an optical force sensor based on a diffractive linear encoder, consisting of two constant period gratings. As depicted in Fig. 7.1, the scale grating is fixed to the substrate, while the index grating is suspended above the fixed grating by four beams. The two gratings are in alignment when no external force is applied to the index grating. For the purpose of characterizing microinjection forces, a microinjection needle is monolithically incorporated into the index grating. Any force applied to the needle causes the index grating to displace. Given a minimum detectable displacement, defined by the optics and photodiodes, the spring constant determines the sensitivity of the sensor. The more compliant the sensor, the smaller the minimum changes in force it may register.

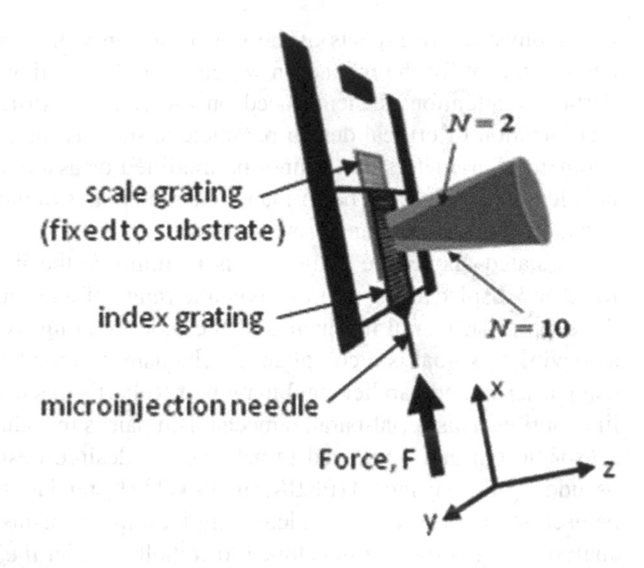

Fig. 7.1 Depiction of optical force sensor, illustrating 2 and 10 grating periods (N) illuminations

7.2.1.1 Optical Displacement Detection

For diffractive linear encoders, the magnitude of displacement is determined by the intensity distribution of the first diffraction order. Using Fraunhofer diffraction theory, a relationship between first diffraction mode intensity and injector displacement can be developed as follows:

$$I_1(d) = I_0 N^2 \left(\frac{\mathrm{sinc}^2(Nd/2L)}{\mathrm{sinc}^2(d/2L)} \right) \left[(L-d)\,\mathrm{sinc}\frac{(L-d)}{4L} \right]^2 \sin^2 \phi_0 G(d) \tag{7.1}$$

$$G(d) = \begin{cases} \sin^2 \dfrac{\pi(L+d)}{4L} & d \in [0,L] \\ \sin^2 \dfrac{\pi(3L-d)}{4L} & d \in [L,2L] \end{cases}$$

$$\mathrm{sinc}(x) = \begin{cases} 1 & \text{for } x=0 \\ \dfrac{\sin(x)}{x} & \text{otherwise} \end{cases}$$

Equation (7.1) shows that the first diffraction mode intensity (I_1) is a periodic function of grating displacement (d), where I_0 is the laser source intensity, N the number of grating periods under illumination, ϕ_0 is the phase delay over the thickness of grating finger, and $2L$ the period of the grating. The direction of translation for the device is considered the x-direction, shown in Fig. 7.1. The sensor is designed to be compliant in the x-direction while resisting motion in the other five degrees of freedom (y- and z-directions and rotations about x-, y-, and z-axes). Of those five, the sensor is most sensitive to z-axis rotation. To minimize this effect, the device uses maximally separated suspension beams to maximize the rigidity in that direction. This particular optical encoder design is less susceptible to errors induced via motion from the remaining four degrees of freedom because they have a lesser effect on the optical readout.

7.2.1.2 Material Selection

Silicon nitride (Si_3N_4) was chosen as the material due to its good optical and stress qualities. To enhance optical transmission, the nitride layer is deposited under high NH_3 conditions. The thickness of the nitride layer is chosen to minimize reflections from the grating elements. The estimated refractive index of nitride is 1.9. The HeNe laser used to illuminate the gratings has a wavelength of 633 nm. To achieve destructive interference, a phase shift that is close to an integer multiple of 2π must be created for light that is being reflected off the fronts and backs. A nitride layer thickness of 1.5 μm yields a phase shift of nine times 2π.

7.2.1.3 Dynamic Range: Sensitivity and Linearity

The most significant performance characteristics for this device are sensitivity and dynamic range. Sensitivity is defined as the change in the first-order diffraction intensity for a given amount of displacement. Dynamic range is the range in which the position (displacement) of the index grating can be unambiguously determined. There is a direct trade off between these performance characteristics, that is by increasing one the other is sacrificed. These performance characteristics are a function of the pitch of the grating. A finer pitch yields better resolution over a smaller usable range when compared to a more coarse pitch. The pitch design variable is fixed once the device is fabricated.

Another manner in which to tune the performance characteristics, sensitivity, and dynamic range is by adjusting the number of individual grating periods illuminated by the laser (N). The illumination can be changed using optics to vary the diameter of the laser spot size. An increased number of illuminated grating periods would have the similar effect as changing the grating to exhibit a finer pitch, improving the sensitivity. The trade off in illuminating more periods is a reduction in dynamic range. This trade off is seen in Figs. 7.1 and 7.2. The curve generated with ten periods illuminated ($N = 10$) has dead zones in between the peaks. While operating in this portion of the curve, a change in displacement does not cause a change in intensity. However, the slope of the peaks is steep, and while operating in this portion of the curve, a small change of displacement will induce a large change in intensity. Examination of the curve with two periods illuminated ($N = 2$) reveals that the dead zone is effectively eliminated at the cost of reducing the slope of the peaks.

Although the effects of changing the spot size are not as dramatic compared to changing the pitch on the device, a more compliant device would have a larger displacement for the same applied force. This would allow a device with a course pitch to compensate for its loss of sensitivity while maintaining an increased

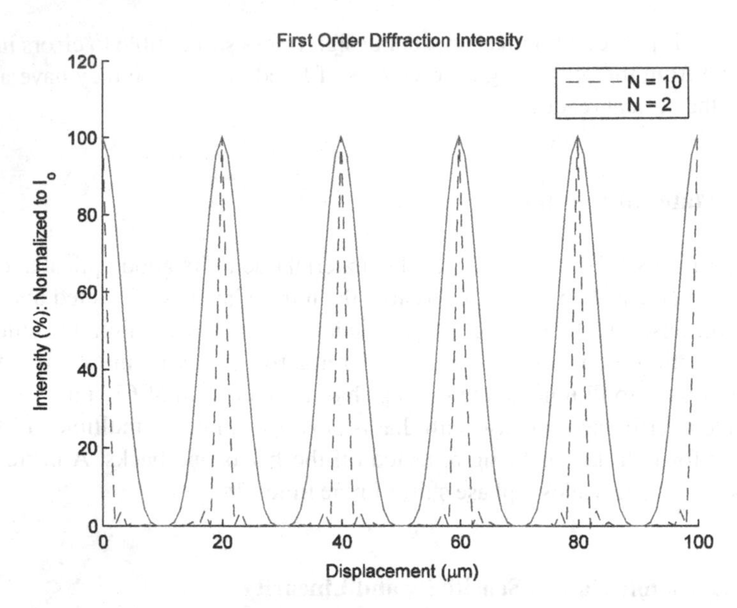

Fig. 7.2 Effects of changing the number of periods under illumination (N) on the sensitivity and dynamic range of the optical detection scheme

dynamic range. Whereas the sensor in [8] would obtain more compliance by narrowing the width of the supporting beams (*w*). The sensor unfortunately loses linearity as *w* decreases. Important in this detection scheme is maintaining a linear relationship between force and displacement.

7.2.1.4 Sensor Support Structure: Theory and Analysis

The index grating supported by four beams was modeled as having two beams that pass through the index grating. The beams act as springs in parallel. Therefore, the equivalent spring constant of one beam may be determined and multiplied two times to give the overall spring constant of the force sensor. The boundary conditions of the beam are set up as both ends fixed and center loaded. The fixed ends are anchored to the substrate, and the index grating displacement (*d*) is equivalent to the center applied load displacement. Equation (7.2) shows the spring constant or stiffness of a beam with these boundary conditions [18].

$$k = \frac{192EI}{l^3} \tag{7.2}$$

The stiffness is a function of material and geometric properties. The defining material property is Young's Modulus (*E*). The cross-sectional area moment of inertia (*I*) is a function of the beam thickness (*h*) and width (*w*), and *l* is the length of the beam.

$$I = \frac{hw^3}{12} \tag{7.3}$$

For the designs to be considered, Young's Modulus ranges from 240 to 270 GPa are typical for Si_3N_4 [19–21]. Choosing a value of 270 GPa and using (7.2) and (7.3) yields a spring constant of 10.8 N/m for the four beam sensor design of dimensions stated in reference [8] (a beam thickness of 1.5 μm, a beam width of 8 μm, and a beam length of 850 μm). The selection of a Poisson's ratio of 0.24 is valid for various deposition parameters of nitride [19–21].

7.2.1.5 Beam Nonlinearity Impact on Linearity Assumptions

Using the optics to precisely determine displacement and the calculated spring constant, the magnitude of the force acting on the sensor can be determined. For small tip deflections, bending is the dominating mode and the relationship between force and displacement can be considered linear. This relationship comes from assumptions made in solving the Bernoulli–Euler equation stated as:

$$M - EI\frac{d^2y/dx^2}{[1+(dy/dx)^2]^{3/2}} \tag{7.4}$$

where M is the bending moment, y is the transverse deflection, and x is the coordinate along the undeformed beam. For small deflections, the square of the slope (dy/dx) is close to zero and the denominator goes unity. When this occurs, the bending moment becomes proportional to the curvature (d^2y/dx^2). When beam deflections become large, the slope can no longer be approximated by a small number. This causes the denominator to deviate from unity which is a factor for the inaccuracies of the linear assumption [17]. For a fixed beam that undergoes a large displacement, axial stretching induces a nonlinear bending relationship.

For the four-beam sensor design of reference [8] using the dimensions, Young's Modulus and Poisson's ratio values stated at the end of Sect. 7.2.1.4, the linear operating range is found to be approximately 12 μm. Referring to the force-displacement curves shown in Fig. 7.3, the solid curve shows that displacements beyond what is considered "small" do not maintain the analytical model's linear relationship. The nonlinear results were obtained using a 2D nonlinear static finite element analysis (FEA) model with ten equivalently spaced displacement load steps, generated using ANSYS® software. Due to the extremely small scale of MEMS, beam elements that do not account for mass properties were used, ignoring inertial effects at this time. Other assumptions made in the FEA include no shear in the z-direction and no pre-strain. Furthermore, all beam elements not representing compliant members were given a much larger width so that they would act rigid compared to the flexible links. As observed in Fig. 7.3, the initial slope of the curve is linear; however, it turns cubic as the magnitude of the deflections increases. Although varying the value of Young's Modulus would change the scale of the curve, it has no effect to general nonlinear trend. Therefore, the stiffness of the device in question as well as the sensitivity can be altered via manipulating the Young's modulus; however, the nonlinear trends will remain.

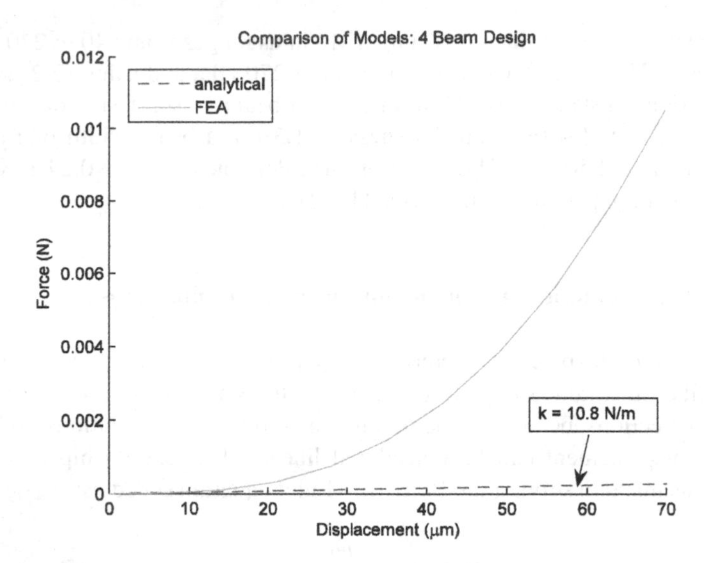

Fig. 7.3 Plots of force-displacement curves for analytical and numerical results of sensor with four-beam configuration [22]

7.2.2 Compliant Mechanism Enhancement

The main thrust for a redesign is to improve the linear relationship between the device's input force and displacement while at the same time minimizing the stiffness for improving the sensitivity of the device. In addition, a similar overall dimensional size of the sensor is maintained. To illustrate the need, consider the embryo microinjection experiments from [8] which reveal an average penetration force of 52.5 μN and embryo deformation of 58 μm. An embryo deformation of this magnitude requires the sensor to displace beyond its linear range, as computed above. If this is not accounted for in the calibration, the force reading will be grossly inaccurate. A force sensor with a much larger linear range will alleviate this error. This scenario of desirable sensor characteristics in dynamic range and linearity will govern the selection of design targets that will be used in selecting parameters for the mechanism enhanced design approach detailed below. Herein, a compliant mechanism is used in place of the simple beams to support the index grating of the prior optical force sensor design.

7.2.2.1 Robert's Mechanism Principles

To address the need for improved linearity characteristics of the force sensor over the dynamic range, this section introduces a compliant mechanism design based on a design that Hubbard et al. [16] conceived of to form a new special mechanism

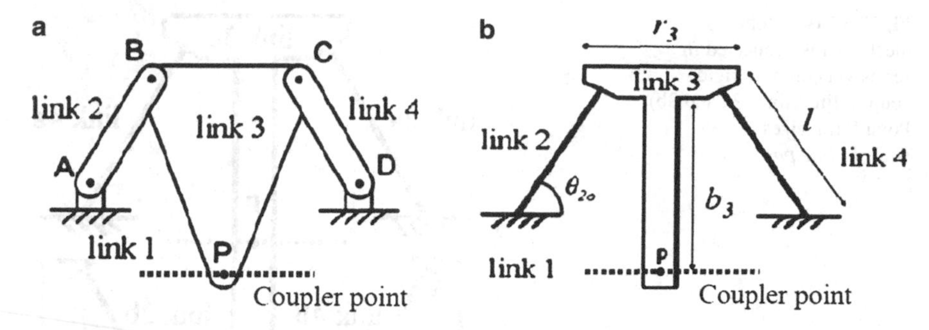

Fig. 7.4 Robert's mechanism used for straight line motion: (**a**) rigid-body model and (**b**) equivalent complaint mechanism model

with straight-line motion characteristics. The compliant mechanism is based on a Robert's four-bar mechanism, shown in Fig. 7.4. The one degree of freedom Robert's mechanism provides a coupler point (P) that translates with an approximate straight-line motion as denoted by the dotted lines in Fig. 7.4. Certain geometric constraints exist that define a Robert's mechanism, such as $\overline{AB} = \overline{DC}$ and $\overline{BP} = \overline{CP}$. The compliant version of the Robert's mechanism has two links that are compliant (links 2 and 4) and two that are rigid, with one of those being the ground (link 1) as illustrated in Fig. 7.4b.

To eliminate the need for a revolute joint to connect the index grating to the mechanism at point P, two Robert's mechanisms are rigidly connected in series at the shared coupler point (P). In doing so, the two mechanisms in series must also have identical dimensions so that the attached coupler links (links 3a and 3b) will rotate through the same arc and can therefore be attached rigidly. Thus, links 3a and 3b form a single rigid coupler link common between the two Robert's mechanisms. For implementing into the sensor, one Robert's mechanism is anchored to the substrate, while the other is rigidly connected to the moving index grating. This mechanism configuration is illustrated in Fig. 7.5.

In the configuration shown in Fig. 7.5, the index grating now has two degrees of freedom; it can displace linearly and can rotate. To eliminate the rotation, another set of Robert's mechanisms is added in parallel. As shown in Fig. 7.6, this setup is then mirrored about the center x-axis of the index grating to yield eight total Robert's mechanisms being used. The mirroring allows for the elimination of errors in the straight-line motion caused by structural errors in the mechanism [16].

7.2.2.2 PRBM Modeling

The PRBM is a concept used to model the force-displacement relationships of a compliant member using an equivalent rigid-body mechanism with the compliance modeled as springs in the joints [16]. For each of the eight Robert's mechanisms within the design, the compliant model's flexible segments are considered to have

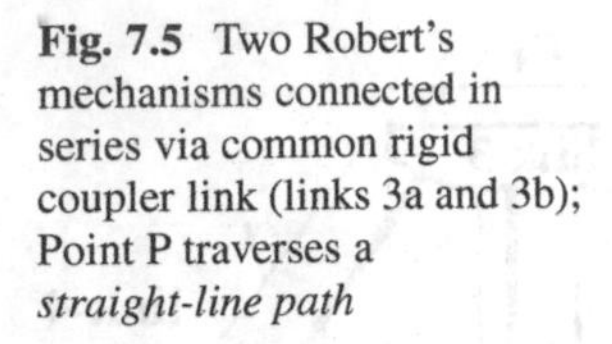

Fig. 7.5 Two Robert's mechanisms connected in series via common rigid coupler link (links 3a and 3b); Point P traverses a *straight-line path*

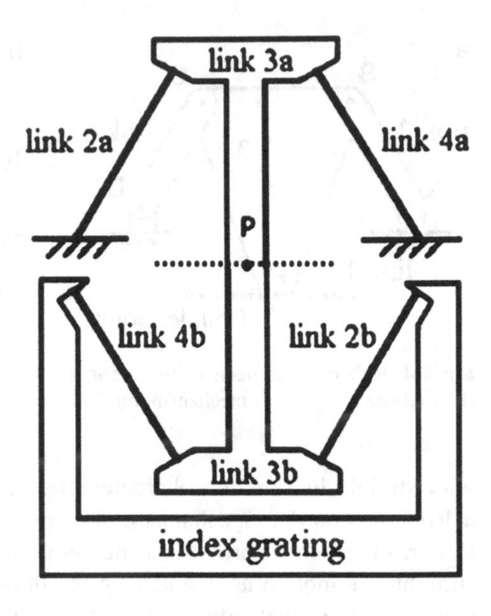

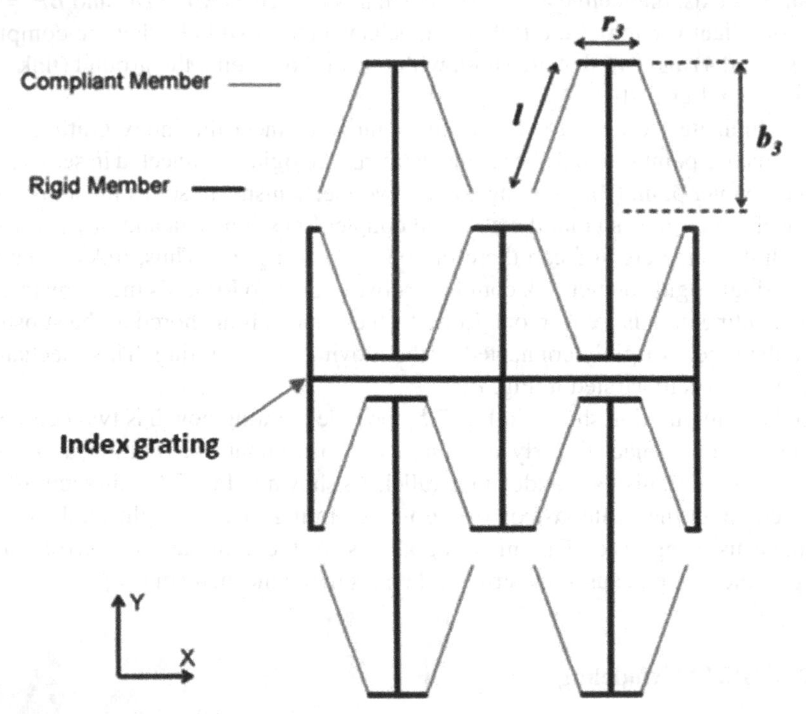

Fig. 7.6 Nonlinear finite element 2D model of serial-parallel pairing of eight compliant Robert's mechanisms, symmetry about center-line of index grating (x-axis, displacement axis of sensor) [22]

boundary conditions: fixed where attached to the ground or index grating and fixed-guided where attached to the middle common rigid coupler links. The PRBM for each compliant Robert's mechanism consists of three rigid links connected via pin joints (A, B, C, and D) with torsional springs at each of the joints (refer to Fig. 7.7). The spring constant (K) is:

$$K = 2\gamma K_{\Theta} \frac{EI}{l} \tag{7.5}$$

where γ is the characteristic radius factor, K_{Θ} is the stiffness coefficient, E is the Young's modulus, I is the cross-sectional area moment of inertia, and l is the length of the flexible member. For this PRBM model approximation, the values for γ and K_{Θ} are 0.8517 and 2.65, respectively [17]. Using the PRBM and principles of virtual work, the kinematic relation of the index grating displacement is:

$$x = N_S \left[r_2(\cos\theta_2 - \cos\theta_{2o}) + b_3 \sin\theta_3 + \frac{r_3}{2}(\cos\theta_3 - 1) \right] \tag{7.6}$$

where N_S is the number of Robert's mechanism in series (2 for this case), b_3 is the perpendicular distance from the center of link 3 to the coupler point, r_i, θ_i, and θ_{io} are the length, final angle, and initial angle of link i, respectively. The value for r_2 (and r_4) is defined as γl. The force required to displace the index grating is:

$$F = -N_T K \frac{(2 - h_{32})\Delta\theta_2 + (2h_{42} - h_{32})\Delta\theta_4 - (1 + h_{42} - 2h_{32})\Delta\theta_3}{N_S \left[r_2 \sin\theta_2 + h_{32} \left(\frac{r_3}{2} \sin\theta_3 - b_3 \cos\theta_3 \right) \right]} \tag{7.7}$$

where N_T is the total number of Robert's mechanisms used (8 for this case), and h_{32} and h_{42} are kinematic coefficients defined in (7.8) and (7.9)

$$h_{32} = \frac{r_2 \sin(\theta_4 - \theta_2)}{r_3 \sin(\theta_3 - \theta_4)} \tag{7.8}$$

$$h_{42} = \frac{r_2 \sin(\theta_3 - \theta_2)}{r_4 \sin(\theta_3 - \theta_4)} \tag{7.9}$$

The $\Delta\theta_2$, $\Delta\theta_3$, and $\Delta\theta_4$ terms are the change in link angle from their initial position. In order to determine the force for a given displacement, the instantaneous angles of the three links must be known. Using closed-loop kinematics, two scalar equations may be obtained from the vector loop equation for the PRBM of the single four-bar mechanism, refer to Fig. 7.7. The r_1 is the distance between joint centers A and D.

$$r_1 = r_2 \cos\theta_2 + r_3 \cos\theta_3 - r_4 \cos\theta_4 \tag{7.10}$$

$$0 = r_2 \sin\theta_2 + r_3 \sin\theta_3 - r_4 \sin\theta_4 \tag{7.11}$$

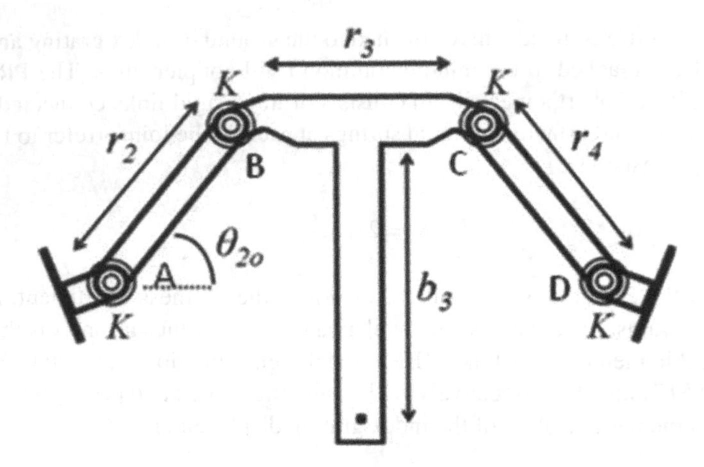

Fig. 7.7 Pseudo-rigid-body model equivalent of single Robert's mechanism

Equations (7.6), (7.10), and (7.11) are solved simultaneously for the three unknown angles, given an input displacement. These angles can then be substituted into (7.7) to find the force. The system of equations was evaluated in Matlab© using a nonlinear solver. The Matlab© software was also used to generate analytical force-displacement plots using the above equations. These plots demonstrate the linear relationship that is desired. Therefore, the resulting spring constant of the sensor is determined from the slope of the force-displacement plot.

7.2.2.3 FEA Modeling

In addition to the above analytical equations, an FEA is conducted to validate the PRBM model. The FEA also captures any nonlinear phenomenon that may occur, which is not possible using the PRBM alone. The model used in the FEA is shown in Figs. 7.6 and 7.8. All beam elements not representing compliant members were given a larger width (20 μm) than the flexible links so that they would exhibit rigid behavior compared to the flexible links. The same value of 270 GPa was used for the Young's modulus and 0.24 for Poisson's ratio. All force-displacement curves were generated by a static analysis of ten equivalently spaced displacement load steps. In order to verify that ten load steps provided enough data points for accurate results, a test case was run. The test case held everything constant and ran the simulation with 50 load steps and 10 load steps. It was found that the selection of 10 data points did not omit any significant trends when compared to the 50 load step case. The displacement input was applied to the center of the rigid section, which represents the index grating. The reaction force at all fixed nodes was tabulated for each load step.

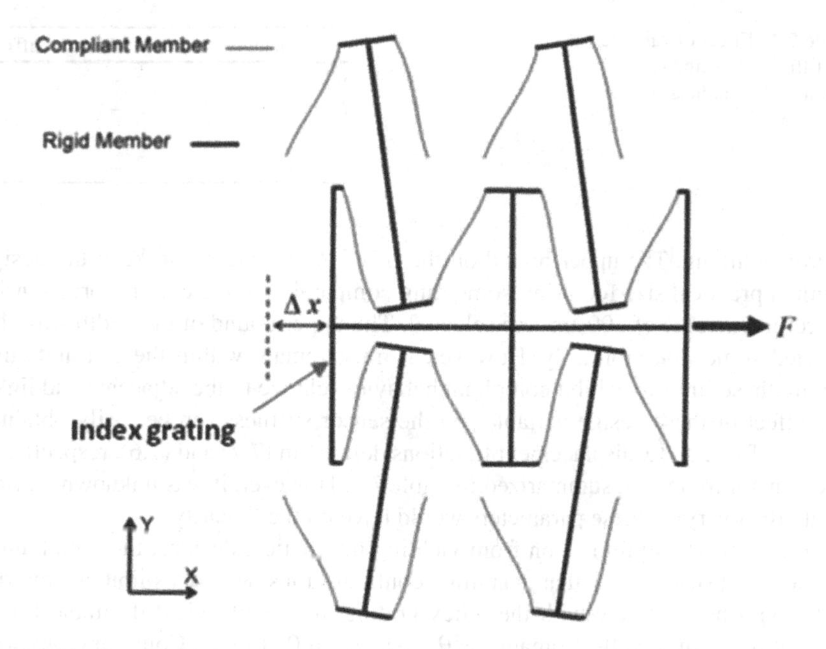

Fig. 7.8 Nonlinear finite element 2D model of compliant mechanism with load applied [22]

7.3 Enhancement of Optical Force Sensor: Design Optimization

For optical force sensor, nitride is the material of choice due to its optical properties. The thickness of the nitride is also preselected to achieve desirable optical properties of a phase shift times 2π. This corresponds to a beam thickness of 1.5 μm. This leaves the length of the links, the initial angle of the link 2, and the width of the compliant member as available design variables for parameter optimization. For the design optimization, the objective function was selected to maximize full scale linearity and minimize the overall stiffness, subject to constraints and boundaries. This is a dual objective problem where initially each objective was weighted equally ($\alpha = 0.5$). This can be stated mathematically as follows:

$$\begin{gathered}(\alpha)\min\ k_{\text{sensor}} \text{ and } (1-\alpha)\max \text{ linearity} :\rightarrow f(w, l, r_3, b_3, \theta_{2o}) \\ l, r_3, b_3 \in [1\,\mu\text{m}, 300\,\mu\text{m}] \\ w \in [1\,\mu\text{m}, 10\,\mu\text{m}] \\ \theta_{2o} \in [20^\circ, 90^\circ]\end{gathered} \tag{7.12}$$

The lower bound of 1 μm was placed on all length and width variables because this is a minimum feature size that can be fabricated using most common surface

Table 7.1 Effect of variables on stiffness: "+" indicates increase, "−" indicates decrease [22]

	Change variable	Change stiffness
w	−	−
l	+	−
r_3	−	−
b_3	+	−

micromachining. The upper bound on the link lengths was set to keep the design within a practical size footprint: something comparable to the current force sensor. Therefore, a value of 300 μm was selected. The upper bound of the width was also selected somewhat arbitrarily. However, it must remain within the restraints that require these links to exhibit compliant behavior relative to the adjacent rigid links. The effect of these design variables on the sensor stiffness can be easily obtained from the force and x displacement functions defined in (7.7) and (7.6), respectively. The general trends are summarized in Table 7.1. However, it was unknown a priori what effect varying these parameters would have on the linearity.

To prevent the optimization from yielding impractical designs, the initial angle θ_{2o} was constrained so that resulting configurations would exhibit geometries that would not interfere with the index grating and would yield desirable link 2 orientations. Initially, the domain for θ_{2o} was from 0° to 90°. Configurations with angles below 20° did not violate any of the specified constraints, but did cause errors for the FEA simulation. Configurations with angles in this range resulted in mechanisms with force-displacement relationships that could not be handled by the FEA nonlinear solver. For this reason, the lower bound on θ_{2o} was raised to 20°.

The metric used for linearity was the adjusted R-square (ARS) value. Mathematically, the ARS is expressed as:

$$\mathrm{ARS} = 1 - (1 - \mathrm{RS})\left[\frac{n-1}{n-j-1}\right] \tag{7.13}$$

where RS is the R-square (coefficient of determination), n the number of observations, and j the number of unknown coefficients (2 for linear fit). This number describes the quality of fit of data points to a polynomial, in this case a linear curve. The ARS statistical metric can take on any value less than or equal to 1, with a value closer to 1 indicating a better fit. The stiffness was calculated as the slope of the best-fit curve resulting from a linear regression of the data points. The resulting objective is stated more formally as:

$$\begin{aligned} f_{\mathrm{obj}}(\overline{X}) &= \alpha \times \mathrm{slope} + (1-\alpha) \times (1 - \mathrm{ARS}) \\ \overline{X} &= [w, l, r_3, b_3, \theta_{2o}] \end{aligned} \tag{7.14}$$

where the value of f_{obj} is being minimized. The combination of parameters in the vector $\overline{X}$ that provide the lowest f_{obj} value within the feasible domain is considered the optimal design.

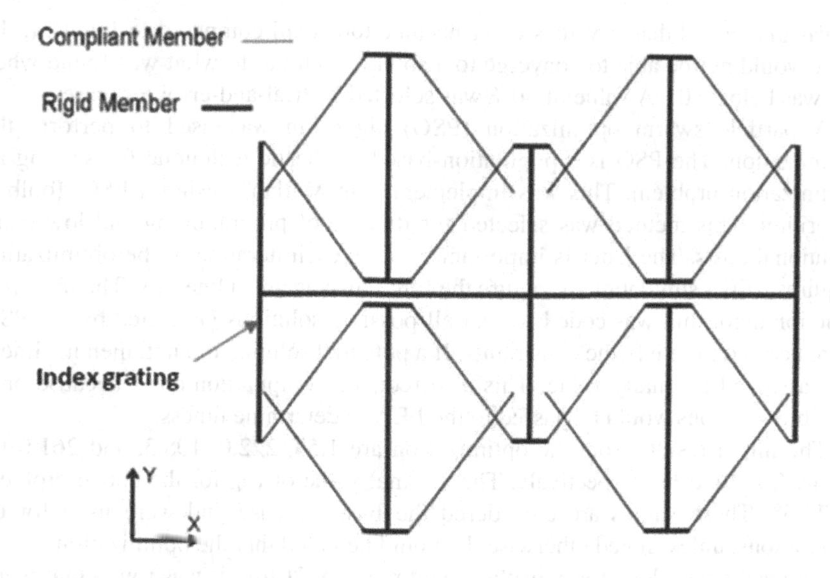

Fig. 7.9 Nonlinear finite element 2D model of infeasible design configuration; compliant members cross each other [22]

Another constraint introduced into the optimization is that the maximum stress is to not exceed the failure stress throughout the entire range of displacement. This is checked analytically using the following equation:

$$\sigma_{max} = \frac{Pac}{2I} \tag{7.15}$$

where P is the load applied to the guided end of the flexible link, a is the perpendicular distance to that load from the fixed end, and c is the distance from the neutral axis to the extreme outer surface. Since the flexible member has a rectangular cross section, c is simply half the value of w.

Another constraint declares that the configurations must not violate any physical kinematic constraints. In particular, a constraint is introduced that prevents links from occupying the same space at the same time and prevents any binding or interference. For example, links 2 and 4 for each set of Robert's mechanisms have the potential to cross each other if they extend too far (e.g., link 4a and 2b from Fig. 7.5). This is prevented by constraining the location of the fixed end for the upper links so that they are always above the location of the fixed end for the lower links. An example of an infeasible design is shown in Fig. 7.9.

After initial optimization trials, the following constraint was implemented:

$$r_3 \geq 0.30 \times \max(l, b_3) \tag{7.16}$$

It was uncovered that if values of r_3 became too small compared to l and b_3, the FEA would not be able to converge to a solution, similar to what was found when θ_{2o} was below 20°. A value of 30% was selected by trial-and-error test cases.

A particle swarm optimization (PSO) algorithm was used to perform the optimization. The PSO is a population-based stochastic technique for solving an optimization problem. This was implemented in Matlab© using a PSO Toolbox algorithm. This method was selected for its ease of programming and low computational costs. The latter is important because each iteration of the optimization requires a FEA simulation to acquire the force-displacement linearity. The objective function algorithm was coded so that all possible solutions generated by the PSO were verified to satisfy the constraints. If a potential solution did not, then its fitness was assigned a penalty value. This also reduced computation costs because only feasible solutions would be passed to the FEA to determine fitness.

The initial results from the optimization are 1.53, 222.0, 108.3, and 261.0 μm for w, l, r_3, and b_3, respectively. The optimal value of θ_{2o} for the above problem is 72.5°. These values are considered the baseline case and were used for all simulations, unless noted otherwise. It should be noted that the optimization for the above results was done for a displacement range of 70 μm. It was found that when the sensor's displacement range considered in the design process is increased, the above combination of parameters no longer yields the optimum design. It was found that a value of 1 μm for w provides the best combination of stiffness and linearity for displacements greater than 70 μm, when using the above combination of link lengths. Referring to the constraint limits, only the lower constraint on the width parameter would become active. A value of 1 μm for w was also expected because this would yield the lowest stiffness within the bounds specified. This finding is covered further in the following section.

7.4 Enhancement of Optical Force Sensor: Design Results

7.4.1 *Preliminary Results*

The results obtained by PRBM and FEA of the above baseline case are compared in Fig. 7.10. Spring constants for the FEA results are determined from a linear regression of the data points using least squares approximation. The analytical model predicts a spring constant of 0.25 N/m [slope of the plot of (7.7) versus (7.6)] and the FEA predicts 0.26 N/m. The ARS for the PRBM and FEA is 1 and 0.9997, respectively. Typically, sensors are characterized by their nonlinearity. This nonlinearity is the maximum deviation of output from the "best-fit" straight line through the data points. The equation for calculating nonlinearity is:

$$\mathrm{NL} = \frac{\text{Maximum deviation}}{\text{Full scale output}} \times 100\% \tag{7.17}$$

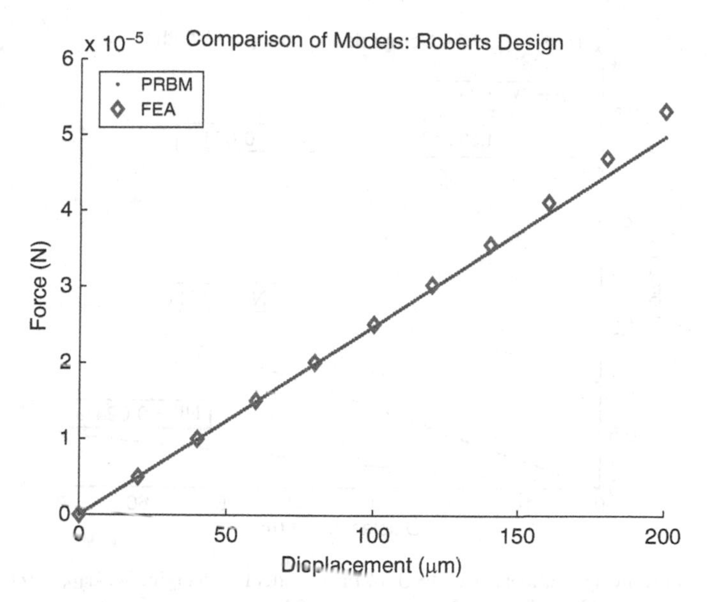

Fig. 7.10 Force-displacement plots from both analytical and numerical models: $k_{\text{PRBM}} = 0.25\,\text{N/m}$ and $k_{\text{FEA}} = 0.26\,\text{N/m}$ [22]

The nonlinearity (NL) measure of the above FEA results is 0.02% of full scale output (FSO). Thus, the values of both the analytical and numerical models are in close agreement. The analytical model, however, has no way to account for nonlinear phenomenon; thus the FEA provides the validation that the optimization was successful in acquiring a high degree of linearity in the design.

The effect of w on the linearity for a displacement range of 70 μm was examined by running additional FEA with values of w at 1 and 2 μm (Fig. 7.11). The resulting FEA spring constants for the force sensor with values of w at 1 and 2 μm are 0.07 and 0.56 N/m, respectively. The same simulations were run with the final displacement set to 100 μm. As seen in Fig. 7.12 in comparison with Fig. 7.11, the effect that w has on the NL measure changed where the w effect on the stiffness remained unchanged. Although the values of NL measure vary by changing w or the displacement range, they are all well below 3%FSO, a typical metric for sensors. Furthermore, the 100 μm displacement range is well below the predicted 170 μm maximum allowable displacement before failure for all widths examined. These values are summarized in Table 7.2.

The initial results of the optimization were obtained for equal weighting of the objective functions. If a higher weight was placed on the stiffness objective, the optimizer would probably have selected a value of 1 μm for w for the 70 μm displacement case. The fact that the nonlinearity remains well below the tolerable limit within this range of w; the width can be selected for a desired stiffness. The nonlinearity is also dependent on the displacement range over which it is being calculated. To further investigate, it was decided that this range should be set to

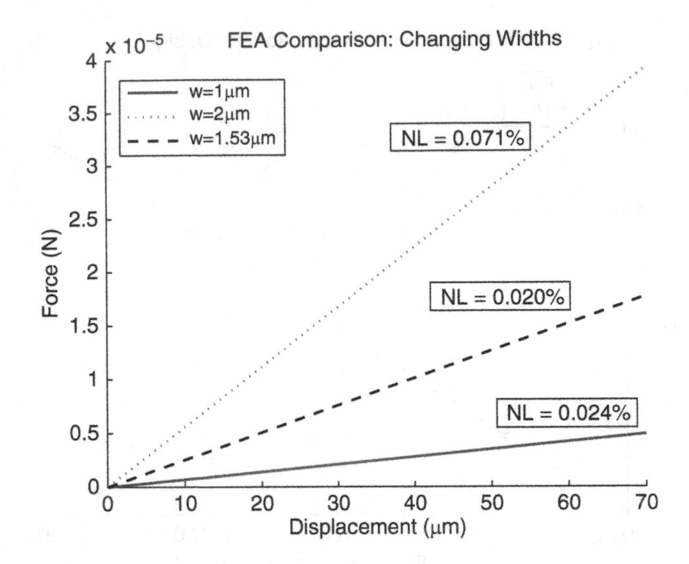

Fig. 7.11 Plot showing nonlinearity (%FSO) of FEA model for changing w: $k_{\mathrm{FEA}} = 0.07\,\mathrm{N/m}$ for $w = 1\,\mu\mathrm{m}$; $k_{\mathrm{FEA}} = 0.56\,\mathrm{N/m}$ for $w = 2\,\mu\mathrm{m}$; $k_{\mathrm{FEA}} = 0.25\,\mathrm{N/m}$ for $w = 1.53\,\mu\mathrm{m}$

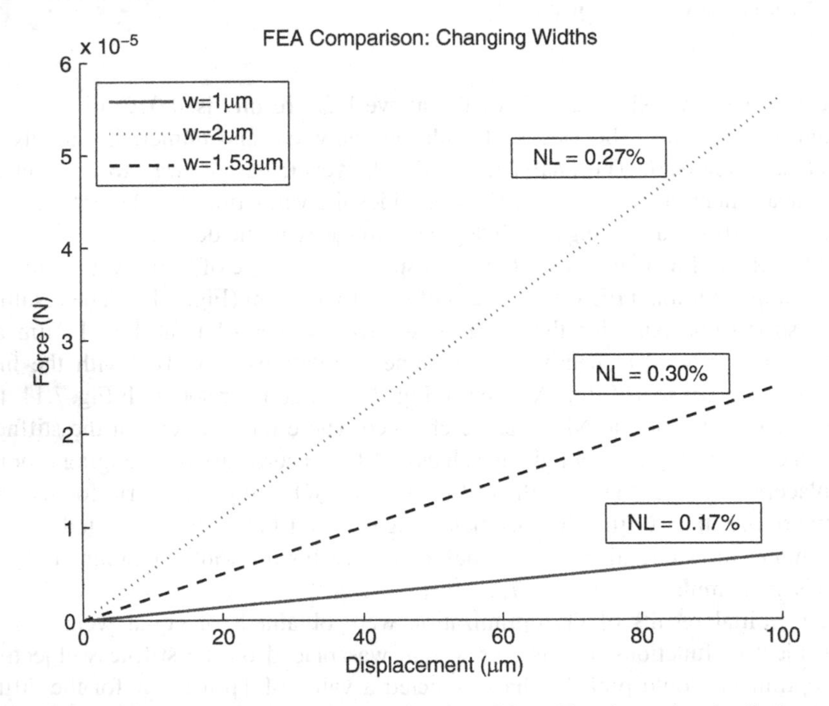

Fig. 7.12 Plot showing nonlinearity (%FSO) of FEA models and how increasing the displacement range changes the NL value: spring constants are the same as in Fig. 7.11

Table 7.2 Stiffness and NL for baseline configuration of link lengths with various values of w for two displacement ranges [22]

w		1 μm	1.53 μm	2 μm
70 μm disp. range	K (N/m)	0.07	0.25	0.56
	NL (%FSO)	0.024	0.007	0.071
100 μm disp. range	K (N/m)	0.07	0.25	0.56
	NL (%FSO)	0.17	0.30	0.27

Table 7.3 Sensor performance characteristics for varying compliant member width [22]

w (μm)	Max displacement (μm)	Max force (μN)	FEA stiffness (N/m)	NL (%FSO)
1.53	200	53.4	0.26	2.8
2.0	170	98.6	0.57	1.73
2.5	138	153	1.10	0.98
3.0	118	222	1.88	0.74

the maximum displacement achievable for each respective value of w examined, in place of a fixed range across the board.

The FEA was also used to determine the maximum displacement of the device. Since nitride is a brittle material, it will not plastically yield, but rather fail catastrophically. The failure or ultimate stress for nitride was set at 2.7 GPa [20]. The baseline configuration was displaced until the maximum calculated stress equals the ultimate stress. The maximum displacement is defined as 75% of the total displacement before failure to provide some factor of safety. For the baseline configuration this value is 200 μm. The maximum displacement was determined for other values of w and summarized in Table 7.3 along with corresponding maximum force, stiffness, and NL measures. The value of 1.53 μm was selected as the lower threshold, because lower values would be pushing the limits of fabrication capabilities. Values chosen for investigation beyond the baseline were 2.0, 2.5, and 3.0 μm. Table 7.3 reveals that increasing the thickness reduces the nonlinearity; however, this also decreases the maximum displacement. There is also a similar inverse relationship found between the sensing range and the sensitivity. If a device was needed for μN applications with high resolution, a smaller width would provide a solution yielding lower stiffness. On the other hand, the width could be increased and allow the device to sense in the mN range with a reduction in resolution.

7.4.2 Parametric Sensitivity Study and Design Robustness

The general effects of the link lengths on the spring constant are summarized in Table 7.1. The magnitude of these effects can be efficiently calculated by the PRBM equations. To see the effects of link lengths on linearity, a sensitivity study was done for l, r_3, and b_3. The parameter of interest for each simulation was varied while all other values remained at the baseline.

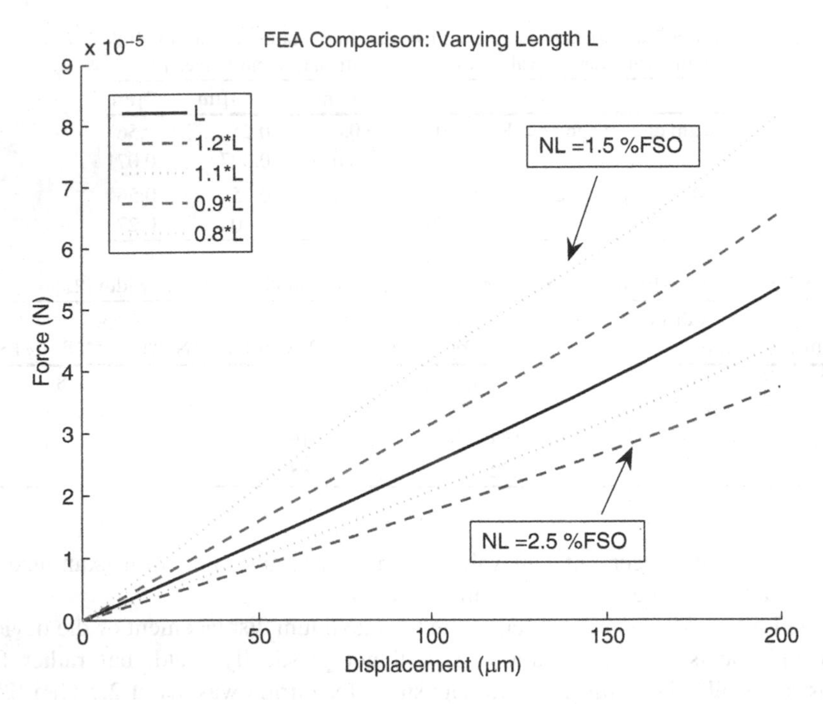

Fig. 7.13 Plot showing parameter study on effects of l on linearity and stiffness

Table 7.4 Effects of changing the l parameter on stiffness and NL; bold numbers indicate optimal design from initial optimization using (7.14); underlined numbers are potential Pareto optimal solutions when considering 200 μm displacement

	Stiffness (N/m)		NL (%FSO)	
Parameter variation from baseline config.	200 μm displacement	70 μm displacement	200 μm displacement	70 μm displacement
$1.2 \times l$	<u>0.19</u>	0.21	<u>2.5</u>	0.11
$1.1 \times l$	<u>0.22</u>	0.17	<u>2.7</u>	0.25
l	<u>0.26</u>	**0.25**	<u>2.8</u>	**<0.01**
$0.9 \times l$	0.31	0.32	3.6	0.13
$0.8 \times l$	<u>0.40</u>	0.43	<u>1.5</u>	0.60

The first variable studied was l. Per Fig. 7.13 and Table 7.4, the linearity of the curve improves as l increases, with $1.2 \times l$ having a NL value of 2.5%FSO. Contrary to this trend, a decrease in the length to $0.8 \times l$ shows a good improvement with an NL of 1.5%FSO. This does not appear to agree with the plots in Fig. 7.13. The curve appears wavy when compared to the baseline, yet has a lower NL value. This deviation in the trend is due to the manner in which the NL is calculated (7.17). Although there are many deviations from a straight line, the maximum deviation is

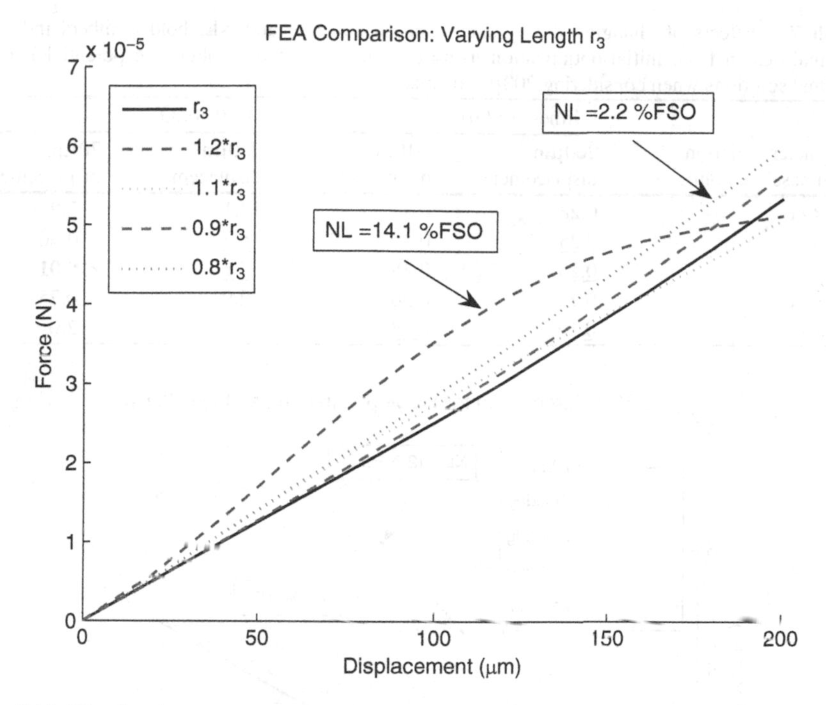

Fig. 7.14 Plot showing parameter study on effects of r_3 on linearity and stiffness

smaller than that of the baseline. Another factor is that the denominator of (7.17) is larger for the $0.8 \times l$ case because the full scale value is greater than it is for the baseline case. Furthermore, the maximum deviation of the baseline case is occurring around 200 μm of displacement. The values in bold indicate the best design as defined by the optimization objective specified in (7.14). The $1.2 \times l$ configuration yields a device with lower stiffness and slightly better linearity than the baseline configuration. In the initial parameter optimization, NL was computed over a range of 70 μm. When calculating NL over a 200 μm displacement range, the parameters obtained from the initial optimization no longer yield the optimal configuration. The underlined values in Table 7.4 indicate potential optimal solutions for the 200 μm displacement case. The nonlinearity remains under 3% for the change of lengths studied. Within a tolerance of 20% of the original length, the device remains robust with respect to nonlinearity.

The next variable studied was r_3. Increasing the value by just 20% destroys the linearity almost completely, seen in Fig. 7.14. Decreasing the value improves the linearity over the baseline configuration, but increases the stiffness. An interesting note is that this trend goes against what is reported Table 7.1. This trend deviation is considered to be caused by the resulting r_3 value nearing its physical limit, requiring it to be longer than 30% of the other lengths l and b_3. That is, (7.6) and (7.7), which were used to determine the trends in Table 7.1, do not hold up when the

Table 7.5 Effects of changing the r_3 parameter on stiffness and NL; bold numbers indicate optimal design from initial optimization using (7.14); underlined numbers are potential Pareto optimal solutions when considering 200 μm displacement

	Stiffness (N/m)		NL (%FSO)	
Parameter variation from baseline config.	200 μm displacement	70 μm displacement	200 μm displacement	70 μm displacement
$1.2 \times r_3$	0.28	0.35	14.1	2.9
$1.1 \times r_3$	<u>0.25</u>	0.28	<u>6.3</u>	0.90
r_3	<u>0.26</u>	**0.25**	<u>2.8</u>	**< 0.01**
$0.9 \times r_3$	0.28	0.26	2.9	0.75
$0.8 \times r_3$	<u>0.30</u>	<u>0.28</u>	<u>2.2</u>	2.0

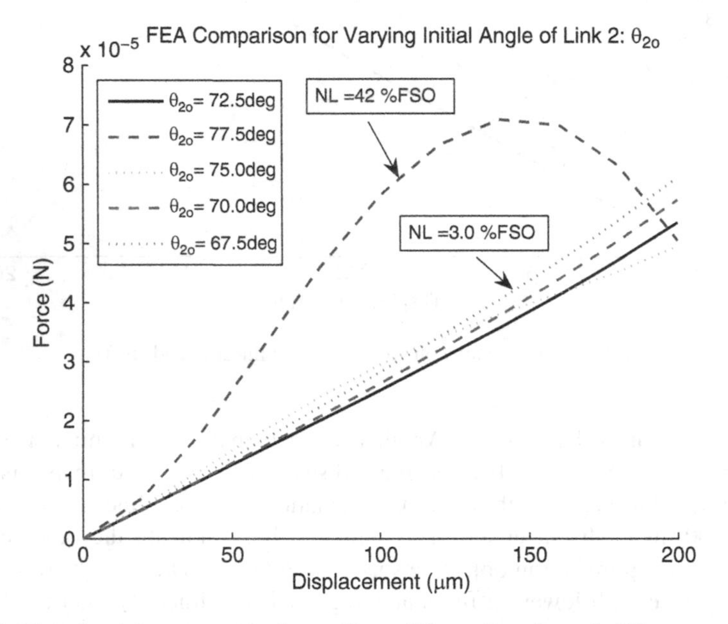

Fig. 7.15 Plots showing parameter study on effects of θ_{2o} on linearity and stiffness

length deviates from the other too much. The results of these variations are listed in Table 7.5; again underlined values indicate potential optimal solutions for the 200 μm displacement case.

The effect of the initial link angle for link 2 was also investigated. Varying the value of θ_{2o} in either direction results in reduced linearity of the device. An extreme example of this occurs when θ_{2o} is increased by only 5°; this results in a huge deviation from a straight line, seen in Fig. 7.15. The θ_{2o} value obtained from the optimization provides the best design whether the displacement range is 70 or 200 μm. This is supported by Table 7.6, where bold values indicate the lowest stiffness and nonlinearity, respectively.

The last length variable studied was b_3. Although the changes in stiffness are as expected from Table 7.1, the nonlinearity and the stiffness of the $1.2 \times b_3$ curve is lower than the baseline configuration (Fig. 7.16). Increasing the length of b_3

Table 7.6 Effects of changing the θ_{2o} parameter on stiffness and NL; bold numbers indicate optimal design from initial optimization using (7.14); this value remains the optimal setting when the displacement is increased to 200 μm

	Stiffness (N/m)		NL (%FSO)	
Parameter variation from baseline config.	200 μm displacement	70 μm displacement	200 μm displacement	70 μm displacement
$\theta_{2o} + 5°$	0.36	0.54	42.0	6.1
$\theta_{2o} + 2.5°$	0.26	0.30	4.8	1.7
θ_{2o}	**0.26**	**0.25**	**2.8**	**<0.01**
$\theta_{2o} - 2.5°$	0.28	0.26	3.2	0.77
$\theta_{2o} - 5°$	0.30	0.28	3.0	1.9

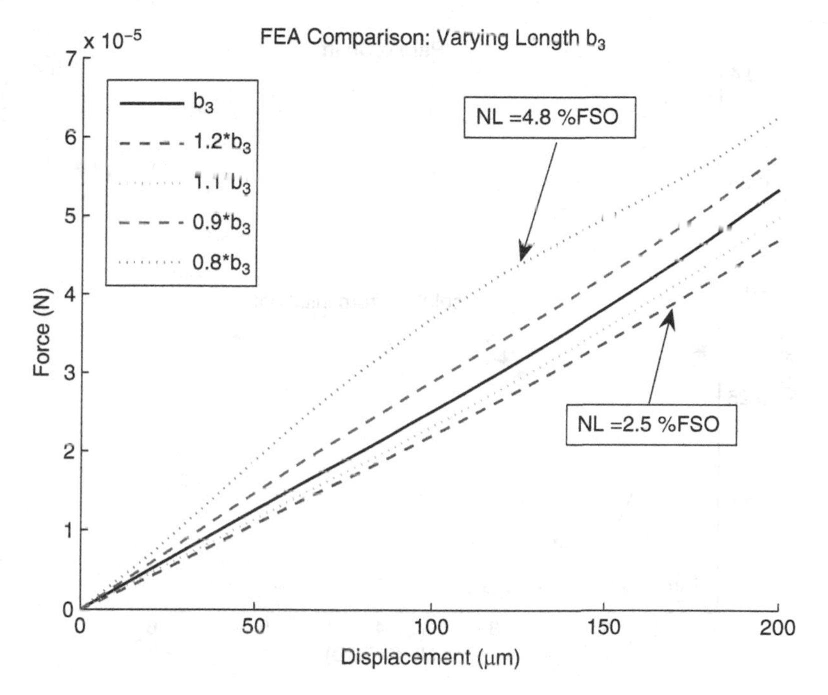

Fig. 7.16 Plots showing parameter study on effects of b_3 on linearity and stiffness

by 20% would make it larger than 300 μm, violating the upper bound constraint from the optimization, this case is shaded gray in Table 7.7. Referring to the same table, increasing and decreasing the length of b_3 by 10% also improves optimality (underlined values). Although the results obtained from the initial optimization still produce acceptable designs with respect to stiffness and nonlinearity, the optimization is refined using a larger displacement range setting (i.e., 200 μm).

The underlined values in Tables 7.4–7.7 are possible Pareto optimal solutions. These values are plotted to form a Pareto chart, shown in Fig. 7.17. The definition of Pareto optimal points are that there can be no improvement in one objective

Table 7.7 Effects of changing the b_3 parameter on stiffness and NL; bold numbers indicate optimal design from initial optimization using (7.14); underlined numbers are potential Pareto optimal solutions when considering 200 μm displacement; gray shading indicates a length that exceeds the upper bound placed on that link length

	Stiffness (N/m)		NL (%FSO)	
Parameter variation from baseline config.	200 μm displacement	70 μm displacement	200 μm displacement	70 μm displacement
$1.2 \times b_3$	0.23	0.22	2.4	1.0
$1.1 \times b_3$	<u>0.25</u>	0.23	<u>2.8</u>	0.30
b_3	<u>0.26</u>	**0.25**	<u>2.8</u>	**<0.01**
$0.9 \times b_3$	<u>0.28</u>	0.29	<u>1.1</u>	0.40
$0.8 \times b_3$	0.31	0.38	4.8	1.8

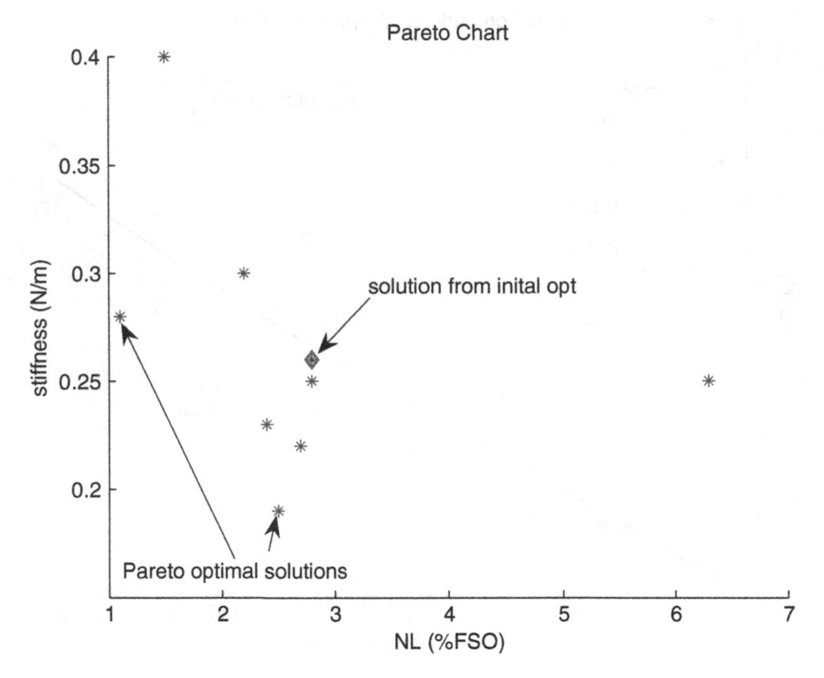

Fig. 7.17 Pareto chart for designs showing optimality with respect to stiffness and nonlinearity calculated over 200 μm; *highlighted* is the solution from the initial 70 μm displacement optimization

without a sacrifice in another objective. These points will form what is known as a Pareto front when plotted. Per this definition, only two configurations in Fig. 7.17 were found to lie on the Pareto front. The notion of optimum changes because the selection of a configuration that lies on this front is dependent on which objective the user values more (e.g., α value), reducing stiffness or NL. Since below 3%FSO is considered acceptable for most sensors, the new optimization could be done to minimize the stiffness and the NL could be added as a constraint.

Before more analysis of this type could be conducted, it was necessary to decouple the FEA from the optimization algorithm. The FEA is very expensive and the optimization can go through thousands of iterations, resulting in extremely long simulation times. A design of experiments was used to derive a polynomial that could predict the NL of a device without resorting to a FEA simulation. This will vastly reduce the cost of a simulation and greatly reduce the required time. For this design of experiments, only the width and link lengths were used as variables while the initial angle of link 2 is held constant. This was done to reduce the degrees of freedom and the complexity of the polynomial. A cubic polynomial was selected a candidate model. A cubic polynomial with four independent variables has 35 coefficients. The fact that the independent variables lay within a design space that has a box shape allows Latin hypercube sampling (LHS) to be used. LHS populates the design space with an ensemble of random numbers that are a good representative of the real variability. Twice as many sample points are needed as there are coefficients; therefore, this design of experiments used 70 sample points. The FEA was used to calculate the NL at these 70 data points. Then a multiple polynomial regression was used to derive the 35 coefficients. The regression produced a polynomial that has ARS of 0.95. The polynomial was used to replace the FEA in the optimization algorithm. The new optimization was conducted with the nonlinearity as a constraint. The tolerance of NL was varied from 3%FSO down to 0.5%FSO in 0.5 increments. The optimization produced the same solutions for every case between 1 and 3%FSO, 1, 300, 80, and 300 μm for w, l, r_3, and b_3, respectively. The resulting stiffness from this combination of dimensions is 0.029 N/m. All geometric constraints are active for this solution, which indicates that the NL and maximum stress constraints are not active. To achieve NL below 1%FSO, the optimization yields a solution of 1, 295, 116, and 300 μm for w, l, r_3, and b_3, respectively. The resulting stiffness from this combination of dimensions is 0.033 N/m. Only the w and b_3 geometric constraints are active in this solution. This suggests that the NL constraint is now active.

The FEA was used to test the devices robustness with respect toward fabrication errors. A device with the following geometric parameters was used: 1, 295, 116, and 300 μm for w, l, r_3, and b_3, respectively and 70° for θ_{2o}. This device will be fabricated using surface micromachining technologies. This process is subject to tolerances and errors similar to those of conventional machining. The relative tolerance for surface micromachining is about 10^{-1} μm. These fabrication errors will have the greatest impact on the width of the compliant members. The width is a cubed term in the calculation of the stiffness, per (7.2) and (7.3). In addition, the width is the smallest feature size, about 100 times smaller than lengths of the links and 20 times smaller than the width the rigid links. The width of the complaint links is 1 μm; therefore, the predicted tolerance of this feature size is ±0.1 μm. The baseline stiffness for this device is 0.033 N/m. The stiffness for the extreme case of over-etching is 0.024 N/m. The stiffness for the extreme case of under-etching is 0.043 N/m. There is approximately 30% change in either direction. This can be compensated for by experimental calibration of a completed device.

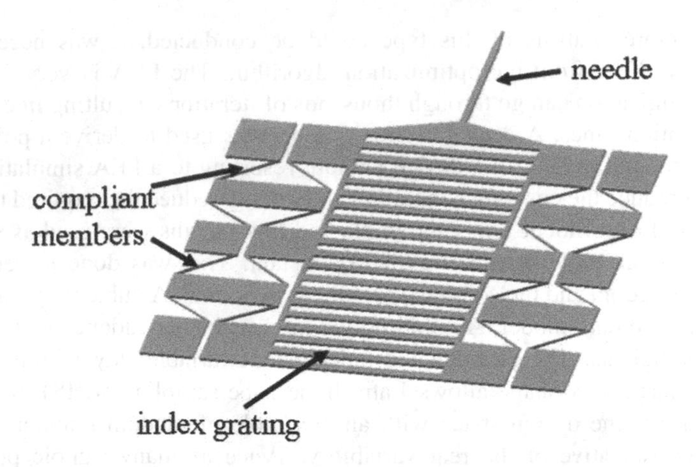

Fig. 7.18 3D model of final design: top nitride layer shown (not to scale) [22]

7.4.3 Comparison of New Design to Original

If a device was intended for the experiments described in [8, 23], the new force sensor design will use link length values from the baseline configuration and a width of 2 μm. This will provide force measurements in the desired range of forces with vastly improved linearity and close to four times the sensitivity. A 3D model of the top layer of nitride of the MEMS device is shown in Fig. 7.18. This device would have a spring constant of 0.57 N/m, almost a quarter of what was published for the four-beam design [8, 23]. The sensor would have large enough displacement range to handle the average 58 μm embryo deformation before penetration. The vast improvement becomes apparent when the force-displacement curves are plotted together, as in Fig. 7.19. In the design of the four-beam sensor, nonlinearities begin to dominate after only 15 μm of displacement. The new Robert's design remains virtually linear throughout the entire displacement range (in the plot of Fig. 7.19, it appears to be a horizontal line due to the scale of the nonlinearities of the four-bar sensor). Therefore, the force can be simply calculated using the predetermined spring constant and computed displacement. Based on the analysis presented in this chapter, the force values obtained with the mechanism enhanced design will be accurate over the specified full range of motion. The four-beam design will only exhibit a linear relationship if the sensor operates up to 10 μm, after which the sensor will be producing erroneous force values or would require nonlinear calibration curves.

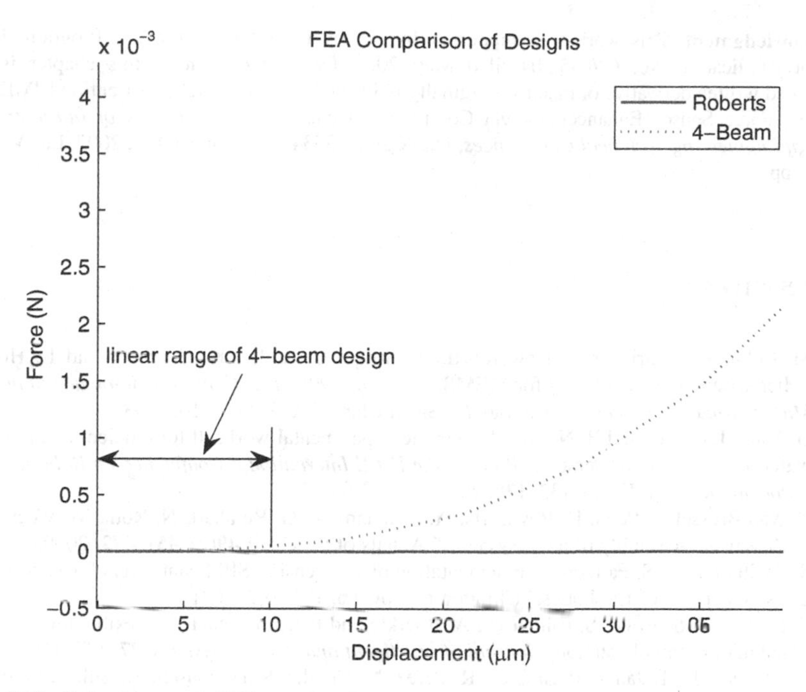

Fig. 7.19 Comparison of FEA results for mechanism enhanced (Robert's) and four-beam designs [22]

7.5 Summary and Conclusions

This work demonstrated the effectiveness of using a mechanism approach for enhancing the performance of a MEMS sensor. The work entailed the redesign of a surface micromachined optical force sensor. The sensor uses an optical diffraction sensing scheme to resolve forces induced from input displacements. This device is geared toward cell manipulation and microinjection. Focusing on the structural support elements of the sensor, a Robert's mechanism was selected to replace the current simple beam structure. The Robert's mechanism was chosen because of its linear motion and force characteristics. This mechanism was combined in series and parallel to form another mechanism with desirable traits. The geometric parameters of the Robert's mechanism were optimized to obtain a sensor with improved linearity and sensitivity. The presented techniques from this research could also be used to pursue designs for other applications. The Robert's mechanism is designed to be implemented as a compliant mechanism. This allows the sensor to be monolithically fabricated via surface micromachining and bulk etching technologies.

Acknowledgment This work was supported in part by the GEM Fellowship Program. U.S. Patent Application No. 60/885,304 filed May 20, 2009. The content of this chapter is an expanded written derivative of material originally published in the authors' paper entitled "MEMS Optical Force Sensor Enhancement via Compliant Mechanism", *Proceedings of International Design Engineering Technical Conferences*, DETC2007–35345, September 4–7, 2007, Las Vegas, NV, 9 pp.

References

1. M. Cohn, K. Böhringer, J. Noworolski, A. Singh, C. Keller, K. Goldberg and L. Howe, "Microassembly technologies for MEMS," *Proc. of 1998 SPIE Conf. on Micromachining and Microfabrication Process Technology IV*, Santa Clara, CA, 3511, 2–16, 1998.
2. G. Yang, J. Gaines and B. Nelson, "A flexible experimental workcell for efficient and reliable wafer-level 3D microassembly," *Proc. of the IEEE International Conference on Robotics and Automation*, Seoul, Korea, 133–138, 2001.
3. H. Van Brussel, J. Peirs, D. Reynaerts, A. Delchambre, G. Reinhart, N. Roth, M. Weck and E. Zussman, "Assembly of microsystems," Annals of the CIRP, **49**(2), 451–472, 2000.
4. S. Fahlbusch and S. Fatikow, "Implementation of self-sensing SPM cantilevers for nano-force measurement in microrobotics," Ultramicroscopy, **86**, 181–190, 2001.
5. S. Fatikow, J. Seyfried, S. Fahlbusch, A. Buerkle and F. Schmoeckel, "A flexible microrobot-based microassembly station," *Journal of Intelligent and Robotic Systems*, **27**, 135–169, 2000.
6. K. Domanaski, P. Janus, P. Grabiec, R. Perez, N. Chailet, S. Fahlbusch, A. Sill, S. Fatikow, "Design, fabrication and characterization of force sensors for nanorobot," Microelectronic Engineering, **78–79**, 171–177, 2005.
7. S. Fahlbusch and S. Fatikow, "Force sensing in microrobotic systems – an overview," *Proc. of the IEEE International conference on Electronics, Circuits and Systems*, 259–262, 1998.
8. X. Zhang, S. Zappe, R. Berstein, O. Sahin, C. Chen, M. Fish, M. Scott, and O. Solgaard, "Micromachined silicon force sensor based on diffractive optical encoders for characterization of microinjection," *Sensors and Actuators: A Physical*, 114(2–3), 197–203, 2004.
9. Y. Sun, B. Nelson, B. Potasek, and E. Enikov, "A bulk-fabricated multi-axis capacitive cellular force sensor using transverse comb drives," *J. Micromechanics and Microengineering*, **12**, 832–840, 2002.
10. D. Dao, T. Toriyama, J. Wells and S. Sugiyama, "Silicon piezoresistive six-degree of freedom force-moment micro sensor", *Sensors and Materials*, **15**(3), 113–135, 2002.
11. G. Roman, J. Bronson, G. Wiens, J. Jones and J. Allen, "Design of a Piezoresistive Surface Micromachined Three-axis Force Transducer for Microassembly," *Proc. of International Mechanical Engineering Conference and Exposition*, Orlando, Florida, No. IMECE2005–81672, 9 pp., Nov 2005.
12. A. Sexena and G. Ananthasuresh, "Optimal property of compliant topologies," *Struct. Multidiscip. Optimization*, **19**(1), 36–49, 2000.
13. C. Pedersen and A. Sechia, "On the optimization of compliant force amplifier mechanisms for surface micromachined resonant accelerometers," *J. Micromech. Microeng.*, **14**, 1281–1293, 2004.
14. J. Wittwer, T. Gomm and L. Howell, "Surface micromachined force gauges: uncertainty and reliability," *J. Micromech. Microeng.* **12**(1), 13–20, 2002.
15. X. Wang and G. Ananthasuresh, "Vision-based sensing of forces in elastic objects," *Sensors and Actuators: A Physical*, **94**(3), 142–156, 2001.
16. N. Hubbard, J. Wittwer, J. Kennedy, D. Wilcox and L. Howell, "A novel fully compliant planar linear-motion mechanism." *Proc. of the 2004 ASME Design Engineering Technical Conferences*, Salt Lake City, Utah, DETC2004–57008, 2004.
17. L. Howell, *Compliant Mechanisms*, Wiley, New York, 2001.

18. J. Przemieniecki, *Theory of Matrix Structural Analysis*, McGraw-Hill, New York, 1985.
19. MEMS and Nanotechnology Clearinghouse, 1895 Preston White Drive, Suite 100, Reston, Virginia 20191 www.memsnet.org, 2006.
20. J. Allen, *Micro Electro Mechanical System Design*, CRC Press, Boca Raton, 2005.
21. M. Madou, *Fundamentals of Microfabrication: The Science of Miniaturization*, CRC Press, Boca Raton, 2002.
22. G. Roman and G. Wiens, "MEMS Optical Force Sensor Enhancement via Compliant Mechanism", *Proceedings of International Design Engineering Technical Conferences*, DETC2007–35345, September 4–7, 2007, Las Vegas, NV, 9 pp.
23. X. Zhang, S. Zappe, R. Berstein, O. Sahin, C. Chen, M. Fish, M. Scott and O. Solgaard, "Micro-optical characterization and modeling of positioning forces on drosophila embryos self-assembled in two-dimensional arrays," *J. Microelectromechanical Systems*, **14**(5), 1187–1197, 2005.

Chapter 8
Observer Approach for Parameter and Force Estimation in Scanning Probe Microscopy

Gildas Besançon and Alina Voda

Abstract This chapter discusses a possible *state-observer* approach for various estimation problems arising in the context of so-called Scanning Probe Microscopes. The discussion is based on the example of the *Electric Force Microscope*. It is first emphasized how a typical force measurement purpose can be formulated as a model-based state observer problem, for which some standard *Kalman observer* can for instance be designed. The notion of *parametric amplification* sometimes used in order to enhance force measurement accuracy is then interpreted in the light of the here discussed observer approach. Finally, the issue of *parameter estimation* for this model-based approach is also discussed. All of those items are illustrated with the considered example and corresponding simulation results.

Keywords Scanning probe microscopy • Electric force microscope • Observer techniques • Force estimation • Parametric amplification • Parameter identification

8.1 Introduction

Various types of microscopes are developed with the purpose of inspecting material properties of various natures, in the context of so-called *Scanning Probe Microscopy* SPM (see e.g. [12]). Most of them are based on a sensitive element carried by a cantilever, which is approached to the surface under study, following the original idea of Scanning Tunneling Microscope [7]. As a consequence a simple dynamical description of the involved physics can be provided by the dynamics of the cantilever itself under the effect of the actuation as well as all additional external effects,

G. Besançon (✉)
Control Systems Department, GIPSA-lab, Grenoble Institute of Technology and Institut Universitaire de France, Ense[3] BP 46, 38402 Saint-Martin d'Hères, France
e-mail: gildas.besancon@grenoble-inp.fr

C. Clévy et al. (eds.), *Signal Measurement and Estimation Techniques for Micro and Nanotechnology*, DOI 10.1007/978-1-4419-9946-7_8,

either due to interaction forces or noises. On this basis, estimation problems can be handled via so-called *observer techniques* as they are developed in the control community.

Hence, the purpose of this chapter is to provide some general guidelines for an observer approach to solve for estimation problems arising in SPM frameworks, on the basis of the typical example of Electric Force Microscopy (EFM) [14], either as a tool towards force measurement or as a help for parameter estimation.

In fact, this chapter summarizes and extends various former studies of the authors in the same spirit [2–5].

For readers interested in physicists methods to determine tip-sample interaction forces in SPM, like the one called *inversion procedure* in Atomic Force Microscopy, there is a good review in [10] and the references therein. Another recent review on force measurement with Atomic Force Microscopy based on force-versus-distance curves is [8]. In the spirit of observers, one can also mention [17], where an observer-based technique is considered for the purpose of tip-sample interaction estimation, basically using likelihood ratio tests to infer the presence of the sample.

The present chapter adopts a different viewpoint, based on direct force estimation, and is organized as follows:

The proposed general statement of SPM estimation problems together with the corresponding observer approach are first presented in Sect. 8.2. The particular case of EFM is then chosen as an illustrative example in Sect. 8.3, where corresponding simulation results are gathered, and some conclusions end the chapter in Sect. 8.4.

8.2 General Problem Statement and Observer Approach

8.2.1 State Space Modeling for Kalman Observer and Force Estimation

The principle of SPM measurement can in short be depicted by Fig. 8.1 below: the sensitive tip is carried by a cantilever, the dynamics of which can roughly be described through its first mode, corresponding to some equivalent mass m, stiffness k and friction f.

Those dynamics typically characterize the motion described by the position variation z w.r.t. some steady-state value, and are assumed to be subject to some known forces F_a on the one hand, which can even be driven via some appropriate actuation u (generally $F_a = F_a(t,z,u)$), and some unknown forces F on the other hand, which can be considered as the forces to be measured by the device. Of course the dynamics are also subject to various types of noises, which can be denoted by v_z.

In addition, the device is equipped with some sensor, typically delivering an information upon z, again possibly disturbed by measurement noises v_y.

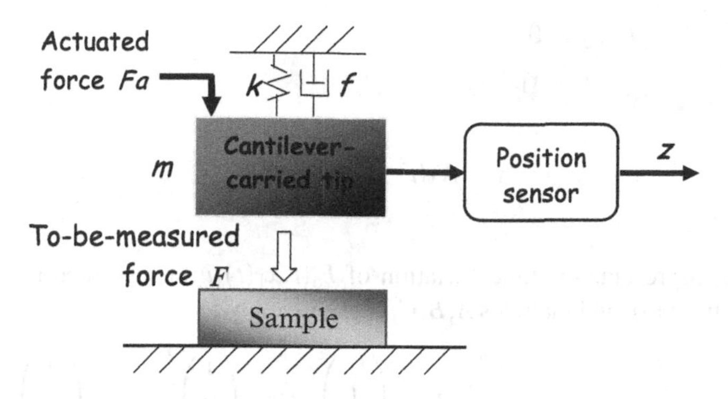

Fig. 8.1 SPM-like measurement principle

As a summary, the whole dynamics can be described by a differential equation of the form:

$$m\ddot{z}(t) = -kz(t) - f\dot{z}(t) + F_{\mathrm{a}}(t,z,u) + F(t) + v_z \tag{8.1}$$

with a measurement:

$$y(t) = z(t) + v_y, \tag{8.2}$$

where the sensor dynamics are neglected and normalized to get a unitary DC gain.

Equivalently, this can be re-written under the form of a classical state-space representation w.r.t. a couple of state variables $x_1 = z$, $x_2 = \dot{z}$, as follows:

$$\begin{aligned}
\dot{x}_1 &= x_2 \\
\dot{x}_2 &= -\frac{k}{m}x_1 - \frac{f}{m}x_2 + \frac{F_{\mathrm{a}}(x_1,u)}{m} + \frac{F}{m} + \frac{v_z}{m} \\
y &= x_1 + v_y.
\end{aligned} \tag{8.3}$$

Notice that this model can be extended with a more precise description of the modes or effects of external variables, without changing the methodology which will be presented hereafter.

In short, the idea is that the purpose of force measurement on the basis of model (8.3) translates into a problem of estimation of F.

In the absence of any further knowledge about F, it can be simply assumed to be modeled as:

$$\dot{F} = 0$$

(if a more complex model is known, it can obviously be used in the same way).

Then, by extending the original state vector with this additional variable $x_3 := F$, an extended state-space representation is obtained as follows:

$$\begin{aligned}
\dot{x}_1 &= x_2 \\
\dot{x}_2 &= -\frac{k}{m}x_1 - \frac{f}{m}x_2 + \frac{F_{\mathrm{a}}(x_1,u)}{m} + \frac{x_3}{m} + \frac{v_z}{m}
\end{aligned}$$

$$\dot{x}_3 = 0$$
$$y = x_1 + v_y. \tag{8.4}$$

namely:

$$\dot{x} = Ax(t) + Bv(t) + Gv_z$$
$$y = Cx + v_y, \tag{8.5}$$

where $v(t)$ represents the time variation of $F_a(t, x_1(t), u(t))$ assumed to be known (or reconstructed), and matrices A, B, C, G are given by:

$$A = \begin{pmatrix} 0 & 1 & 0 \\ -\frac{k}{m} & -\frac{f}{m} & \frac{1}{m} \\ 0 & 0 & 0 \end{pmatrix}; B = \begin{pmatrix} 0 \\ \frac{1}{m} \\ 0 \end{pmatrix}; C = \begin{pmatrix} 1 \\ 0 \\ 0 \end{pmatrix}^T; G = \begin{pmatrix} 0 \\ \frac{1}{m} \\ 0 \end{pmatrix}.$$

From this, a direct estimation of F can be obtained from a state observer design of the classical form:

$$\dot{\hat{x}}(t) = A\hat{x}(t) + Bv(t) - K(C\hat{x}(t) - y), \tag{8.6}$$

where K is to be chosen according to the desired estimation performances.

In particular, a classical *Kalman observer* can provide a solution for this problem, in addition yielding the optimal state estimate in the sense of least mean square of the estimation error, whenever covariances matrices are available for state and output noises v_z, v_y [11]. This just needs a slight modification of the form $\dot{F} = v_F$ for some noise v_F in the model, so that conditions for the Kalman design to be indeed appropriate are satisfied (for instance as in [6] for the case of Atomic Force Microscopy). In the absence of any information about noises, a deterministic approach can be considered omitting any noise, the system being obviously observable, and a Luenberger-type observer can be designed.

8.2.2 About Parametric Amplification and Force Estimation

The proposed approach for the estimation of the force from the directly available measurements is guaranteed to asymptotically provide the right (steady-state) value of the unknown force (at least in average). But various studies in force measurement have proposed some estimation scheme with improved accuracy by a specific *excitation* of the system (known as parametric amplification, e.g. as in [9, 15, 16]).

Such an approach is here discussed in the light of observer-based force measurement, in the spirit of [3].

In particular, it is shown how using this idea of signal amplification can indeed result in a better accuracy of the estimation results given an estimation time, or, conversely, a better estimation time, given an accuracy.

To do so, let us consider the situation when a constant-like unknown force is to be estimated and a constant actuation $u = u_0$ is applied.

Then a corresponding steady state behaviour can be computed for x_1, x_2 (when $F = 0$) as:

$$x_{20} = 0;\ x_{10} \text{ such that } f x_{10} = F_a(x_{10}, u_0).$$

Then, a first order approximation of model (8.4) around this steady state is given by:

$$\begin{aligned}
\dot{\delta x_1} &= \delta x_2 \\
\dot{\delta x_2} &= -\frac{(k-\kappa_{u_0})}{m}\delta x_1 - \frac{f}{m}\delta x_2 + \frac{\delta x_3}{m} + \frac{b}{m}\delta u + \frac{v_z}{m} \\
\dot{\delta x_3} &= 0 \\
y &= \delta x_1 + v_y
\end{aligned} \tag{8.7}$$

for $\kappa_{u_0} = \frac{\partial F_a}{\partial z}|_{u_0,x_{10}}$ and $b = \frac{\partial F_a}{\partial u}|_{u_0,x_{10}}$.

Assume that κ_{u_0} can be arbitrarily driven by u_0 (as in the example of EFM considered in next section for instance [3]).

Then it results that one can improve the performance in estimating $F = x_3$ w.r.t. output noise v_y by appropriately choosing u_0. This corresponds to a particular case of the so-called *parametric amplification* approach as proposed and experimented in [16] for instance.

The result can be formally stated by transforming the considered system (8.7) into a canonical observable form:

$$\begin{aligned}
\dot{\zeta} &= \begin{pmatrix} -a_1 & 1 & 0 \\ -a_2 & 0 & 1 \\ 0 & 0 & 0 \end{pmatrix} \zeta + B_{co}\delta u + G_{co}v_z = A_{co}\zeta + B_{co}\delta u + G_{co}v_z \\
y &= \begin{pmatrix} 1 & 0 & 0 \end{pmatrix} \xi + v_y = C_{co}\zeta + v_y,
\end{aligned} \tag{8.8}$$

where $a_1 = \frac{f}{m}$ and $a_2 = \frac{(k-\kappa_{u_0})}{m}$, which can be driven as desired by u_0.

Then we have [3]:

Proposition 8.2.1. *Given system (8.8) where v_y is a gaussian white noise with variance W, and v_z is omitted, and given any observer:*

$$\dot{\hat{\zeta}} = A_{co}\hat{\zeta} + B_{co}\delta u - K(C_{co}\hat{\zeta} - y)$$

for a convergence rate arbitrarily chosen by $K = \begin{pmatrix} k_1 & k_2 & k_3 \end{pmatrix}^T$, then the third component e_3 of the estimation error $e := \hat{\xi} - \xi$ decays to zero in means with a rate given by K, and its variance $v_3(t) := E[e_3(t)^2]$ satisfies:

- *v_3 asymptotically goes to:*

$$v_{3\infty} = \frac{k_3 W}{2[(k_1+a_1)(k_2+a_2)-k_3]} \times \left[k_3 a_1^2 + k_2^2(k_1+a_1) + (a_2-k_2)[(k_1+a_1)(k_2+a_2)-k_3]\right]. \quad (8.9)$$

- *$v_{3\infty}$ decreases with a_2, as long as:*

$$a_2 \geq \frac{k_3}{k_1+a_1} =: a_{20}. \quad (8.10)$$

- *$v_{3\infty}$ admits a minimum for $a_2 = a_{20}$ of the form:*

$$v_{3\infty 0} := \frac{k_3^2}{2k_2(k_1+a_1)}\left(k_2+a_1^2\right) W. \quad (8.11)$$

This means that given a desired rate of estimation (characterized by the observer gain K), one can get an optimal variance on the estimation error for x_3 by appropriate tuning of a_2 (via u_0).

Conversely, it is also clear that given a tuning of a_2, one can play on the accuracy by appropriate choice of K.

Notice that a similar result can be carried out for varying excitation $u(t)$ as in more classical parametric amplification approaches (e.g. as in [13]).

8.2.3 *About Model Identification and Force Estimation*

All the results discussed so far being based on a state observer approach, they strongly rely on the model. This means that the corresponding parameters are assumed to be known, or estimated a priori.

In the present section, the purpose is to emphasize how estimating such parameters can also be done with observer techniques, allowing to adapt the parameters at the same time as the force is itself estimated.

This is also known to be feasible depending on a specific excitation of the system.

Here, let us show that the structure of the model is a priori appropriate for this simultaneous state and parameter purpose.

To that end, the key-point is to notice that system (8.4) can be re-written in terms of new variables:

$$\begin{aligned} \xi_1 &= x_1 \\ \xi_2 &= x_2 + \frac{f}{m} x_1 \\ \xi_3 &= \frac{1}{m} x_3. \end{aligned} \quad (8.12)$$

The new representation then reads:

$$\begin{aligned}
\dot{\xi}_1 &= \xi_2 - \frac{f}{m}\xi_1 \\
\dot{\xi}_2 &= -\frac{k}{m}\xi_1 + \frac{v}{m} + \xi_3 + \frac{v_z}{m} \\
\dot{\xi}_3 &= 0 \\
y &= \xi_1 + v_y.
\end{aligned} \tag{8.13}$$

Notice that this possible transformation is a special case of some approach holding for any so-called *Liénard* system [2].

Notice also that the new representation can also be written in a more compact form:

$$\begin{aligned}
\dot{\xi} &= A_0\xi + \Phi(\xi_1, v)\theta + Gv_z \\
y &= C_0\xi + v_y,
\end{aligned} \tag{8.14}$$

where A_0, C_0 are under some Brunowsky canonical form, θ is a set of parameters equivalent to k, f, m ($\theta = \left(\frac{f}{m}\ -\frac{k}{m}\ \frac{1}{m}\right)^T$) and Φ follows from (8.13).

By neglecting the noises, we get a form allowing for a possible so-called *adaptive observer* which can be used as a candidate for estimation of both ξ *and* θ in (8.14), and thus from (8.12), of both model parameters and unknown force, for instance as in [1], with an observer given by:

$$\begin{aligned}
\dot{\hat{\xi}} &= A_0\hat{\xi} + \Phi(y, v)\hat{\theta} + (K + \Lambda\Gamma\Lambda^T C_0^T)\left(y - C_0\hat{\xi}\right) \\
\dot{\Lambda} &= (A_0 - KC_0)\Lambda + \Phi(y, v) \\
\dot{\hat{\theta}} &= \Gamma\Lambda^T C^T (y - C_0\hat{\xi}),
\end{aligned} \tag{8.15}$$

where K is chosen such that $A_0 - KC_0$ is stable and $\Gamma = \Gamma^T > 0$.

8.3 Illustrative Simulation Results for EFM Example

The purpose in this section is to illustrate the three possible observer approaches for estimation problems in SPM techniques discussed in the former section, on the basis of the specific example of EFM [14].

8.3.1 EFM Model

A typical example of SPM can be that of an Electric Force Microscope (EFM), which can indeed be described as in former section by a dynamical model (8.3), where F_a is an electrostatic force provided by a polarizing voltage u applied between the cantilever and the sample.

This force can thus be represented here by:

$$F_a(z,u) = \frac{\varepsilon_0 S \cdot u^2(t)}{(D - z(t))^2}, \tag{8.16}$$

where S denotes the surface of the polarized part of the cantilever in front of the sample, D their distance in the absence of any position variation, and ε_0 the permittivity in vacuum.

In the sequel, let us consider numerical values taken from [16], namely:

$$m = 0.22e - 12\,\text{kg},\; f = 4.7e - 11\,\text{Nsm}^{-1},\; k = 1\,\text{Nm}^{-1},$$
$$S = 3.4e - 8\,\text{m}^2,\; D = 20e - 6\,\text{m}.$$

All simulations will be provided in the presence of measurement noise as well as state equation noise, and with the unknown force to be estimated with a step-like profile of a magnitude equal to 1 nN.

The measurement noise is of a magnitude larger or equal to the cantilever motion magnitude under the effect of the $1 - nN$-force which is simulated, and the state equation noise about ten times lower.

With those values, a typical noise-free cantilever position step response, as well as an example of corresponding noisy measurement, are illustrated by Figs. 8.2 and 8.3 below.

8.3.2 EFM Force Estimation Results

Assuming the full model known (made of (8.3)–(8.16)), the force reconstruction approach based on a Kalman observer for model (8.4) is first illustrated here.

The possible improvement on the estimation results obtained with *signal amplification* is then illustrated.

In the first case of the Kalman approach, the observer is here tuned so as to achieve a good trade-off between estimation time (less than 0.5 s) and accuracy (less than 0.5 nN error), while the applied voltage is set to $u = 1$ V.

The corresponding estimation result is provided by Fig. 8.4. Notice that this is given as a basic illustration, and that some better accuracy could of course be achieved, but at the expense of the estimation time (according to standard features of Kalman observers).

Our purpose then is to illustrate the effect of parametric-like amplification on the estimation performance by playing on u: to that end, let us for instance consider the purpose of a reconstruction time of about 1 ms instead of the 0.5 s case previously considered.

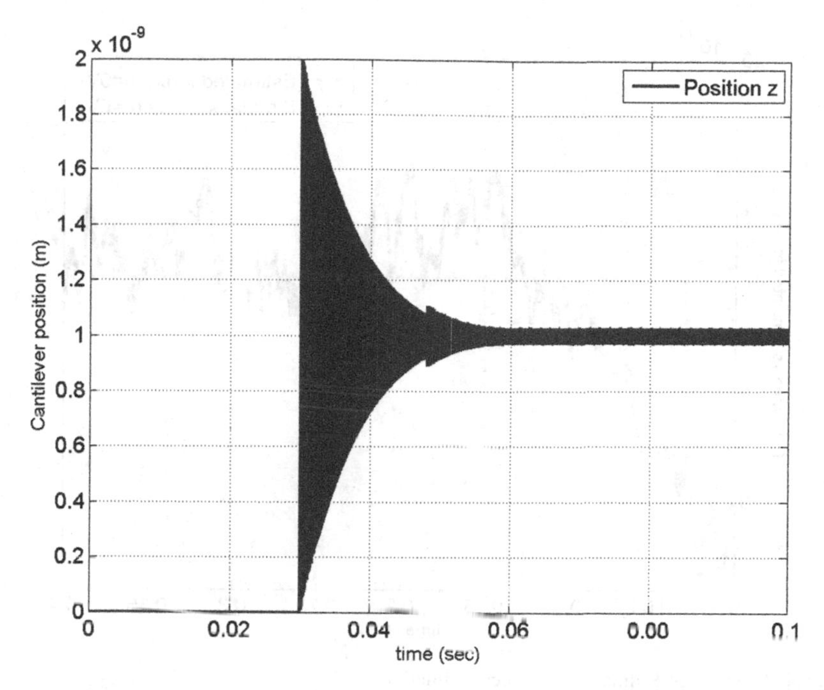

Fig. 8.2 Position response to a step-force F

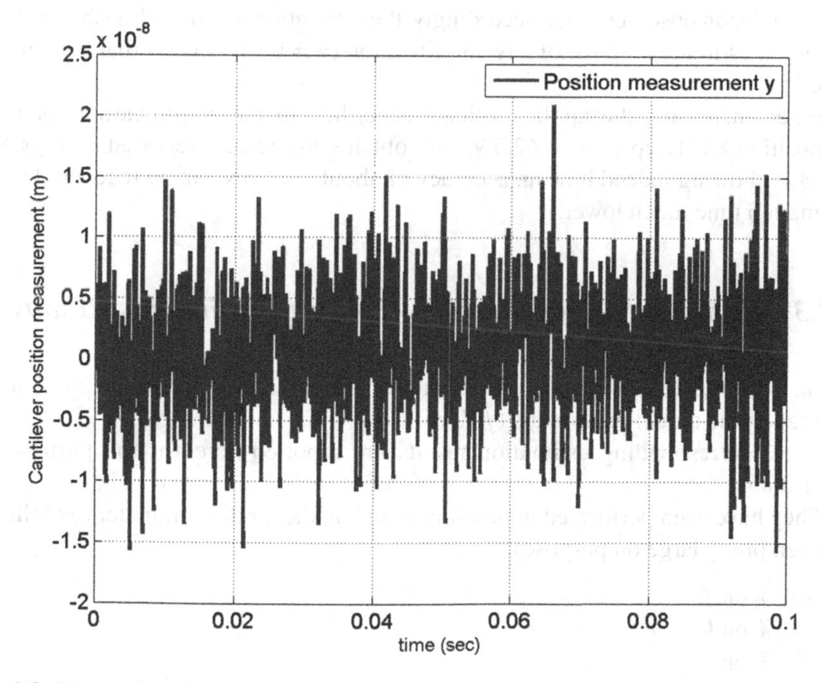

Fig. 8.3 Measured position response to a step-force F

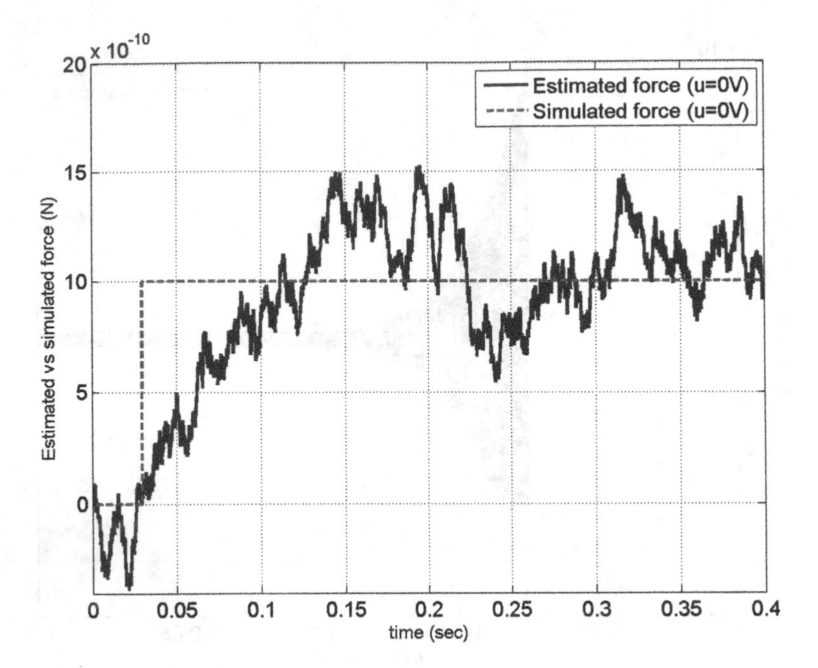

Fig. 8.4 Example of Kalman-based force estimation

Then, for an observer tuned accordingly, the estimation results of Figs.8.5 and 8.6 are obtained in the absence of any amplification ($u = 0$): it is clear that they are of poor interest.

When increasing the applied voltage according to the amplification result of proposition 8.2.1, up to $u = 62.5$ V, one obtains the results reported in Figs. 8.7 and 8.8, showing indeed how an accuracy of about ± 0.5 nN can be recovered in an estimation time much lower.

8.3.3 EFM Model Identification with Force Estimation Results

In this section, the use of observer techniques is further extended to the problem of estimating the model parameters (f, k, m).

Some corresponding estimation results are reported here for the purpose of illustration.

They have been performed with initial errors on the model parameters as follows (chosen pretty large on purpose):

- $+90\%$ on f
- $+50\%$ on k
- -20% on m

and with an observer of the form (8.15).

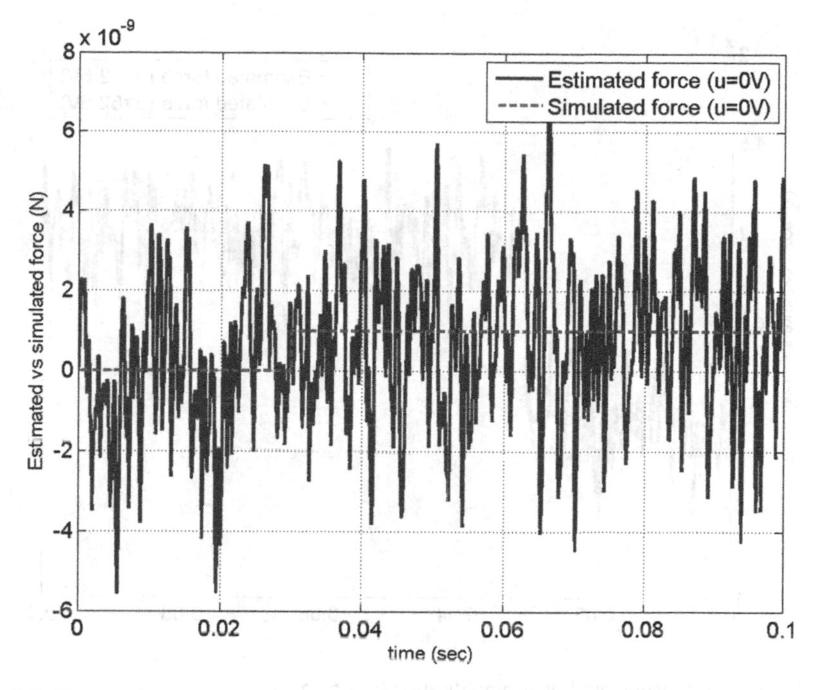

Fig. 8.5 Force estimation result without amplification ($u = 0\,\text{V}$)

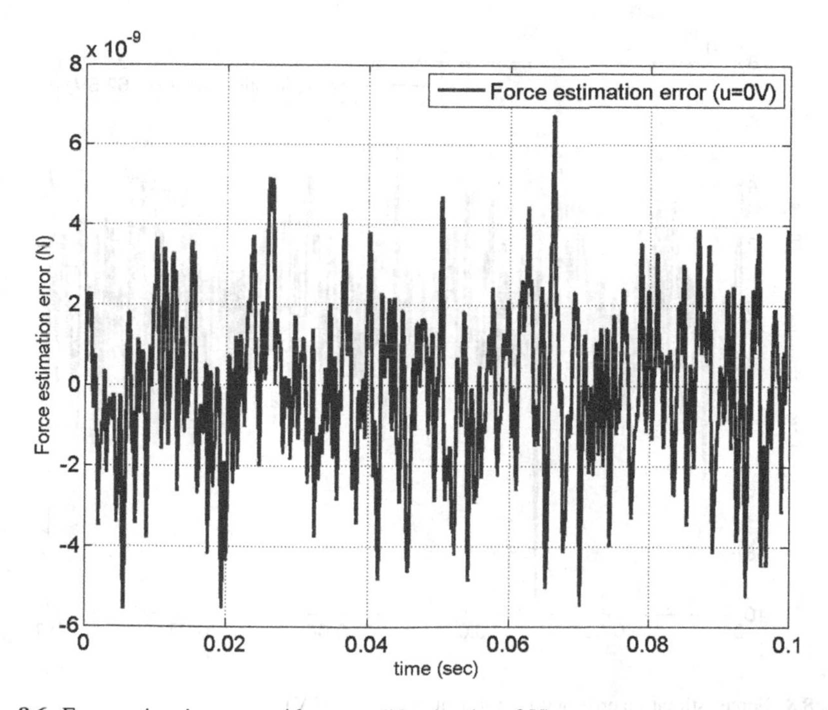

Fig. 8.6 Force estimation error without amplification ($u = 0\,\text{V}$)

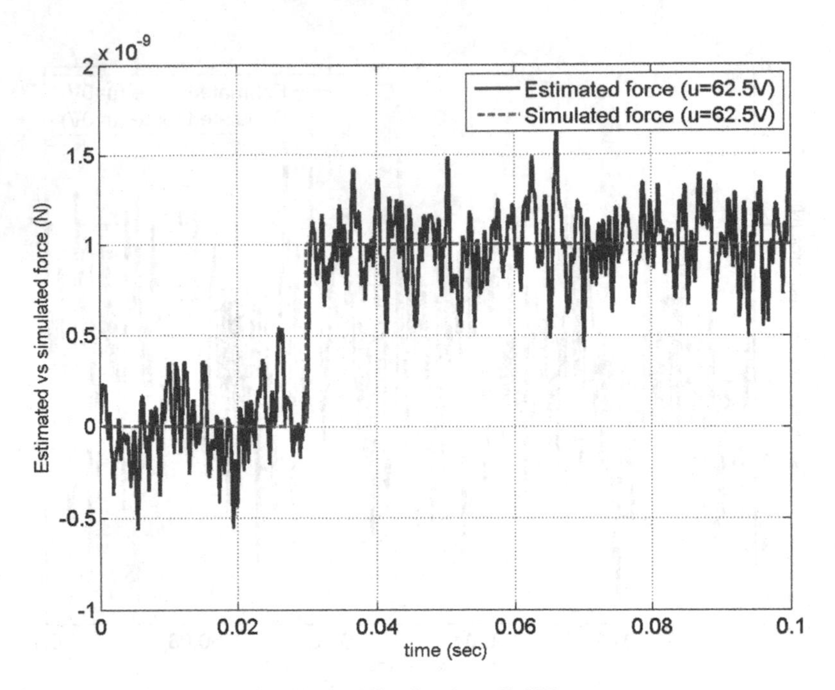

Fig. 8.7 Force estimation result with amplification ($u = 62.5\,\mathrm{V}$)

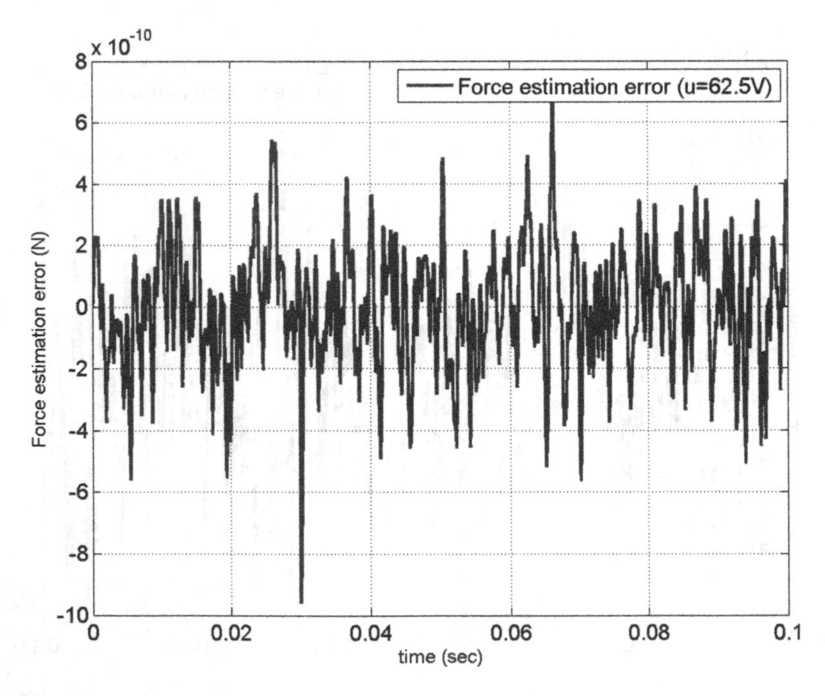

Fig. 8.8 Force estimation error with amplification ($u = 62.5\,\mathrm{V}$)

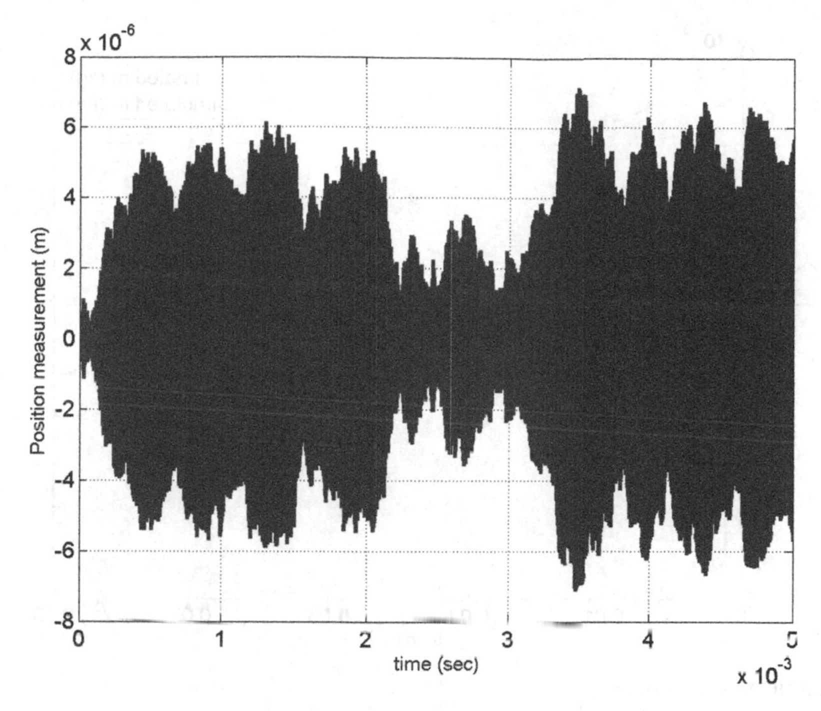

Fig. 8.9 Cantilever motion for parameter and force estimation

For the results which are presented hereafter, the system has been excited with a random-like signal, generating a cantilever motion as described in Fig. 8.9.

Notice that this large excitation is needed to counteract the presence of noise, in a similar way as in former subsection, and that the results provided here were obtained with a measurement noise level comparable to the cantilever motion due to the unknown force action.

Notice also that for larger noises, under the same excitation level (restricted by the physically limited range of cantilever motion), small steady-state errors appear.

Under the conditions described above, the obtained estimation results for the model parameters can be seen on Figs. 8.10–8.12.

It can be checked that m and k are indeed rapidly well identified, and f in a bit longer time and with a much larger transient error (see Fig. 8.13 for a view of the final accuracy).

The force is anyway very well estimated, as illustrated by Fig. 8.14.

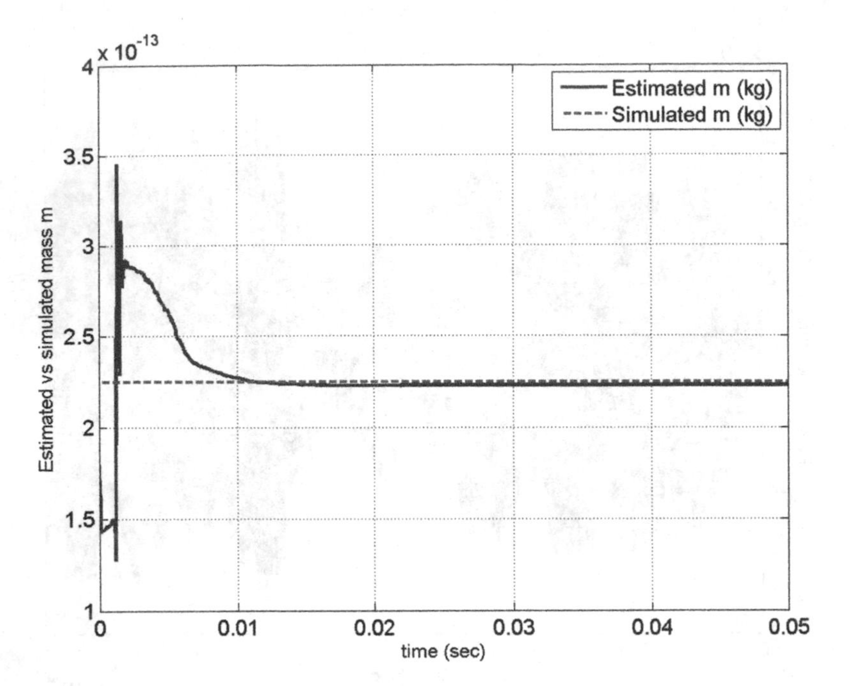

Fig. 8.10 Mass estimation

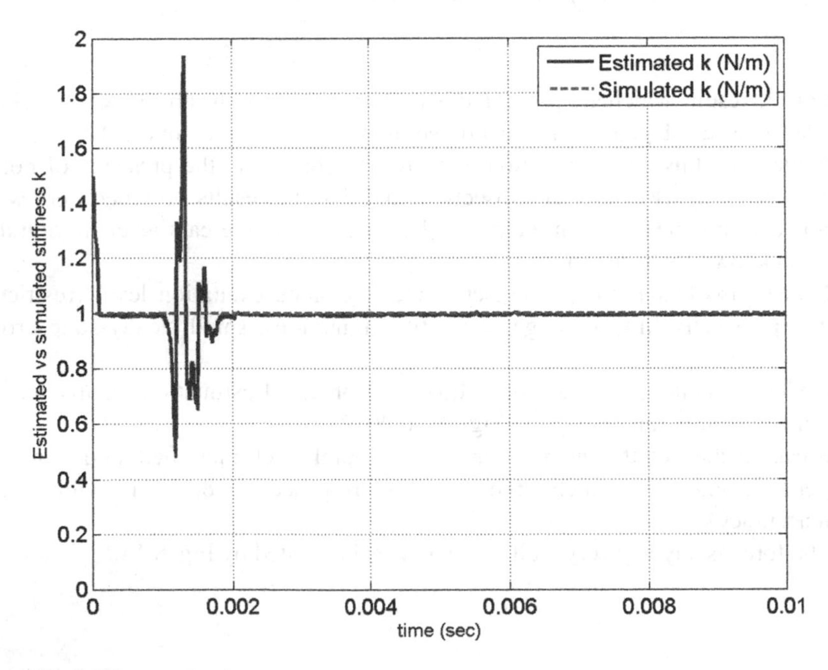

Fig. 8.11 Stiffness estimation

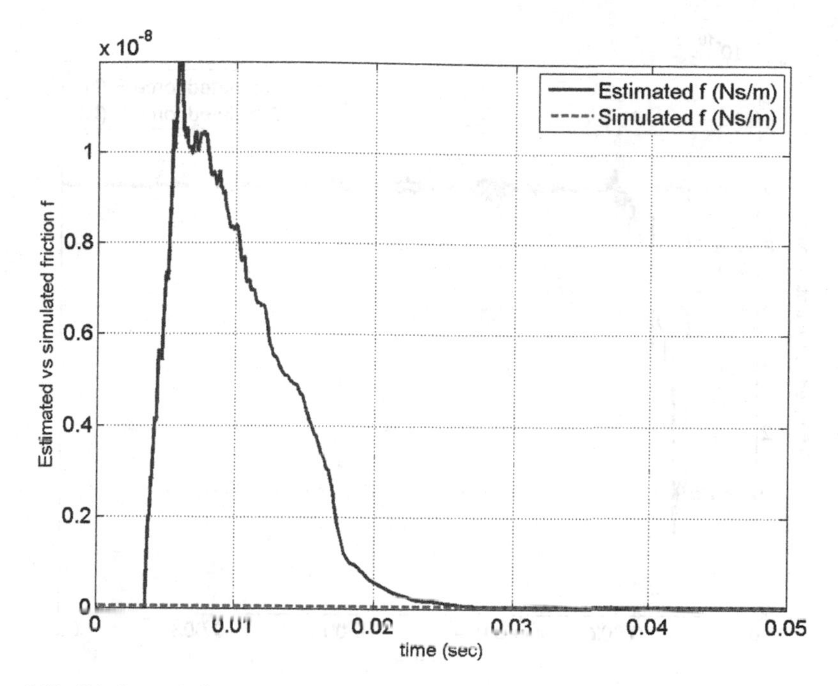

Fig. 8.12 Friction estimation

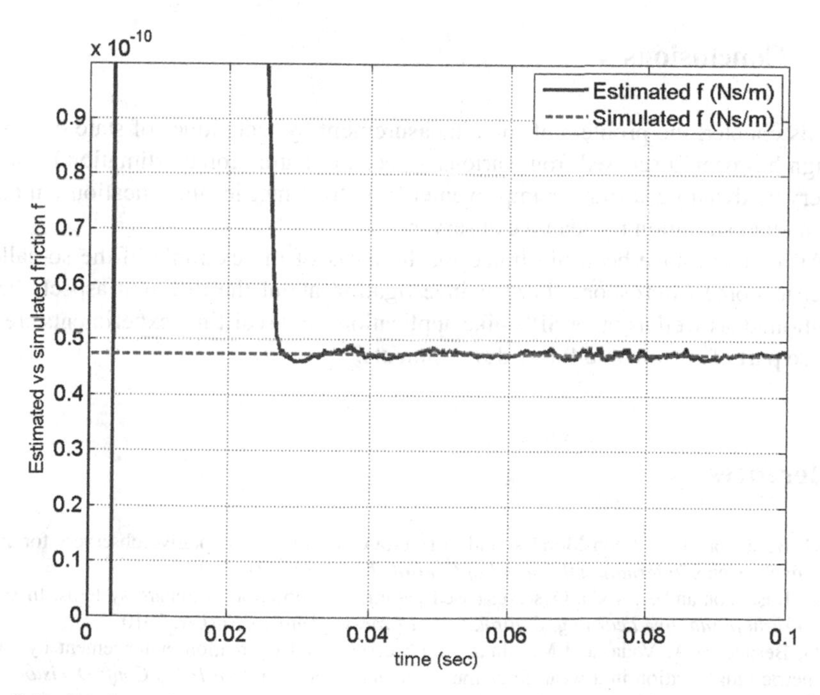

Fig. 8.13 Friction estimation (zoom)

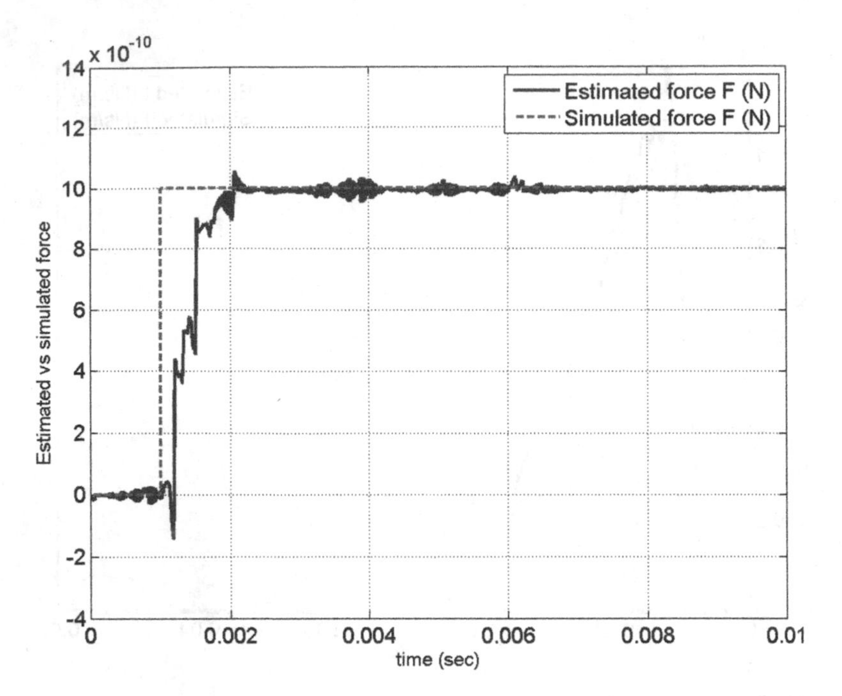

Fig. 8.14 Force estimation

8.4 Conclusions

In this chapter, the problem of force measurement by techniques of state observer design has been discussed from various aspects: dynamic force estimation by state observer, dynamic accuracy improvement by 'parametric amplification', model parameter estimation by adaptive observer.

All of them have been illustrated on the basis of the example of the so-called Electric Force Microscope. Further investigation about the different aspects here mentioned, as well as other SPM-like applications and real-time experiments are of course part of current and future developments.

References

1. G. Besançon, J. de Leon-Morales, and O Huerta-Guevara. On adaptative observers for state affine systems. *International Journal of Control*, 79(6):581–591, 2006.
2. G. Besançon and A. Voda. Observer-based parameter estimation in liénard systems. In *IFAC Workshop Adaptive Learning, Control, Signal Process. Antalya, Turkey*, 2010.
3. G. Besançon, A. Voda, and M. Alma. On observer-based estimation enhancement by parametric amplification in a weak force measurement device. In *47th IEEE Conf. Decision and Control, Cancun, Mexico*, 2008.

4. G. Besançon, A. Voda, and J. Chevrier. Measurements in fundamental physics: an observer application. In *2nd Symp. on System Structure and Control, Oaxaca, Mexico*, 2004.
5. G. Besançon, A. Voda, and G. Jourdan. Kalman observer approach towards force reconstruction from experimental afm measurements. In *15th IFAC Symposium on System Identification, St Malo, France*, 2009.
6. G. Besançon, A. Voda, and G. Jourdan. *Micro, nanosystems & systems on chips - Modeling, control and estimation*, chapter Observer-based estimation of weak forces in a nanosystem measurement device. J. Wiley & Sons, 2010.
7. G. Binnig and H. Rohrer. Scanning tunneling microscopy. *Surface science*, 126:236–244, 1983.
8. H-J. Butt, B. Cappella, and M. Kappl. Force measurement with atomic force microscope: Technique, interpretation and applications. *Surface Science Reports*, 59:1–152, 2005.
9. A.N. Cleland. *Foundations of nanomechanics*. Springer, 2003.
10. R. García and R. Pérez. Dynamic atomic force microscopy methods. *Surface science reports*, 47:197–301, 2002.
11. H. Kwakernaak and R. Sivan. *Linear Optimal Control Systems*. Wiley-Interscience, New York, 1972.
12. E. Meyer, H.J. Hug, and R. Bennewitz. *Scanning Probe Microscopy - The Lab on a tip*. Springer, 2004.
13. M. Napoli, B. Bamieh, and K. Turner. Mathematical modeling, experimental validation and observer design for a capacitively actuated microcantilever. In *American Control Conf., Denver, Colorado, USA*, 2003.
14. T. Ouisse, M. Stark, F. Rodrigues-Martins, B. Bercu, S. Huant, and J. Chevrier. Theory of electric force microscopy in the parametric amplification regime. *Physical Review B*, 71, 2005.
15. J.F. Rhoads, S.W. Shaw, K.L. Turner, and R. Baskaran. Tunable microelectromechanical filters that exploit parametric resonance. *ASME J. of Vibration and Acoustics*, 127(5):423–30, 2005.
16. D. Rugar and P. Grütter. Mechanical parametric amplification and thermomechanical noise squeezing. *Physical Review Letters*, 67:699–702, 1991.
17. D.R. Sahoo, A. Sebastian, and M.V. Salapaka. Harnessing the transient signals atomic force microscopy. *International Journal of Robust and Nonlinear Control*, 15:805–820, 2005.

Index

C. Clévy et al. (eds.), *Signal Measurement and Estimation Techniques for Micro and Nanotechnology*, DOI 10.1007/978-1-4419-9946-7,